TETRAHEDRON ORGANIC CHEMISTRY SERIES
Series Editors: J E Baldwin, FRS & R M Williams

VOLUME 17

Solid-Supported Combinatorial and Parallel Synthesis of Small-Molecular-Weight Compound Libraries

Related Pergamon Titles of Interest

BOOKS

Tetrahedron Organic Chemistry Series:
CARRUTHERS: Cycloaddition Reactions in Organic Synthesis
DEROME: Modern NMR Techniques for Chemistry Research
FINET: Ligand Coupling Reactions with Heteroatomic Compounds
GAWLEY & AUBE: Principles of Asymmetric Synthesis
HASSNER & STUMER: Organic Syntheses Based on Name Reactions
and Un-named Reactions
McKILLOP: Advanced Problems in Organic Reaction Mechanisms
PERLMUTTER: Conjugate Addition Reactions in Organic Synthesis
SESSLER & WEGHORN: Expanded, Contracted & Isomeric Porphyrins
SIMPKINS: Sulphones in Organic Synthesis
TANG & LEVY: Chemistry of C-Glycosides
WILLIAMS: Synthesis of Optically Active Alpha-Amino Acids 2nd Edition*
WONG & WHITESIDES: Enzymes in Synthetic Organic Chemistry

JOURNALS

BIOORGANIC & MEDICINAL CHEMISTRY
BIOORGANIC & MEDICINAL CHEMISTRY LETTERS
TETRAHEDRON
TETRAHEDRON: ASYMMETRY
TETRAHEDRON LETTERS

Full details of all Elsevier Science publications/free specimen copy of any Elsevier Science journal are available on request from your nearest Elsevier Science office

* In Preparation

Solid-Supported Combinatorial and Parallel Synthesis of Small-Molecular-Weight Compound Libraries

DANIEL OBRECHT
Polyphor Ltd, Zurich

and

JOSÉ M. VILLALGORDO
University of Girona

PERGAMON

An Imprint of Elsevier Science

ELSEVIER SCIENCE Ltd
The Boulevard, Langford Lane
Kidlington, Oxford OX5 1GB, UK

Library of Congress Cataloging-in-Publication Data
A catalog record from the Library of Congress has been applied for.

British Library Cataloguing in Publication Data
A catalogue record from the British Library has been applied for.

First Edition 1998

ISBN 0 08 043258 1 Hardcover
ISBN 0 08 043257 3 Flexicover

Transferred to digital printing 2005
Printed and bound by Antony Rowe Ltd, Eastbourne

This book is dedicated to

Sir Jack E. Baldwin
(Waynflete Professor of Chemistry, Oxford UK)

and

Prof. Dr. Heinz Heimgartner
(University of Zürich, CH)

List of Contributors

Boopathy Dhanapal (BD)
Polyphor Ltd.
Winterthurerstr. 190
CH-8057 Zürich

Werner Doebelin
Prolab/LabSource
Zihlackerstr. 4
CH-4153 Reinach

Philipp Ermert (PE)
Polyphor Ltd.
Winterthurerstr. 190
CH-8057 Zürich

Patrick Jeger (PJ)
Lipomed AG
Fabrikmattenweg
CH-4144 Arlesheim

Liza Koltay
Koltay Design
Gustackerstr. 20
CH-4103 Bottmingen

Thierry Masquelin (TM)
Pharma Research
F.Hoffmann-La Roche Ltd.
CH-4070 Basel

Daniel Obrecht (DO)
Polyphor Ltd.
Winterthurerstr. 190
CH-8057 Zürich

Klara Sekanina (KS)
Polyphor Ltd.
Winterthurerstr. 190
CH-8057 Zürich

José M. Villalgordo (JMV)
Departamento de Quimica
Facultad de Ciencias
Universitad de Girona
Campus di Montilivi
E-17071 Girona

Cornelia Zumbrunn (CZ)
Pharma Research
F.Hoffmann-La Roche Ltd.
CH-4070 Basel

Table of Contents

FOREWORD

The words "Combinatorial Chemistry" have different meanings to different people, ranging from split and mix strategies to parallel synthesis using robots, and embracing the whole range of preparative chemistry from organic molecules, to catalyst ligands, and even inorganic solids. All of these activities have in common an attempt to expand the diversity of structure available to the chemist as well as the access to this diversity, permitting the discovery of new and valuable biological acid material properties. In this outstanding survey of combinatorial organic chemistry the authors Obrecht, who has established a new combinatorial chemistry company called *Polyphor,* and Villalgardo have brought together the literature, including that from 1998, and have concisely analysed the applications and achievements of this new field. This work will be of value to all chemists engaged in preparative work, both in industry and academe.

J E Baldwin, FRS

List of Abbreviations

Ac	acetyl
acac	acetylactone
ACD	available chemicals directory
ACE	angiotensin converting enzyme
ADP	adenosine diphosphate
AIBN	azoisobutyronitrile
Ala	L-alanine
Arg	L-arginine
Asn	L-asparagine
Asp	L-aspartate
ATP	adenosine triphosphate
9-BBN	9-borabicyclo[3.3.1]nonane
BHA	benzhydrylamine
Bn	benzyl
Boc	*tert*-butoxycarbonyl
Bpoc	2-(4-biphenylyl)isopropyloxycarbonyl
Bu	butyl
Bz	benzoyl
CAE	carbonyl alkyne exchange reaction
CAN	ceric ammonium nitrate
Cbz	benzyloxycarbonyl
CNS	central nervous system
COD	1,5-cyclooctadiene
COS	combinatorial organic synthesis
Cys	L-cysteine
DAMGO	[D-Ala, MePh^4, Gly-01^5]enkephalin
DBU	1,8-diazabicyclo[5.4.0]-undec-7-ene
DCC	*N,N'*-dicyclohexylcarbodiimide
DDQ	2,3-dichloro-5,6-dicyanobenzoquinone
DEAD	diethyl azodicarboxylate
DIAD	diisopropyl azodicarboxylate
DIBAH	diisobutylaluminium hydride
DIC	*N,N'*-diisopropylcarbodiimide
DIEA	*N,N*-diisopropyl-*N*-ethylamine
DMA	*N,N*-dimethylacetamide
DMAD	dimethyl acetylenedicarboxylate

DMAP	*N,N*-dimethylaminopyridine
DME	ethylene glycol dimethyl ether
DMF	*N,N*-dimethylformamide
DMS	dimethylsulfide
DMSO	dimethylsulfoxide
DMT	dimethoxytrityl
DNA	desoxyribonucleic acid
DTBP	2,6-di-(*tert*-butyl)pyridine
DVB	divinylbenzene
EDCI	1-ethyl-3-[3-(dimethylamino)propyl]carbodiimide
ELISA	enzyme linked immunosorbant assay
Et	ethyl
FAB	fast atom bombardment
Fmoc	9-fluorenylmethoxycarbonyl
FT	Fourier transform
GABA	γ-aminobutyric acid
GC	gas chromatography
Gln	L-glutamine
Glu	L-glutamate
Gly	L-glycine
HATU	azabenzotriazolyl-*N,N,N′,N′*-tetramethyluronium hexafluorophosphate
HBTU	2-(1H-benzotriazole-1-yl)-1,1,3,3-tetramethyluronium hexafluorophosphate
His	L-histidine
HIV	human immunodeficiency virus
HMB	*p*-(hydroxymethyl)benzoyloxy-methyl
HMDS	hexamethyldisilazane
HMP	(hydroxymethyl)phenoxy)acetic acid
HOBt	1-hydroxybenzotriazole
HPLC	high performance liquid chromatography
HTS	high throughput screening
Ile	L-isoleucine
IR	infrared spectroscopy
LAH	lithium aluminium hydride
LDA	lithium *N,N*-diisopropylamide
Leu	L-leucine
LPCS	liquid-phase combinatorial synthesis
LPE	liquid-phase extraction
Lys	L-lysine

m-CPBA	*m*-chloroperbenzoic acid
MALDI-TOF	matrix-assisted, laser desorption ionisation time-of-flight MS
MAS-NMR	magic-angle spinning NMR
MBHA	*p*-methylbenzhydrylamine
Me	methyl
Met	L-methionine
MMP	matrix metalloproteinase
Mpc	2-(4-methylphenyl)isopropyloxycarbonyl
MS	mass spectroscopy
Mtr	methoxytrityl
NBS	*N*-bromosuccinimide
NGF	nerve growth factor
NMM	*N*-methylmorpholine
NMO	*N*-methylmorpholine *N*-oxide
NMP	*N*-methylpyrrolidinone
NMR	nuclear magnetic resonance
NVOC	*N*-nitroveratryloxycarbonyl
PAL	5-(4-Fmoc-aminomethyl-3,5-dimethoxyphenoxy)valeric acid
PAM	phenylacetamidomethyl
PCD	pyridinium dichromate
PCR	polymerase chain reaction
PEG	polyethylene glycol
Pfp	pentafluorophenyl
Ph	phenyl
Phe	L-phenylalanine
PNA	peptide-nucleic acid
PPOA	4-(2-bromopropionyl)-phenoxyacetic acid
PPTS	pyridinium *p*-toluene sulfonic acid
Pro	L-proline
PS	polystyrene
PSP	polymer-supported perruthenate
PSQ	polymer-supported quenching procedure
PTBD	polymer-supported 1,5,7-triazabicyclo[4.4.0]dec-5-ene
PVPCC	poly(4-vinylpyridinium chlorochromate)
py	pyridine
PyBOP	(benzotriazole-1-yl)-oxy-tris-pyrrolidino-phosphonium hexafluorophosphate
PyBrOP	bromo-tris-pyrrolidino-phosphonium hexafluorophosphate
RAM	*Rink* amide resin

RNA	ribonucleic acid
SCAL	safety catch amide linkage
Ser	L-serine
SPE	solid-phase extraction
SPOS	solid phase organic synthesis
Su	succinimidyl
TADDOL	α, α, α', α'-tetraaryl-1,3-dioxolane-4,5-dimethanol
TBAF	tetrabutylammonium fluoride
TBTU	*O*-(benzotriazole-1-yl)tetramethyluronium tetrafluoroborate
TCEP	tris-(2-carboxyethyl)phosphine
TES	triethylsilyl
Tf	trifluoromethanesulfonyl (triflyl)
TFA	trifluoroacetic acid
TFE	2,2,2-trifluoroethanol
THF	tetrahydrofuran
Thr	L-threonine
TIPS	triisopropylsilyl
TLC	thin layer chromatography
TMEDA	tetramethylethylenediamine
TMG	tetramethylguanidine
TMS	trimethylsilyl
Trp	L-tryptophane
Ts	*p*-toluolsulfonyl
Tyr	L-tyrosine
UV	Ultraviolet spectroscopy
Val	L-valine
XAL	[[9-[(9-fluorenylmethyloxycarbonyl)amino]xanthen-2(or 3)-yl]oxy]alkanoic acid

Polystyrene resin (cf. *Chapter 1.7, Table 1.7.1*)

Wang resin (cf. *Chapter 1.7, Table 1.7.1*)

Sasrin resin (cf. *Chapter 1.7, Table 1.7.1*)

Rink resin (cf. *Chapter 1.7, Table 1.7.1*)

Introduction, Basic Concepts and Strategies

1.1. Opening remarks

Friedrich Woehler's discovery of synthetic urea in 1828 was not only the start of synthetic organic chemistry, but was ultimately also the beginning of a change of paradigm in biology, which led finally to molecular biology.[1] Although biopolymers such as proteins, DNA and RNA play a dominant role in living matter, and synthetic polymers derived from polystyrene have been known since 1839,[2] it was only in 1963 when *R. B. Merrifield* published his seminal paper on solid-supported peptide synthesis[3] that organic synthetic chemists started to consider polymers as valuable tools for synthesis. It is not surprising that polymer-supported synthesis found wide applications in the field of polypeptide and oligonucleotide chemistry as their synthesis consists of linear repetitive cycles of deprotection-, coupling- and washing steps, which ultimately could be fully automated and carried out on robotic systems.

In the two decades following *Merrifield's* pioneering work many other synthetically useful reactions were successfully transferred to mainly polystyrene-derived resins by scientists such as *Leznoff*[4] and *Fréchet*.[5] It was also recognised by several authors that polymer-bound reagents and catalysts[2,5] such as boronic acids, peracids and phosphines might be extremely valuable tools to carry out solution chemistry while keeping all the benefits of solid supports such as simple work up and purification procedures. Additional exciting aspects initiated by the use of polymer-bound reagents such as employing oxidising and reducing agents in the same reaction compartment, were recognised early on[2] but their scope was not fully exploited. Twelve years after *Merrifield's* work on polypeptides *Rapoport*[6] critically reviewed the field of solid-phase organic chemistry by asking: "Solid-phase organic synthesis: Novelty or fundamental concept?"

Maybe due to the lack of industrial applications, the interest in solid-supported chemistry somehow slowed down for the next decade and was essentially concentrated on polypeptide and oligonucleotide chemistry. The field of organic synthesis was focusing on the total synthesis of complex natural products, on the elucidation of their biosynthesis and on the development of novel synthetic methodologies. Synthetic organic chemistry quickly reached the level of a "mature" science with the general perception that chemists were now able to synthesise any given small-molecular-weight natural and synthetic compound no matter how complex the structure would be.[7] A lot of the motivation to synthesise complex molecules such as taxol (*Figure 1*) has been initiated by the fact that these molecules exhibit interesting biological activities. Thus, the significant progress made in synthetic organic chemistry had direct repercussions in drug discovery in that more and more complex molecules with increasing potency and selectivity could be synthesised, tested and ultimately manufactured successfully.

Figure 1

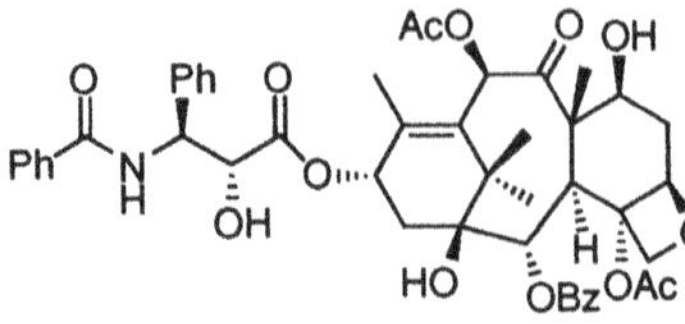

1: taxol

The main stream of synthetic organic chemists most often did not incorporate solid supports and solid-supported reagents in their synthetic concepts. For this reason, it was not a surprise that the first ideas for the parallel and combinatorial synthesis of compound libraries emerged from scientists like *Geysen*[8] and *Houghten*[9] working in the polypeptide area. While this area offered a lot of potential applications for the use of peptide libraries it was really the dramatic changes in the methods and throughput of biological screening in pharmaceutical and agrochemical research that initiated a literal burst of interest for novel and diverse compound collections.[10,11] Alongside molecular biology and genomic sciences, which are creating an ever-growing number of new and biologically relevant targets, and providing deeper insights into biological mechanisms and processes, novel high-throughput screening methods offer also the possibility to screen thousands of molecules per day. Hence, there was a real need for a large number of novel small-molecular-weight compounds and synthetic chemists were challenged to develop efficient high-throughput synthesis methods. For obvious reasons solid-supported chemistry became, from the nineties on, a fashionable area of research.

At the outset chemists focused on known solid-phase methodologies available from peptide- and oligonucleotide chemistry but the shift towards small-molecular-weight compounds necessitated new synthetic strategies. In the early nineties several research groups in academia and industry started extensive programs on transferring solution chemistry to solid supports,[12] developing novel linker and cleavage procedures in order to broaden the scope of molecules amenable on solid supports. Once again polymer-bound reagents and catalysts enjoyed a renaissance because they offered the possibility to carry out reactions in solution while maintaining the advantages of solid supports such as ease of isolation, work-up and purification.[13] While peptide- and oligonucleotide chemists created the prospect of synthesising libraries of millions of molecules, the shift towards small-molecular-weight "drug-like" compounds needed by the agrochemical and pharmaceutical companies initiated a reassessment and redimensioning of the library sizes. From initially mixtures of thousands of compounds possible in the field of peptide- and oligonucleotide libraries, necessitating complex tagging and deconvolution strategies, there was a gradual shift towards mixtures of 10-20 members[14] of a given library observable. As solid-supported synthetic strategies became more sophisticated and convergent and thus the resulting products more structurally complex and "drug-like", there was again a shift towards the parallel synthesis of single compounds with a defined chemical purity and confirmed structure. The progress made in the development of efficient workstations and robotic systems for parallel synthesis guaranteeing a high synthetic throughput, the call for small-molecular-weight compounds together with the severe problems resulting from screening of mixtures, were responsible for this dramatic shift in paradigm. Novel synthetic strategies, polymer supports and polymer-supported reagents and catalysts, new linker- and cleavage strategies and new reactive building blocks are now being discovered at high speed, and it is becoming more and more accepted that solid supports offer an new dimension when incorporated into synthetic design strategies.

In this book *"Solid-Supported Combinatorial and Parallel Synthesis of Small-Molecular-Weight Compound Libraries"* we focus on the synthetic aspects of polymer-supported chemistry without neglecting related areas such as tagging- and deconvolution aspects, analytical methods and spectroscopy. After some historical landmarks we present a short survey of applications of solid supports followed by the four elements of lead finding and aspects of the drug development process. We then focus on polymer supports and polymer-supported reagents and catalysts used for the synthesis of small-molecular-weight compounds. Forming the connection between the solid support and the actual molecules that are synthesised, the linkers play a crucial role in the synthetic concepts of solid supported synthesis (*Chapter 1.7*). While in the fields of peptide- and oligonucleotide chemistry the linkers were mainly designed to release one functional group, carbocyclic- and heterocyclic compounds require new resin-releasing techniques such as cleavage by cyclisation and multidirectional cleavage procedures which introduce new elements of diversity in the very last step and allow the production of very pure products. We discuss in *Chapter 1.8* the different synthetic strategies amenable to solid-phase chemistry, briefly mention aspects of solution- and solid-phase synthesis, mixtures versus single compound collections and explain the key technologies of parallel and combinatorial chemistry such as the "split-mixed technology".[15] Focusing on convergent strategies we will highlight the importance of multicomponent one-pot and multigeneration reactions as efficient strategies to create small molecule libraries. In all these approaches, reactive building blocks ("*reactophores*") which are essentially constituted of a reactive group to which are attached orthogonally protected functional groups ready for subsequent easy transformations, play a crucial role. The reactophores are designed to react in simple high yielding one-step transformations to give a large array of interesting core structures ("*pharmacophores*") which can subsequently be easily transformed into the final molecules. We have classified these reactophores into three categories: *donors, acceptors and donor-acceptors*. We will discuss the possibilities to combine these reactophores in multicomponent and multigeneration reactions and comment on the intriguing result that the same set of building blocks can give rise to a multitude of novel core structures depending on the sequence of mixing. Amongst others, building blocks derived from natural products will increasingly be developed with a view to synthesising focused libraries. The *"combinatorics of reactophores"* thus constitutes a novel interesting dimension in combinatorial and parallel chemistry. After explaining the terms and strategies of tagging and deconvolution, we will talk about aspects of diversity assessment and planning, library formats, analytical methods used in solid-supported organic chemistry and finally very briefly touch on aspects of robotics and automation.

After having explained the rules and the methods used in combinatorial and parallel chemistry we will turn our attention to examples and applications which have appeared in the recent literature, starting with a brief account of linear strategies towards polypetides, peptidomimetics, peptoids, oligonucleotides and oligosaccharides. Focusing much more on convergent approaches towards small molecules, we will first analyse in *Chapter 3* examples which have been performed in solution using polymer-supported scavengers, and solid-supported reagents and catalysts, in order to speed up the extraction, work-up and purification procedures. In *Chapter 4* we will then concentrate on examples which have been completely performed on solid-phase. In analysing the examples we will always highlight the reactive building blocks, the solid supports, the linkers and the synthetic strategies that have been employed. We have attempted to increase the complexity of the combinatorial elements as we go through the examples presented. Thus, we will conclude with examples combining essentially all useful strategies of combinatorial and parallel chemistry, such as "safety-catch" linkers, highly reactive and versatile building blocks, diversification using multicomponent one-pot and multigeneration approaches with multidirectional resin cleavage procedures. By following this strategy we hope that the reader will get a flavour for where this exciting field is heading and what technologies might emerge in the future.

Those people who have tried here to summarise solid-supported organic synthesis in a hopefully clear and comprehensive way agree:

"Solid-supported organic chemistry has only just started to become a new dimension in organic synthesis"

D. Obrecht, April 1998

1.2. Historical landmarks in solid-supported organic synthesis

Although natural polymers such as cellulose nitrate (1838) and cellulose acetate (1870) were discovered in the last century, and also synthetic polymers derived from styrene (1839), isoprene (1879) and methacrylic acid (1880) have been known for quite some time, it was really *H. Staudingers* macromolecular theory (1930) that set the stage for a better understanding of polymers.[2] Concerning organic chemistry on polymers, it was definitely *R. B. Merrifield* and his fundamental contribution about solid-supported peptide synthesis in 1963[3] and *Letsinger*[16] and *Khorana's* work on oligonucleotide synthesis that inspired and initiated a lot of further investigations. Some of the historical landmarks are mentioned below (this list is by no means complete or representative).

1963: Seminal paper by *R. B. Merrifield* describing for the first time the successful synthesis of a short peptide on a polystyrene resin (*J. Am. Chem. Soc.* **1963**, *85*, 2149)

1965: *Letsinger* and *Khorana* applied solid supports for the synthesis of oligonucleosides (*J. Am. Chem. Soc.* **1965**, *87*, 3526; ibid. **1966**, *88*, 3181)

1967: *J. Fréchet* described a highly loaded trityl resin (2.0 mmol/g)

1967: *Wilkinson et al.* described polymer-bound tris(triphenylphosphine) chlororhodium as hydrogenation catalyst (*J. Am. Chem. Soc.* **1967**, 1574)

1969: Solid-phase synthesis of Ribonuclease (*J. Am. Chem. Soc.* **1969**, *91*, 501)

1970: *H. Rapoport* introduced the term *hyperentropic efficacy* (effect of high dilution) on solid supports (*J. Am. Chem. Soc.* **1970**, *92*, 6363)

1971: *Fréchet et al.* pioneered solid-phase chemistry in the field of carbohydrate research. (*J. Am. Chem. Soc.* **1971**, *93*, 492)

1973: Application of intramolecular *Dieckmann*-condensation for the solid-phase synthesis of lactones by *Rapoport et al.* (*J. Macromol. Sci. Chem.* **1973**, 1117)

1973: *Leznoff et al.* described the use of polymer supports for the mono-protection of symmetrical dialdehydes, describing oxime formation, *Wittig*-reaction, crossed aldol-condensation, benzoin-condensation and *Grignard*- reaction on solid support. (*Can. J. Chem.* **1973**, *51*, 3756)

1974: *F. Camps et al.* described the first synthesis of benzodiazepines on solid support
(*Ann. Chim.* **1974**, *70*, 1117)

1976: *Leznoff* and *Fyles* described bromination and lithiation of insoluble polystyrene, thus
pioneering the synthesis of functionalised resins
(*Can. J. Chem.* **1976**, *54*, 935)

1976: *Rapoport* and *Crowley* published a review entitled: "Solid phase organic synthesis:
Novelty or fundamental concept?" and raised three important questions
(*Acc. Chem. Res.* **1976**, *9*, 135)
- degree of separation of resin-bound functionalities
- analytical methods to follow reactions on solid support
- nature and kinetics of competing side reactions

1976-
1978: *Leznoff et al.* published a series of papers dealing with the synthesis of insect sex
attractants (*Can. J. Chem.* **1977**, *55*, 1143)

1977: *Wulff et al.* synthesised chiral macroporous resins using carbohydrates as templates for
the use of column materials for the separation of enantiomers
(*Makromol. Chem.* **1977**, *178*, 2799)

1979: *Leznoff* employed successfully a chiral linker for the asymmetric synthesis of (*S*)-2-
methylcyclohexanone in 95% e.e. (*Angew. Chem.* **1979**, *91*, 255)

1984: *Geysen et al.* described the multi-pin technology for the multiple peptide synthesis
(*Proc. Natl. Acad. Sci. USA*, **1984**, *81*, 3998)

1985: *Houghten et al.* described the tea-bag method for multiple peptide synthesis
(*Proc. Natl. Acad. Sci. USA*, **1985**, *82*, 5131)

1985: *G. P. Smith* described in a seminal paper the use of filamentous phage for the synthesis
of peptide libraries (phage display method, *Science*, **1985**, *228*, 1315)

1986: Mixtures of activated amino acid monomers were coupled to solid supports for the
synthesis of peptide libraries as mixtures, the product distribution depending on the
relative coupling rates (*Mol. Immunol.* **1986**, *23*, 709)

1991: *Fodor et al.* described the VLSIPS-method (very large scale immobilised polymer synthesis; photolitographic parallel synthesis (*Science*, **1991**, *251*, 767))

1991: *Furka et al.* described the "portioning-mixing" method (*Int. J. Pept. Prot. Res.* **1991**, *37*, 487), which was termed "split synthesis" by *Hruby et al.* (*Nature* **1991**, *354*, 82) and "divide, couple, and recombine" process by *Houghten et al.* (*Nature* **1991**, *354*, 84)

1992: Oligonucleotide-encoded chemical synthesis proposed by *Lerner* and *Brenner* (*Proc. Natl. Acad. Sci. USA*, **1992**, *89*, 5181)

1992: Synthesis of 1,4-benzodiazepins on solid support described independently by *S. Hobbs-DeWitt* (Diversomer® technology, US 5324483, **1993**) and *J. A. Ellman* (*J. Am. Chem. Soc.* **1992**, *114*, 10997)

1993: Binary encoded synthesis using gas chromatographically detectable chemically inert tags by *W. C. Still et al.* (*Proc. Natl. Acad. Sci. USA*, **1993**, *90*, 10922)

1993: Use of multi-cleavable linkers for the synthesis of peptide-type libraries by *M. Lebl et al.* (*Int. J. Protein Res.* **1993**, *41*, 201)

1994: Use of the "safety-catch" linker principle developed by *Kenner et al.* (*J. Chem. Soc., Chem. Commun.* **1973**, 636) by *J. A. Ellman* for multidirectional cleavage from the resin (*J. Am. Chem. Soc.* **1994**, *116*, 11171)

1995: Synthesis of a potent ACE inhibitor by combinatorial organic synthesis on solid support using a 1,3-dipolar cycloaddition reaction by *M. Gallop et al.* (WO 95/35278, **1995**)

1995: Use of a genetic algorithm for the selection of the products of an *Ugi* four component reaction (*Angew. Chem. Int. Ed. Engl.* **1995**, *34*, 2280)

1996: Use of the *Ugi* four component reaction in combination with a 1,3-dipolar cycloaddition reaction of intermediary formed "Münchnones" with electron-poor acetylenes by *R. Armstrong et al.* (*Tetrahedron Lett.* **1996**, *37*, 1149)

1997: Combination of a cyclo-condensation reaction, multicomponent diversification and multidirectional cleavage from the resin using a novel combined "safety-catch"- and traceless linker yielding highly diverse pyrimidines by *D. Obrecht et al.* (*Chimia* **1996**, *11*, 530; *Helv. Chim. Acta* **1997**, *80*, 65) and *L. M. Gayo et al.* (*Tetrahedron Lett.* **1997**, *38*, 211).

1997: Synthesis of a taxoid-library using radiofrequency-encoding
(*J. Org. Chem.* **1997**, *62*, 6092)

Many recent reviews and books summarise the work that has been done in this field:

Functionalized resins and linkers for solid-phase synthesis of small molecules: C. Blackburn, F. Albericio, S. A. Kates, *Drugs of the Future* **1997**, *22*, 1007.

Functionalised polymers: Recent developments and new applications in synthetic organic chemistry: S. J. Shuttleworth, S. M. Allin, P. K. Sharma, *Synthesis* **1997**, 1217.

Recent developments in solid-phase organic synthesis: R. Brown, *Contemporary Organic Synthesis* **1997**, *4*, 216.

Solid-Phase Organic Reactions II. A Review of the Recent Literature: P. H. H. Hermkens, H. C. J. Ottenheijm, D. C. Rees, *Tetrahedron* **1997**, *53*, 5643.

Organic synthesis on soluble polymer supports: Liquid phase methodologies: D. J. Gravert, K. D. Janda, *Chem. Rev.* **1997**, *97*, 489.

The current status of heterocyclic combinatorial libraries: A. Nefzi, J. M. Ostresh, R. A. Houghten, *Chem. Rev.* **1997**, *97*, 449.

Combinatorial Chemistry, Synthesis and Application. S. R. Wilson, A. W. Czarnik, Wiley (1997)

Combinatorial organic synthesis using Parke-Davis's DIVERSOMER® method: S. Hobbs DeWitt, A. W. Czarnik, *Acc. Chem. Res.* **1996**, *29*, 114.

Multiple-component condensation strategies for combinatorial library synthesis: R. W. Armstrong, A. P. Combs, P. A. Tempest, S. D. Brown, T. A. Keating, *Acc. Chem. Res.* **1996**, *29*, 123.

Design, synthesis and evaluation of small-molecule libraries: J. A. Ellman, *Acc. Chem. Res.* **1996**, *29*, 132.

Strategy and tactics in combinatorial organic synthesis. Applications to drug discovery: E. M. Gordon, M. A. Gallop, D. V. Patel, *Acc. Chem. Res.* **1996**, *29*, 144.

Synthesis and application of small molecule libraries: L. A. Thompson, J. A. Ellman, *Chem. Rev.* **1996**, *96*, 555.

Solid-phase organic reactions: a review of the recent literature: P. H. H. Hermkens, H. C. J. Ottenheijm, D. Rees, *Tetrahedron* **1996**, *52*, 4527.

Combinatorial synthesis of small-molecular-weight organic compounds: F. Balkenhohl, C. von dem Bussche-Hünnefeld, A. Lansky, C. Zechel, *Angew. Chem.* **1996**, *108*, 2436.

Organic synthesis on solid phase: J. S. Früchtel, G. Jung, *Angew. Chem. Int. Ed. Engl.* **1996**, *35*, 17.

Kombinatorische Synthese: K. Frobel, T. Krämer, *Chemie in unserer Zeit* **1996**, *30*, 270.

Combinatorial Peptide and Nonpeptide Libraries. G. Jung (Ed.), VCH, Weinheim (1996)

Combinatorial Libraries. Synthesis, Screening and Application Potential. R. Cortese (Ed.), W. de Gruyter, Berlin (1995).

Applications of Combinatorial Technologies to Drug Discovery. 1. Background and Peptide Combinatorial Libraries. M. A. Gallop, R.W. Barret, W. J. Dower, S. P. A. Fodor, E. M. Gordon, *J. Med. Chem.* **1994**, *37*, 1233.

Applications of Combinatorial Technologies to Drug Discovery. 2. Combinatorial Organic Synthesis, Library Screening Strategies, and Future Directions. E. M. Gordon, R. W. Barrett, W. J. Dower, S. P. Fodor, M. A. Gallop, *J. Med. Chem.* **1994**, *37*, 1385

1.3. Applications of solid-supported organic chemistry

Although the combinatorial and parallel synthesis of compound libraries devoted to screening, lead finding and lead optimisation purposes have initiated a real burst of interest for solid-phase organic synthesis (SPOS), there are numerous other applications and opportunities for SPOS emerging from related areas. This chapter will summarise the fields of interest for SPOS and some of those will be discussed more extensively in the following chapters.

Development of novel soluble and insoluble polymer supports

To date still most of the resins that are employed in SPOS are derived from polystyrene (for more details see *Chapter 1.5*). Although polystyrene has certainly many excellent properties there is currently a great interest in finding novel resins with modified mechanical and chemical properties, such as amphiphilic resins consisting of co-polymers of functionalised polystyrene- and polyacrylamide subunits.[17,18] The aims of this field of research can be summarised as follows:

- *increase of the mechanical and chemical stability of the polymers*
- *extension of the scope of reactions that can be performed on solid supports*
- *amphiphilic resins that are compatible with aqueous reaction conditions*
- *shorter and less extensive washing procedures*
- *development of soluble dendrimeric polymers with high loading capacities which can be easily precipitated*
- *development of novel anchors that are compatible with a broad range of reaction conditions and allow a high loading capacity*
- *beneficial effects of high loaded resins on some reactions and on the resin cleavage procedures; drawbacks might be anticipated in intramolecular cyclisations*

Use of solid-supported reagents (cf. *Chapter 1.6*)

One crucial advantage of performing parallel and combinatorial chemistry in solution is the fact that most organic reactions have been studied and developed in solution. Thus, many reaction protocols can be directly applied in a parallel format without having to go through tedious reaction development procedures. On the other side, solid-supported strategies offer the benefit of easy automation and rapid isolation and purification procedures. Insoluble reagents might really constitute the future for solution strategies in that by-products resulting from the reagents can be removed by simple filtrations and thus offer all the benefits of solid-phase and solution chemistry. Many solid-supported reagents have been described in the literature or are commercially available (cf. *Chapter 1.6*) but their use has so far been limited to more academic studies. Novel reagents grafted on highly loaded resins will certainly emerge as important tools in the field of parallel and combinatorial chemistry in solution. Some of the most important features of polymer-bound reagents are demonstrated schematically in *Figure 1.3.1*.

Figure 1.3.1

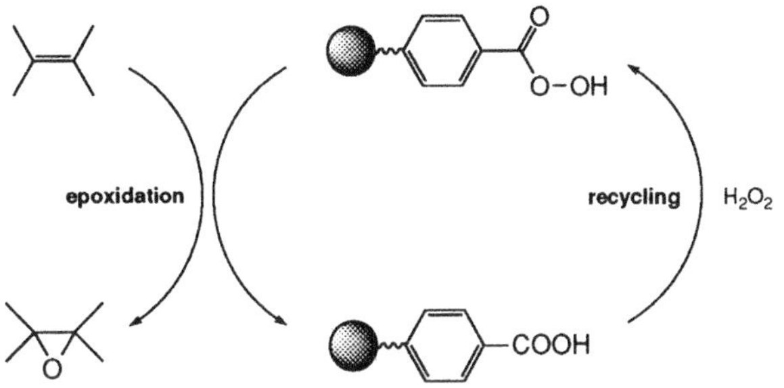

The polymer-bound peracid described by *Fréchet et al.*[5] which was easily obtained by H_2O_2 oxidation of the corresponding benzoic acid can be used to oxidise olefins to epoxides. The purification is achieved by simple filtration of the polymer-bound reagent, which can be recycled and re-used. It has to be pointed out, however, that after a few reaction cycles the reagent looses some of its initial activity. Nevertheless, this example demonstrates nicely the potential of polymer-bound reagents in combinatorial organic synthesis (COS) (more examples will be presented in *Chapter 1.6*).

Supports for linear syntheses

Solid phase chemistry originally started with *Merrifield's* pioneering polypeptide synthesis.[3] Due to the highly repetitive cycles of deblocking-, washing- and coupling procedures, peptide synthesis became a prime target for solid-phase synthesis.[2] For the last thirty years solid-phase peptide synthesis has matured to become a highly automated and powerful technology. Closely related to the linear assembly strategy of polypeptides, the synthesis of oligonucleotides[2,16] and oligosaccharides[2,19] also initiated a lot of interest. The iterative nature of linear assembly strategies is schematically shown in *Figure 1.3.2*. After having grafted the first suitably protected building block (BB1) containing orthogonally protected functional groups G^1 and G^2 (amino acids for peptides, mono-nucleotides for oligonucleotides and mono-saccharides for oligosaccharides) with a suitable linker to the polymeric support, BB2 is coupled after cleavage of the blocking group (BG) followed by the usual washing procedures. This process is repeated n times, the protected functional groups G^1-G^{2n+2} are deprotected and the linear biopolymer is subsequently cleaved from the resin.

11

Figure 1.3.2

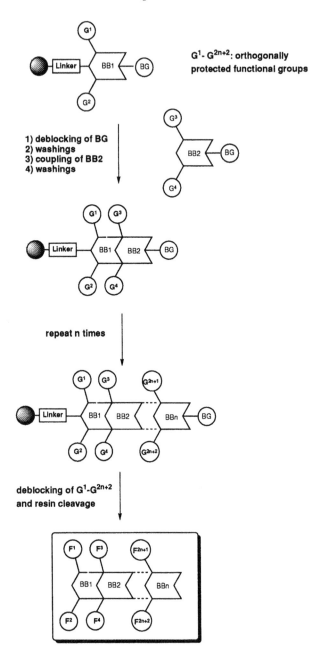

G^1- G^{2n+2}: orthogonally
protected functional groups

1) deblocking of BG
2) washings
3) coupling of BB2
4) washings

repeat n times

deblocking of G^1-G^{2n+2}
and resin cleavage

Following this protocol, it is quite understandable that polypetides, oligonucleotides and oligosaccharides, which belong to the three major classes of biopolymers, constituted the ideal platform for combinatorial chemistry. Extensive studies on the coupling procedures of the corresponding activated amino acid-, mononucleotide- and monosaccharide derivatives and a whole plethora of suitable protective groups and building blocks were available ready to start the combinatorial chemistry.

Traps for toxic and/or odorous molecules

Functionalised polymers have been used to covalently or ionically extract unpleasantly smelling and / or toxic molecules from a solution.[20] A classic example is given in *Figure 1.3.3*.

Figure 1.3.3

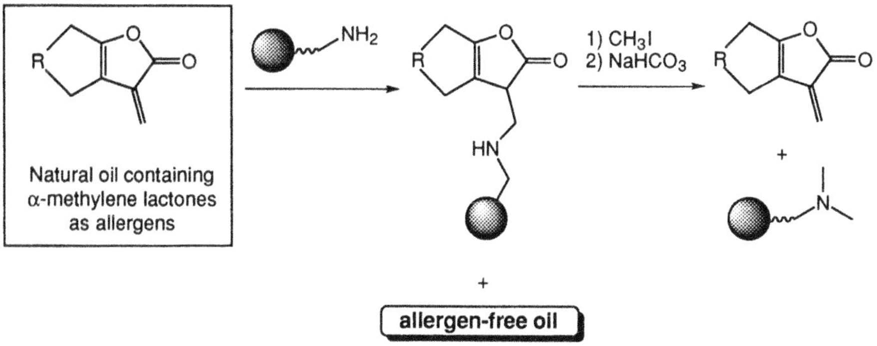

Natural oil containing α-methylene lactones were extracted with a polystyrene derived amino resin to yield the polymer-bound *Michael*-addition product which could be filtered off. It is interesting to note that the allergenic α-methylene lactones could be regenerated by sequential reaction with methyl iodide and NaHCO$_3$ liberating *N,N*-dimethylaminomethyl polystyrene. Following the same idea, notoriously unpleasantly smelling thiols can be captured via disulfide bond formation with a polymer-bound thiol. Further examples will be shown in *Chapter 1.6*.

Support for unstable and explosive reagents

It has been stated numerous times that polymer-bound reactive intermediates are more stable than the corresponding solubilised counterparts. The stabilisation forces exerted by the polymeric matrix are most probably due to hydrophobic clustering. Polymer-bound peracids are believed to be more stable and not explosive.[2,5] The hydrophobic nature of polystyrene resins in combination with the high dilution effect has been used successfully several times to accelerate intramolecular cyclisations by grafting molecules onto a solid support as shown in *Chapter 1.6*.[21,22]

Immobilisation of synthetic receptor molecules

Polymer supports have been used in order to covalently graft enzymes and receptors in order to perform biological tests (ELISA assays). Recently, polystyrene derived resins have been used to covalently link cationic and anionic[23] synthetic receptor molecules. A particularly nice example of a polystyrene-grafted phosphate binding synthetic receptor is shown in *Figure 1.3.4*.

Figure 1.3.4

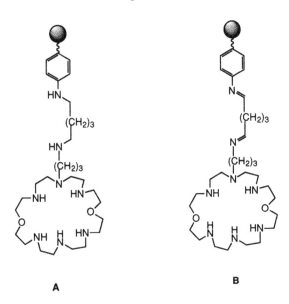

Polymer-supported macrocycles **A** and **B** are able to take up nucleotide phosphates such as adenosine di- and triphosphate (ADP and ATP) at pH 4

Combinatorial and parallel synthesis of small-molecular-weight compound libraries

Among the many applications of solid supports that have emerged recently, the synthesis of compound libraries for biological screening purposes by combinatorial and parallel techniques has certainly created the largest interest. The possibility of creating families of hundreds and thousands of interesting molecules by combinatorial and parallel organic synthesis has attracted with equal intensity scientists from academic and industrial organisations, especially those institutions dealing with biological screening such as pharmaceutical and agrochemical companies.

A library of compounds is defined as an ensemble or a family of compounds that have originated from the same assembly strategy and building blocks. Thus, the members of a library exhibit a common core structure with differing appended substituents. As shown in *Chapter 1.8* the library can consist of single compounds or mixtures depending on whether a parallel or "split-mixed" synthesis strategy was used. The idea of generating ensembles or families of molecules by synthesising all possible combinations of building blocks of type **A**, **B** and **C** in parallel or combinatorial fashion was certainly attractive to many scientists. This account focuses on the solid-supported synthesis of small-molecular-weight compound libraries and we will try to explain the chemical strategies including solid supports, linkers, reactions and reactive building blocks that are especially useful for combinatorial and parallel synthesis. In *Figure 1.3.5* we consider for the sake of simplicity the different combinations of building blocks A^{1-3}, B^{1-3} and C^{1-3} that are possible.

As can be schematically seen in *Figure 1.3.5* the number of resulting products depends highly upon the nature of the reactive building blocks and the assembly strategy. In chart a) where $A=B=C$ ($=A^{1-3}$, *e.g.* amino acids), 27 linear tripeptides with the same basic core structure are the result of all permutations of three amino acid building blocks. In chart b) where building blocks **A**, **B** and **C** are of different nature but can be assembled only in one linear fashion again 27 products with the same basic core structure can be obtained. Conversely, in case c) where again **A**, **B** and **C** are different but can be assembled in all six possible linear ways 162 products with six different basic core structures can be obtained. In chart c) where it is assumed that **A**, **B** and **C** are assembled in a cyclic arrangement with only one directional assembly possible, 27 cyclic products can be obtained which are cyclic analogues of the products obtained in chart b). For the case where all cyclic directed permutations are possible we can obtain 54 cyclic products with two different core structures. This in fact only one third of the linear analogues described in chart c).

This simple sketch shows that the type and the permutability of building blocks **A**, **B** and **C** and the assembly strategy play an equally important role in determining the number and the nature of the compounds that can be obtained. It is striking to note that the permutability of building blocks **A**, **B** and **C** (compare *Figure 1.3.5* charts b) and c)) not only increases the number of final compounds from 27 to 162 but maybe even more importantly increases the number of core

15

structures from one to six. While introducing cyclic core structures the number of products diminishes from 162 to 54 (compare *Figure 1.3.5* charts d) and e)). This reduction in numbers is compensated by the fact that the cyclic compounds are more rigid and the diversity is spatially more condensed.

Figure 1.3.5

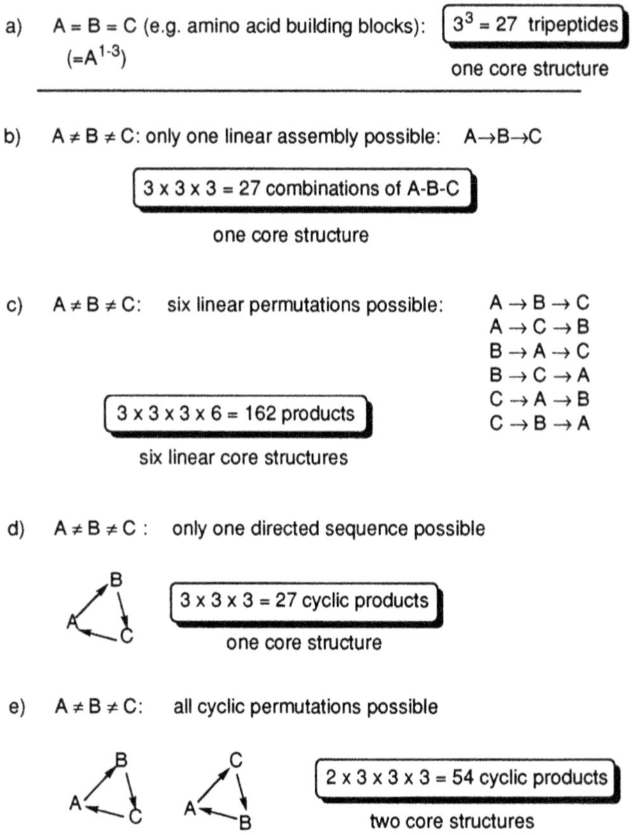

a) A = B = C (e.g. amino acid building blocks): | 3^3 = 27 tripeptides |
    $(=A^{1-3})$
                                                    one core structure

_____

b) A ≠ B ≠ C: only one linear assembly possible: A→B→C

        | 3 x 3 x 3 = 27 combinations of A-B-C |

                one core structure

c) A ≠ B ≠ C: six linear permutations possible: A → B → C
                                                     A → C → B
                                                     B → A → C
        | 3 x 3 x 3 x 6 = 162 products | B → C → A
                                                     C → A → B
                                                     C → B → A
            six linear core structures

d) A ≠ B ≠ C :   only one directed sequence possible

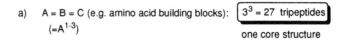

        | 3 x 3 x 3 = 27 cyclic products |

                one core structure

e) A ≠ B ≠ C: all cyclic permutations possible

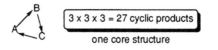

        | 2 x 3 x 3 x 3 = 54 cyclic products |

                two core structures

In *Chapter 1.8* we will present extensively the synthetic strategies used in parallel and combinatorial chemistry and in *Chapter 1.9* we will present a classification of the reactive building blocks of type **A**, **B** and **C** according to their reactivity. In addition to discussing the pivotal role of reactive building blocks and their permutability ("combinatorics of reactive building blocks") we will focus on the important role of solid-phase chemistry for the synthesis of small-molecular-weight compound libraries including linker strategies, synthesis and resin cleavages.

16

1.4. The role of solid-supported chemistry in modern drug discovery

1.4.1. The four key elements of lead finding

The smooth interaction of four key elements has been determined as being crucial for a successful and fast drug development process as schematically shown in *Figure 1.4.1*.

Figure 1.4.1

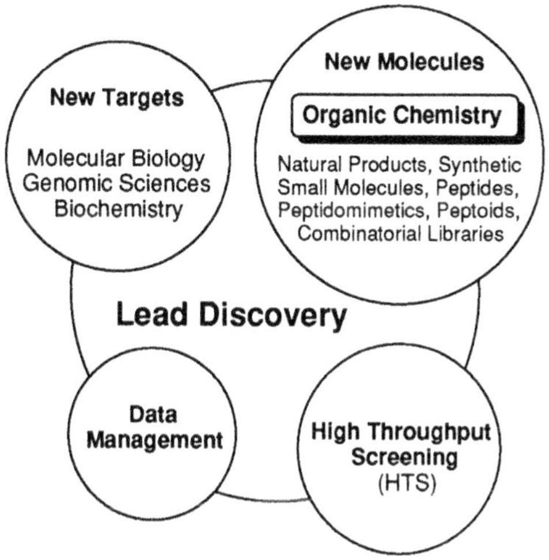

New targets: -Due to the enormous progress made in genomic sciences, molecular biology and biochemistry, an ever-growing number of biologically important target proteins (*e.g.* receptors, enzymes, transcription factors, modulators, chaperones) have become available in pure form allowing to study their structure, function and biological role in living systems. Several crystal structures of pharmacologically relevant proteins (*e.g.* platelet derived growth factor (PDGF),[24,25] nerve growth factor (NGF),[26] HIV-proteinase,[27] collagenase,[28] insulin receptor kinase domain,[29,30] calcineurin/FK BP12-FK 506 complex[31]) have initiated several successful drug development programs. In addition, the rapid elucidation of the entire human genome is expected to further amplify the number of novel biological targets that will become available to drug discovery.

High Throughput Screening (HTS): -The discovery of novel biological targets was paralleled by the development of novel assays and high throughput screening (HTS) technologies including miniaturised screening formats, allowing to test thousands of individual compounds per day.

New molecules: -Since a successful clinical compound originates in average from the evaluation of a pool of 10'000 related analogues, the appearance of novel targets and high throughput screens called for novel sources of structurally and chemically diverse compound collections. Hence, chemists were challenged to develop high throughput synthesis techniques in order to satisfy the increasing demand for screening compounds and allow to maintain a reasonable rate for the detection of valuable hits and leads for subsequent lead optimisation.

Data management: - Integrated data management concepts had to be developed in order to handle the enormous wealth of data arising from chemical and biological data bases.

While up to roughly 1980, drug dicovery was driven by random screening and the chemical and biological intuition of the scientists, the last fifteen years of drug discovery have been heavily influenced by the structural knowledge of the target protein and molecular modelling techniques which allowed in some cases to produce tailor-made ligands.[32,33] This so-called "rational drug design approach" was quite successful in finding potent ligands for well defined enzyme and receptor cavities.[32] Large surface spanning ligand-receptor interactions, however, have been a challenge for drug discovery programs[34] and it has been notoriously difficult to find small molecules as lead compounds.[25] Amino acid derived peptide mimetics,[35,36] corresponding to exposed protein epitopes such as β-turns, loops, 3_{10}- and α-helical structures, which can serve as valuable tools to probe potential ligand-receptor or ligand-enzyme interaction sites have found widespread applications in these structurally more demanding cases. Hence, integrated approaches combining structural knowledge from conformationally constrained small peptides and combinatorial and parallel synthesis of small molecules seem particularly well suited for this purpose.[37] This integrated approach combining rational and random elements is shown in *Figure 1.4.2*.

Figure 1.4.2

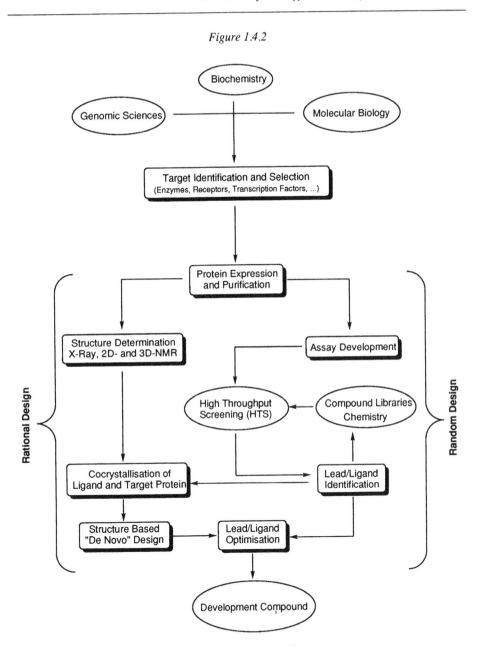

Once a target protein has been identified and selected, it is expressed, amplified and purified using modern biochemical techniques. As soon as the protein is available in pure form and homogeneous form, and in sufficient quantities, the phases of structure determination and assay development usually start simultaneously. Once a sensitive assay has been developed, it is transferred into a format suitable for HTS. Screening of compounds collections as single compounds or as mixtures can then be initiated. Depending on the performance of the test, the compound collections will be screened in a completely random fashion, or, conversely only a subset will be selected based on 2D- and 3D-clustering techniques[13] and screened. After an active ligand molecule has been identified it can serve as a starting point for a lead optimisation program. Alternatively, in the case where crystals suitable for X-ray crystallography are available, the ligand molecule can be co-crystallised and the resulting structural information used for structure-based "de novo" design to find more potent ligands. In the ideal case the rational and random approaches converge into lead compounds showing common structural features which will be ultimately transferable into a development compound.

1.4.2. Sources for novel lead compounds

The different sources for novel lead compounds are schematically represented in *Figure 1.4.3*.

Natural products:

- structurally highly diverse
- molecular weight ~400-1000
- often biologically active ("biological activity maximised through evolution")
- optimisation process is time-consuming
- high manufacturing costs

Natural products isolated from plants, microbes, and animals of terrestrial and marine origin have a long-lasting tradition in medicine.[38] Nature's architecture offers a virtually unlimited plethora of diverse chemical scaffolds going beyond human imagination and hence natural products represent the structurally and chemically most diverse source for finding new lead compounds. Taxol,[7] monensin,[39] avermectin,[40] mitomycin[41-43] and FK506[44] are only a few examples of natural products that served as starting points for intense optimisation programs. Although this process of lead optimisation is usually very time-consuming and manufacturing costs of the final drugs are high, it is expected that natural products will rather gain importance in lead finding.

Figure 1.4.3

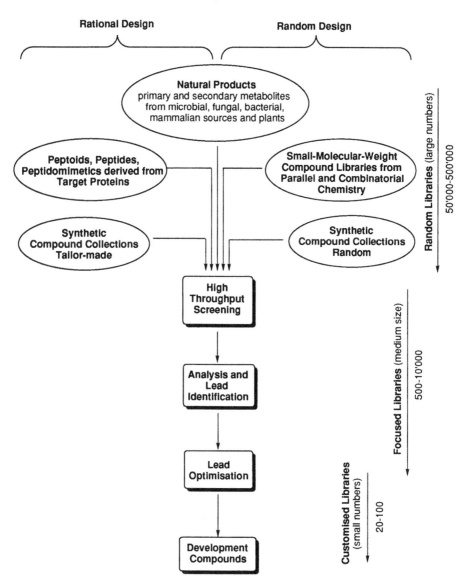

Synthetic compounds collections:

- molecular weight ~300-550
- "drug-like"
- structural diversity is limited
- optimisation process straightforward
- low manufacturing costs

A highly useful source for finding novel lead compounds constitute the compound collections of synthetic small-molecular-weight compounds that have been accumulated over the past decades in industrial companies and academic institutions. Since these compound collections are usually derived from a limited amount of so-called "priviledged" core structures such as benzodiazepines, dihydropyrimidines, phenothiazines and others, they exhibit a more limited chemical and structural diversity than natural products. Due to their ususally simpler structure a given hit or lead can be subsequently optimised in a quite efficient manner resulting in low manufacturing and development costs.

Peptoids, peptides and peptidomimetics derived from a target protein:

- fast assembly through linear assembly of similar building blocks
- limited, "rather flexible" diversity
- molecular weight > 600
- usually not "drug-like"
- optimisation process often long and tedious

As indicated previously, many pharmacologically interesting target receptors exhibit large surface spanning areas where it has been notoriously difficult to find small molecules as inhibitors. For this purpose peptoids, peptides and especially peptidomimetics derived from structural information of the targets can be valuable tools for lead finding (cf. *Chapter 2*).

Small-molecular-weight compound libraries from combinatorial and parallel chemistry:

- fast and convergent assembly using reactive building blocks
- limited chemical and structural diversity
- "drug-like"
- molecular weight < 600
- fast optimisation phase
- low manufactoring costs

These libraries are excellent tools to complement the compound collections already available in terms of numbers and diversity. Hits and leads that have emerged from such a library are especially valuable since the optimisation process is speeded up by the usually fast and convergent assembly strategies that were used to synthesise the compounds. The structural complexity of the resulting compounds is only limited by the *chemical strategies* that are possible, the *reactive building blocks* that can be employed and by the number of available *building blocks* that can be engaged for the final derivatisation of a given library.

As shown in *Figure 1.4.3* the compound collections will be rather large and of random type in the early phases of hit and lead identification. Based on the first lead structures emerging from HTS, more tailor-made compound collections will be synthesised or selected from existing compound libraries. Rational design elements at that stage will help to reduce the size of the compound collections that will be synthesised or selected for further screening. Lead compounds generated by combinatorial and parallel chemistry are of particular value at that stage of development as the optimisation process is accelerated by the modular synthetic approaches used by these techniques. In the lead optimisation phase, focused libraries of reduced size, containing all the structural elements of a given lead compound can have a significant input on a faster development of clinical candidates. In addition, several structural analogues can be followed simultaneously which should result in faster development times and thus effectively help to reduce the overall development costs.

1.5. Solid supports

As indicated in *Chapter 1.3*, polymeric supports have been mainly used to immobilise substrates, reagents and catalysts. Insoluble polymeric matrices allow to fully exploite the ease of solid-liquid separation by filtration. In addition, certain soluble polymers were examined which can be easily precipitated. The polymers which were used for the synthesis of small organic molecules can be divided in the following classes:

-Crosslinked organic polymers:	-insoluble in organic solvents
	-microporous polymers, gels
	(*e.g.* polystyrene, poly-(4-vinylpyridine))
	-macroporous polymers
	(*e.g.* polystyrene derived beads)
-Linear organic polymers:	-usually soluble in organic solvents and can be
	precipitated in certain cases (*e.g.* polyethylene glycol
	(PEG), polystyrene)
-Dendrimers:	-solubility depends on size and shape
-Inorganic supports:	*-porous glass*
	-SiO$_2$
	-Al$_2$O$_3$
	-clays
	-graphite

1.5.1. Immobilisation of substrates

The synthesis of oligomeric molecules such as polypetides, oligonucleotides and oligosaccharides and small molecules have generally been performed either by using *soluble linear, usually PEG-*[45,46] *or polystyrene-derived supports,*[47-50] whereby the isolation and purification of the polymer-bound substrates occurs by precipitation,[45,46,48,49] ultrafiltration,[51] dialysis or gelfiltration.[47] This technology takes formally advantage of combining benefits of well established solution chemistry with simple purification procedures. These purification procedures, however, are often tedious to perform and not generally applicable for automated synthesis.

The vast majority of applications in the field of solid-supported synthesis, however, use *insoluble polymeric matrices*. The advantages of using insoluble polymers can be summarised as follows:

- *purification can be performed by simple washing and filtration cycles*
- *allows the use of large excesses of reagents (typically 3-5 equivalents)*
- *high concentrations of reagents allow slow reactions to be driven to completion*
- *reaction-, washing- and filtration steps can be easily automated*

Among the polymeric supports that have been employed in solid phase synthesis we mostly find polystyrene derived resins crosslinked with varying amounts of divinylbenzene (DVB, typically 1-5%). These resins consist mainly of spherical beads of various sizes. Depending on the polymerisation protocol these resins can be of the *micro-* or *macroporous type*. The resins having found wide applications in organic synthesis are listed below:

- **Micro- and macroporous polystyrene / DVB-crosslinked resin beads** (*Merrifield*-type resins). These resins are fairly cheap, can be easily functionalised with high loading (typically 1-3.5 mmol/g) and are extensively used in solid-supported polypetide- as well as in small molecule synthesis.

- **Polystyrene / DVB-crosslinked polymeric matrix coated with polyethylene-glycol spacers** of various sizes (*e.g.* TentaGel®[52]). These resins are less hydrophobic and show better swelling properties in aqueous solutions. They are usually mechanically less stable, more expensive and show a lower degree of loading than the *Merrifield*-type resins. They find extensive use in solid-supported polypetide and small molecule synthesis.

- **Tea Bags:**[9] polystyrene/DVB-derived resin beads sealed in a porous polypropylene bag developed originally by *Houghten et al*. Each bag contains one type of labeled polymer-bound molecule. Different tea bags can be mixed and are allowed to react with the same building block, can be washed in parallel and then be separated. Since the reactions take place in separate reaction compartments there will be only one type of well specified molecule in each bag. This protocol also allows the synthesis of larger quantities of material. Similar to the "split-mixed" technology the number of coupling steps is reduced while all the benefits of solid-supported chemistry are maintained. This methodology has been mainly used for the combinatorial and parallel synthesis of polypeptide libraries.

- **Diversomer®-technology:**[53-56] Parallel solid-phase synthesis of individual compounds ('diversomers') is achieved in spatially separated compartments using an apparatus allowing the simultaneous handling of an array of resin-bound intermediates. During synthesis the resin is located in glass tubes equipped with frittes at their lower ends, dipping in the wells of a reservoir block containing solutions of reagents.

- **Pins**:[57] polystyrene- or polymethacrylamide-dimethylacrylamide-(copolymerised) derived matrices grafted on polyethylen crowns which are attached on top of pins. The pins usually are assembled in the typical 96-well plate format. This technology originally developed by *Geysen et al.*[8,57,58] for the parallel synthesis of polypetide libraries has found wide applications also for the synthesis of small-molecular-weight compound collections [59-62]

- Other supports such as **foils**[63] and **cellulose disks**[64] should also be mentioned.

1.5.2. Immobilisation of reagents and catalysts

Polymer-bound reagents and catalysts which can be used in large excesses are ideally suited to achieve the complete conversions of reactions in solution. In addition, this approach offers the attractive feature that well established solution protocols can be used while maintaining the easy purification procedures associated with insoluble polymers. This should ultimately result in shorter development times. Many polymer-bound reagents and catalysts, although commercially available (cf. *Chapter 1.6*), show several drawbacks such as low loadings and varying qualities which result in unreproducible yields. The relatively high prices might account for the fact that these polymer-bound reagents and catalysts have not yet found wide applications. As mentioned in *Chapter 1.3*, many research groups have focused on the development of *cheap, recyclable* and *highly loaded* reagents and catalysts suitable for the synthesis of small-molecular-weight compound libraries in solution. The following polymeric matrices have been extensively used to graft reagents and catalysts:

- **Inorganic polymers**:[65,66] SiO_2, silicates, Al_2O_3, charcoal, molecular sieves, porous glass.

- **Organic polymers**:[2,66,67] polystyrene crosslinked with DVB and poly-(4-vinylpyridine) crosslinked with DVB.

1.5.3. Crosslinked polystyrene-derived matrices

Crosslinked polystyrene beads are obtained by free radical initiated copolymerisation of styrene and varying amounts of divinylbenzene (DVB). A suspension of water, styrene and DVB is mixed in the presence of a free radical catalyst such as dibenzoyl peroxide or azoisobutyronitril (AIBN) and heated to the reaction temperature suitable for polymerisation. The aqueous phase initiates a fine dispersion of the mixture of monomers and serves as medium to control the reaction temperature but does not participate in the reaction. Coalescence of monomer droplets leads to association and the formation of droplet conglomerates and larger monomer units by means of fusion of monomers. Efficient mechanical stirring of the reaction mixture results in the formation of a dynamic equilibrium but is not sufficient to suppress the aggregation of viscous droplets during the polymerisation process. Suspension stabilisers such as polyvinylalcohol or derivatives of cellulose are added in order to avoid these aggregations and to ensure a successful and reproducible polymerisation. These aggregation phenomena can also be suppressed by addition of salts to the aqueous phase which results in a change of the surface interface forces. Thus the polymerisation proceeds in each droplet and initiates the formation of a polymer bead whereby the bead size can be controlled by the stirring speed, the relative amounts of aqueous and monomer phases, the amount and nature of the suspension stabilisers (dispergator) and by the reaction temperature. The size of the beads can also be controlled by means of sieving.

The swelling properties of the resin beads depend in essence from the degree of crosslinking and thus from the relative amount of DVB to the core monomer.[68-70] Resins containing a low degree of crosslinking (typically 1-2% DVB) show a more pronounced swelling capacity than those with a high content of DVB[71] (> 5%). The size and the shape of the resin particles can be directed to a certain extent by suspension polymerisation.

Microporous resins.- These resins, also called *gels* or *temporary porous resins*, show a low degree of crosslinking[72] (DVB content of 1-2%) and high swelling properties (see *Table 1.5.3.1*). The swelling properties depend also on the nature of the attached functional groups, *e.g.* chloromethyl polystyrene (crosslinked with 1% DVB) swells four times as much in chloroform as 4-nitro-chloromethyl polystyrene whereas in DMF both resins show the same amount of swelling.[73] The solvent molecules form pores which collapse when the resins are dried.

Table 1.5.3.1. Swelling properties of polymers in different solvents

solvent	swelling capacity[a]	
	crosslinked PS (1% DVB)[b]	Merrifield resin (2% DVB)[c]
MeOH	0.95	
EtOH	1.05	1.0
AcOH		1.0
MeCN	2.0	
pyridine		3.0
DMF	3.5	2.0
THF	5.5	
dioxane	4.9	2.5
Et$_2$O	2.6	
CH$_2$Cl$_2$	5.2	
toluene	5.3	2.8

a) swelling capacity = volume of the swollen resin / original volume

b) values according to Ref.[74]

c) values according to Ref.[75]

The synthesised resin beads show inhomogeneous crosslinking density which can be explained by the fact that DVB is more reactive towards polymerisation than styrene.[72]

Macroporous resins.- These resins, also called *macroreticular* or *permanently porous resins*, can be obtained when the polymerisation process is conducted in the presence of an inert solvent[72,76] ("porogen"). Whether a macroporous or microporous resin is obtained depends largely on the ratio of the content of solubilising and crosslinking agents. Absence or low contents of porogen and low degree of DVB usually yield microporous resins whereas high content of porogen (*e.g.* toluene) and high degree of DVB rather lead to macroporous polymers. Alternatively, macroporous resins are also formed by using a high content of DVB in the presence of a non-solubilising agent such as 2-ethyl capronic acid.[72] Agglomerates of small and smallest resin particles (termed *microspheres* or *nodules*) which depending on their DVB content can significantly swell, form macroreticular beads which keep their porous character in the dried state and show a large pore surface (inner surface).[76,77] The pore surface and the pore size depend again on the reaction conditions and on the nature of the porogen. The functional groups in a microporous resin are homogeneously distributed in the whole bead. Their somewhat restricted solvent accessibility can be modulated by a solvent showing a high degree of swelling. Conversely, the functional groups of macroreticular resins are largely positioned at the outer surface of the bead. Thus, the reactivity and solvent accessibility of the functional groups does not depend on the nature of the solvent and the swelling properties.[71] In this respect macroreticular resins can be compared with inorganic matrices.[66]

1.5.4. Functionalised polystyrene resins

The widespread applications of polystyrene derived resins is due to the fact that styrene consists of a chemically inert alkyl backbone carrying chemically reactive aryl side chains that can be easily modified. As discussed earlier, a wide range of different types of polystyrene resins exhibiting various different physical properties can be easily generated by modification of the crosslinking degree. In addition, many styrene derived monomers are commercially available and fairly cheap.

Polystyrene is chemically stable to many reaction conditions while the benzene moiety, however, can be funtionalised in many ways by electrophilic aromatic substitutions or lithiations. As shown in *Scheme 1.5.4.1* there are principally two different ways to obtain functionalised polystyrene/DVB-copolymers.

Scheme 1.5.4.1

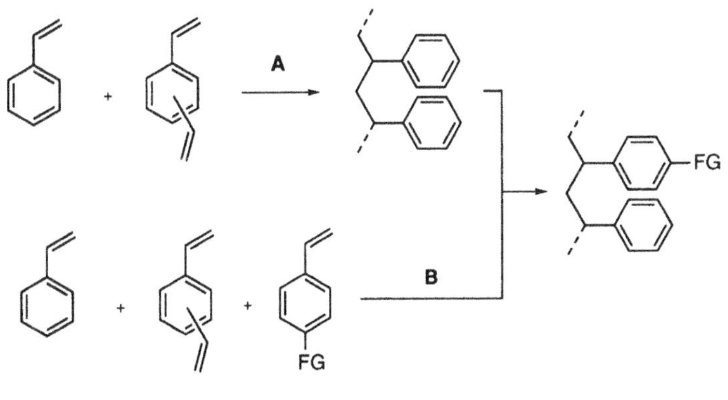

Approach **A** consists of building first the polystyrene/DVB-copolymer followed by subsequent chemical introduction of the functional group, whereas in approach **B** the functional group is already introduced in a modified styrene-derived monomer. The chemical modification of crosslinked polystyrene (approach **A**) offers the advantage that only the accessible aromatic rings and positions on the aromatic rings are functionalised. The drawback of this approach is that the reactions on the polymer are usually slower and difficult to monitor which can significantly affect the yields due to side reactions and thus the degree of loading. Approach **B** assures the exact positioning of the functional groups and usually a high degree of loading but necessitates the synthesis of the appropriate monomers. Since the first functionalisation determines the loading of the resin it is important that these reactions proceed in high yield and with high fidelity.

1.5.4.1. Chloromethylated polystyrenes

Chloromethyl polystyrene was introduced by *Merrifield*[3] into peptide synthesis. It was used, however, first as an intermediate for the synthesis of anion exchange resins.[78] The synthesis of chloromethylated polystyrenes is best achieved by *Friedel-Crafts* alkylation of polystyrene with methoxymethylene chloride in the presence of a *Lewis* acid catalyst such as $SnCl_4$,[3,78] as shown in *Scheme 1.5.4.2*.

Scheme 1.5.4.2

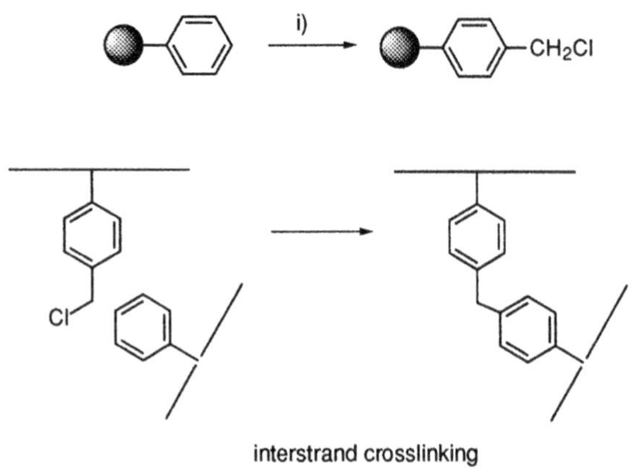

interstrand crosslinking

Reagents and conditions: i) $CICH_2OCH_3$, $SnCl_4$, reflux, 1h ; $CICH_2OCH_3$, $CHCl_3$, $SnCl_4$,0°C, 30 min; $CICH_2OCH_3$, $ZnCl_2$, THF, CH_2Cl_2; $CICH_2OCH_3$, BF_3OEt_2, hexane, 35°C, 3h.

Investigations about the swelling properties of chloromethylated polystyrenes showed that up to a chlorine content of 19% (about 0.75 chlorine atoms per aromatic ring*) there was no significant drop in swelling. An increase of loading, however, led to a markedly less swelling resin indicating a higher degree of crosslinking by 'interstrand' couplings (*Scheme 1.5.4.2*). It could be shown by NMR-measurements that the ratio of *para-* to *ortho*-chloromethylation of soluble polystyrene was 95:5.[80] The same ratio can be expected for resins with a low degree of crosslinking. The use of $ZnCl_2$ instead of $SnCl_4$ for the chloromethylation with methoxymethylene chloride lead to a resin with a low degree of functionalisation.[81] *Sparrow*[82] was able to show that by using BF_3OEt_2 as catalyst and ethoxymethylene chloride as alkylating agent the degree of loading could be controlled depending on the amount of catalyst.

Chloromethylated resins were also synthesised by copolymerisation of styrene, divinylbenzene and chloromethylstyrene [17,83,84] (usually as a 3:2-mixture of *meta* and *para* isomers). *Arshadi et al.*[17] noticed, however, that this approach can lead to substantial losses of chlorine content.

---

* For calculations of loadings and interconversion of different indications see Ref.[79]

Scheme 1.5.4.3

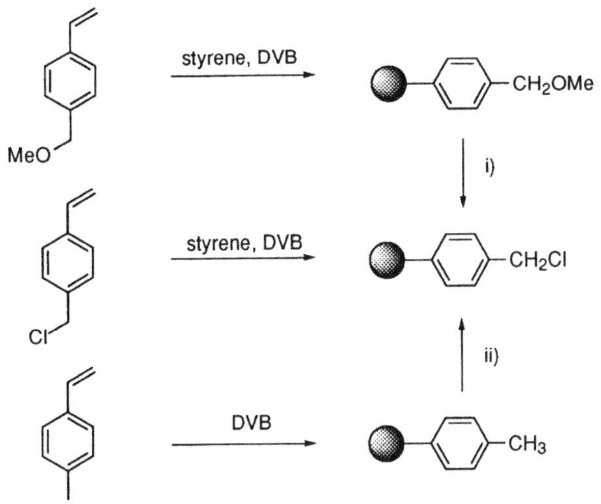

Reagents and conditions: i) BCl$_3$, CCl$_4$, 0°C, 2h; ii) NaOCl, CHCl$_3$
(or ClCH$_2$CH$_2$Cl), BnNEt$_3$$^+Cl^-$; SO$_2Cl_2$, benzene, AIBN, 60°C.

Merrifield resins with a chlorine content up to 22% (corresponding to 1.0 chlorine per aromatic ring) could be obtained by copolymerisation of 4-methoxymethylstyrene with styrene and DVB and subsequent conversion using BCl$_3$ in CCl$_4$[17] as depicted in *Scheme 1.5.4.3*. Chlorination of 4-methylpolystyrene (obtained by copolymerisation of styrene, DVB and 4-methylstyrene) with NaOCl in the presence of a phase transfer catalyst[85] proved a valuable method for the preparation of *microporous* (1% DVB-crosslinking) and *macroporous* (20% DVB-crosslinking) chloromethylpolystyrene. This method was compared with a free radical chlorination using SO$_2$Cl$_2$ and AIBN as shown in *Scheme 1.5.4.3*. Using NMR-techniques it could be shown again that for a loading degree higher than 2.5 mmol/g also dichloromethyl groups were present.[85] Since the elemental combustion analysis gave higher chlorine contents than determined by NMR it was concluded that in both methods also partial chlorination of the backbone occurred.

Guyot[72] investigated in some detail the advantages of the copolymerisation procedures and concluded that a homogenous distribution of the functional groups was only achieved when the monomers showed a comparable reactivity. Some functional groups are usually buried in the interior of the polymer and will not be accessible by the solvents and the reactants. This drawback can be partially overcome by delayed addition of the functionalised monomer.[72] Hence, the maximum loading can be controlled by the amount of the functionalised monomer that is employed in a given polymerisation.

1.5.4.2. *Aminomethylated polystyrene resins*

Aminomethylated, crosslinked polystyrenes constitute very valuable and versatile resins for a variety of applications. The amino group can be easily acylated for the introduction of spacer and linker molecules [86-90] and it was also used for the synthesis of polymer-bound carbodiimides.[91-93] In *Scheme 1.5.4.4*, some of the most common synthetic accesses to aminomethylated polystyrenes are summarised.

Scheme 1.5.4.4

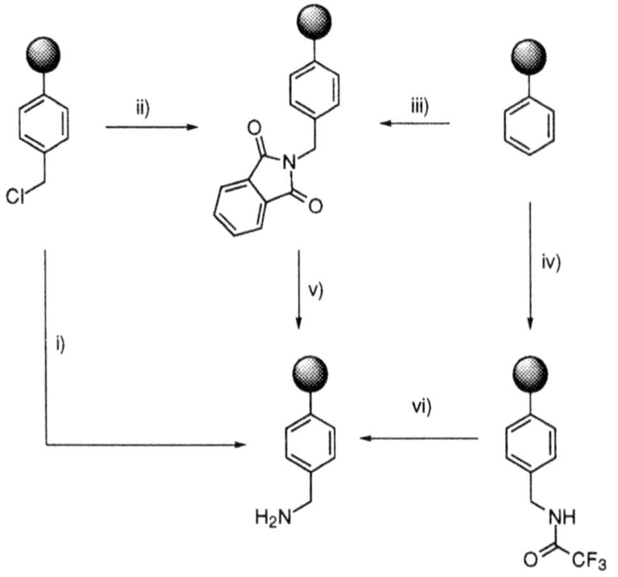

Reagents and conditions: i) NH$_3$, CH$_2$Cl$_2$; ii) potassium phthalimide, DMF; iii) *N*-(hydroxymethyl)- or *N*-(chloromethyl) phthalimide, CH$_2$Cl$_2$, TFA, catalyst (HF, CF$_3$SO$_3$H, SnCl$_4$, FeCl$_3$); iv) *N*-(hydroxymethyl)-trifluoracetamide; v) NH$_2$NH$_2$ H$_2$O, EtOH; vi) KOH, EtOH

These methods comprise the conversion of chloromethylated polystyrene with non-aqueous ammonia described by *Rich et al.*[73] (*Scheme 1.5.4.4*, i)) which is quite slow. Excellent conversions are obtained by using potassium phthalimide[88,89,91-93] followed by treatment with hydrazine (*Scheme 1.5.4.4*, ii) and v)). The same phthalimido intermediate can be obtained by a *Tscherniak-Einhorn* reaction[94] using *N*-(hydroxymethyl)- or *N*-(chloromethyl)phthalimide in the presence of an acid catalyst such as HF, CF$_3$SO$_3$H or SnCl$_4$.[86,87] The degree of functionalisation (typically 0.05 - 3.6 mmol N/g) can be controlled by the amount of reagent and the concentration of the catalyst as recently described by *Zikos et al.*[95] by using *N*-(chloromethyl) phthalimide in CH$_2$Cl$_2$ and FeCl$_3$ as catalyst. Loadings as high as 7.3 mmol N/g can be achieved but those resins show restricted swelling properties. Finally, aminomethylated polystyrene resins have also been obtained via reaction with *N*-(hydroxymethyl)trifluoracetamide and subsequent saponification with KOH in EtOH.[87]

1.5.4.3. *Various functionalised polystyrene resins*

Besides aminomethyl polystyrene, many other functionalised polystyrene resins were prepared from chloromethyl resin by nucleophilic displacement (*Scheme 1.5.4.5*)

Scheme 1.5.4.5

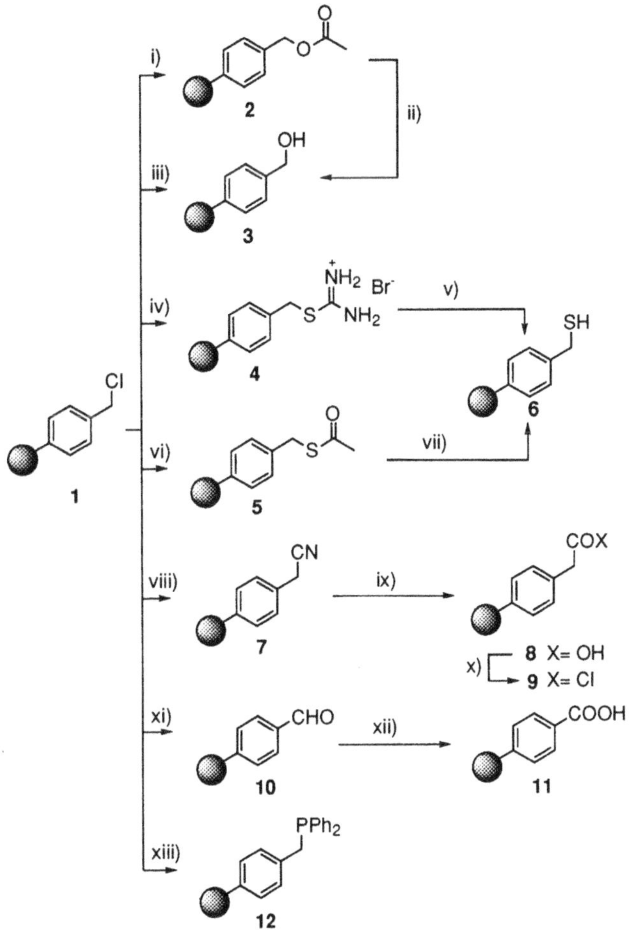

Reagents and conditions: i) **1** (1% DVB, 0.73 mmol Cl/g), KOAc, DMA, 85°C, 24h[96] or **1** (1% DVB, 3.9 mmol Cl/g), KOAc, DMA, 85°C, 24h;[21] ii) LiAlH$_4$, Et$_2$O, r.t., 4h or NH$_2$NH$_2$, DMF, r.t., 76h[96] or NH$_2$NH$_2$, DMF, r.t., 20h;[21] iii) **1** (2% DVB, 2.8 mmol Cl/g), KOH, 1-pentanol, reflux, 24h;[97] iv) **1** (1% DVB, 3.5 mmol Cl/g), thiourea, dioxane/EtOH (4:1), 85°C, 15h;[98] v) KOH, EtOH/dioxane; vi) **1** (1% DVB, 1.15 mmol Cl/g), KSAc, DMF;[99] vii) LiBH$_4$, Et$_2$O, r.t.;[99] viii) **1** (2% DVB, 1.36 mmol Cl/g), NaCN, DMF, H$_2$O, 120°C, 20h;[75] ix) H$_2$SO$_4$, AcOH, H$_2$O, 120°C, 10h;[75] x) SOCl$_2$, toluene, 110°C, 24h;[75] xi) **1** (2% DVB, 1.05 mmol/g), DMSO, NaHCO$_3$, 155°C, 6h;[100,101] xii) 1,2-dimethoxyethane, *m*-CPBA, 55°C, 19h;[100] xiii) **1** (2% DVB, 6.7 mmol Cl/g), ClPPh$_2$, Li, THF.[102]

The synthesis of various functionalised polymers containing spacer or linker units usually starts with a *Friedel-Crafts* alkylation of the basic polystyrene resin as depicted in *Scheme 1.5.4.6*. Treatment of polystyrene with propyleneoxide in the presence of SnCl$_4$ yields a resin containing a non-benzylic hydroxy group (*Scheme 1.5.4.6*). Prior to the addition of the reagent the resin is treated with the *Lewis* acid catalyst SnCl$_4$ forming a complex.[103] *Neckers et al.*[104] indicate that such complexes are also observed with other *Lewis* acids such as AlCl$_3$. The alkylation procedure described by *Tomoi et al.*[105] using microporous polystyrene resins with ω-bromoalkenes in the presence of trifluoromethane sulfonic acid presumably proceeds without additional crosslinking reactions as shown in comparison with similar experiments with soluble polymers.

Scheme 1.5.4.6

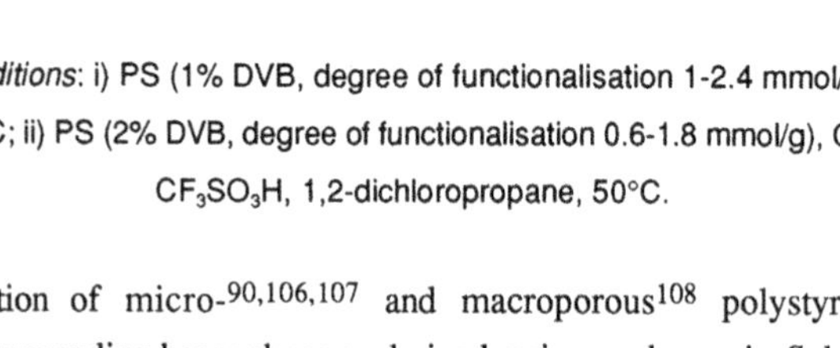

Reagents and conditions: i) PS (1% DVB, degree of functionalisation 1-2.4 mmol/g), propylenoxide, SnCl$_4$, CH$_2$Cl$_2$, 0°C; ii) PS (2% DVB, degree of functionalisation 0.6-1.8 mmol/g), CH$_2$=CH(CH$_2$)$_9$Br, CF$_3$SO$_3$H, 1,2-dichloropropane, 50°C.

Friedel-Crafts acylation of micro-[90,106,107] and macroporous[108] polystyrene resins yielded conveniently the corresponding benzophenone-derived resins as shown in *Scheme 1.5.4.7*. These functionalised resins could be successfully be converted into trityl-,[108] oxim-derived[106] or the photolabile o-nitrobenzhydryl-derived resins[107] which were quite extensively used in peptide and oligonucleotide chemistry. In addition, 2-bromo propionylchloride in the presence of AlCl$_3$[90] or FeCl$_3$[95] was similarly used in acylation reactions with polystyrene resins. The reactions were carried out in solvents like CH$_2$Cl$_2$, 1,2-dichloroethane or nitrobenzene.

Scheme 1.5.4.7

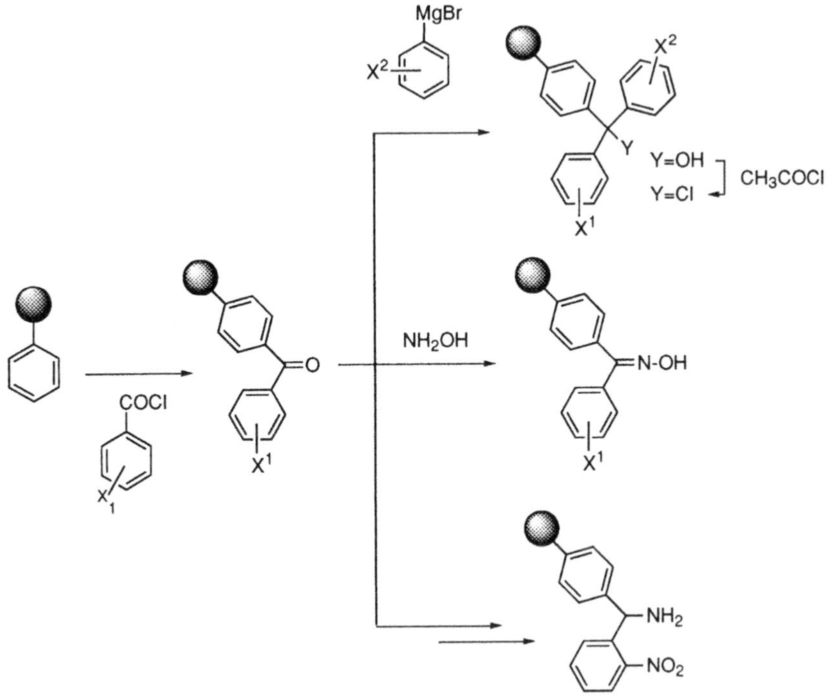

Strongly acidic cation-exchange resins were obtained by sulfonylation of polystyrene using H_2SO_4 or $ClSO_3H^{109}$ as shown in *Scheme 1.5.4.8*.

Scheme 1.5.4.8

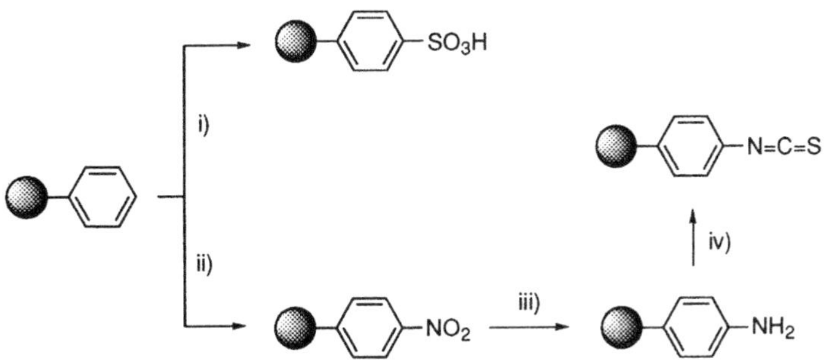

Reagents and conditions: i) H_2SO_4 or $ClSO_3H$; ii) HNO_3, 0°C - 15°C; iii) $SnCl_2$ 2 H_2O, DMF, 100°C, 6 h; iv) a) Et_3N, CS_2, 0°C - r.t., 4h; b) $ClCO_2Et$, 0°C - r.t.

35

Nitration of polystyrene resins followed by reduction with SnCl$_2$ resulted in aniline formation which was further converted into the corresponding isothiocyanate. This polymer was used as an insoluble reagent for *Edman*-degradation.[110]

As shown so far many functionalised resins were obtained by electrophilic aromatic substitution reactions. A very valuable route to other functionalised resins constitute lithiation reactions on polystyrenes as shown by *Fréchet* and *Farrall*[111] and depicted in *Scheme 1.5.4.9*.

Scheme 1.5.4.9

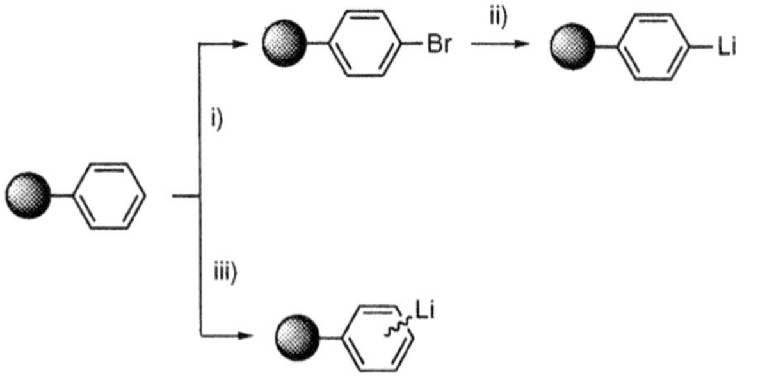

meta and *para* substitution

Reagents and conditions: i) PS (1% DVB), Br$_2$, Tl(OAc)$_3$ or FeCl$_3$, CCl$_4$, Br$_2$, r.t., 1h, reflux, 1.5h;
or Br$_2$, BF$_3$, nitromethane, r.t., 18h; ii) nBuLi, hexane, toluene or benzene, 60°C, 3h;
iii) nBuLi, cyclohexane, TMEDA, 65°C, 4.5h.

Lithiated polystyrene resins can be obtained either via convenient bromine-lithium exchange reaction using nBuLi starting from 4-bromo-substituted polystyrene [102,111-115] or by direct lithiation of polystyrene using nBuLi in cyclohexane in the presence of TMEDA [111,116]. This method, however, yields a mixture of *para*- and *meta* isomers. The bromination of microporous resins in the presence of the *Lewis*-acid catalysts was carried out in the dark whereby the degree of functionalisation could conveniently be controlled by the amount of bromine used in the reaction.[111] Macroreticular resins were brominated using Br$_2$ and FeCl$_3$[71] or stoichiometric amounts of thallium acetate as *Lewis* acid catalysts.[113,114]

The bromine-lithium exchange reaction on macroreticular PS-resins using nBuLi in THF could be driven to completion by repetitive lithiation as described by *Fréchet*.[111] The lithiation of highly loaded microporous resins, however, were successfully carried out in toluene or benzene. The direct lithiation reaction of microporous polystyrene-derived resins (2% DVB) using nBuLi in cyclohexane in the presence of TMEDA is much faster than that of macroreticular resins (20% DVB).[116] It is interesting to note that for this reaction THF and benzene are not the solvents of choice. Using cyclohexane as solvent allows the synthesis of resins with a low or medium degree of loading (up to 2.0 mmol/g). Highly loaded resins have to be synthesised via the bromination-lithiation route. In *Scheme 1.5.4.10* are shown functionalised PS-resins which were derived from lithiated polystyrene.

Scheme 1.5.4.10

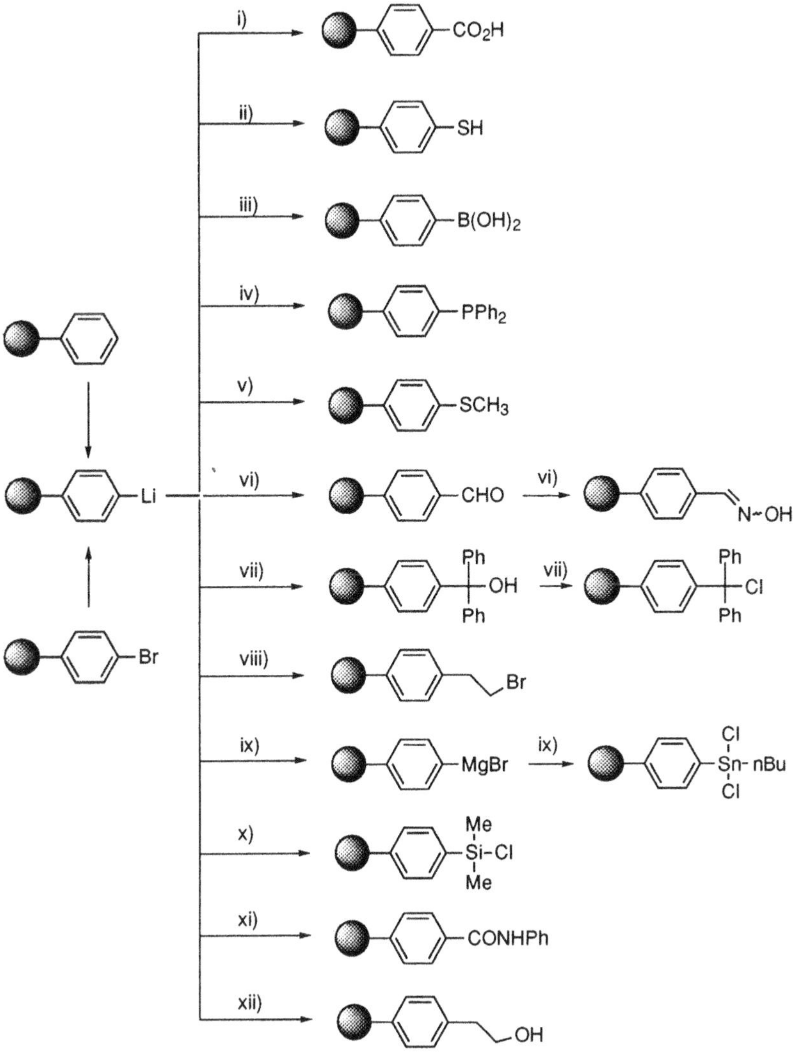

Reagents and conditions: i) CO_2 (dry ice), THF, PS (1% DVB)[111] or CO_2 (gas), THF, PS (2% DVB) or macroreticular PS (Amberlite® AE-305);[116] ii) S_8, THF, 1h then $LiAlH_4$, THF (reduction of S-S bond), PS (1% DVB);[111] iii) $B(OMe)_3$, THF, r.t., 20h, aq. HCl, dioxane, 60°C, 90min., PS (1% DVB);[111] iv) $ClPPh_2$,

THF, r.t., 105min, PS (1% DVB);[111] or ClPPh$_2$, THF, r.t., 4h, PS (2% DVB)[115] or ClPPh$_2$, THF, r.t., 4h, 70°C, 24h, (20% DVB);[115] v) dimethyldisulfide, THF, PS (1% DVB)[111] or dimethyldisulfide, THF, macroreticular resin Amberlite® XE-305);[114] vi) DMF, THF, r.t., 105min. then NH$_2$OH x HCl, pyridine, 90°C, 4h, PS (1% DVB);[111] vii) benzophenone, THF, r.t., 2h, PS (1% DVB)[111] or benzophenone, THF, r.t., 20% AcCl in benzene, reflux, PS (2% DVB);[116] viii) BrCH$_2$CH$_2$Br, benzene, r.t., 150min., PS (1% DVB);[111] ix) MgBr$_2$ x Et$_2$O, THF then nBuSnCl$_3$, r.t., 24h, macroreticular PS (Amberlite® XE-305);[113] x) Cl$_2$SiMe$_2$, benzene, r.t., 45min., PS (1% DVB);[111] xi) PhN=C=O, benzene, PS (1% DVB);[111] xii) ethylene oxide, toluene, PS (2% DVB).[112]

1.5.5. Polyacrylamide resins

The functional groups of microporous resins can only react when the polymers show a high degree of swelling, which depends largely on the nature of the functional group. In the course of a synthesis on a polymer support the swelling properties of the resin can change significantly. In the *Merrifield* peptide synthesis the starting polystyrene derived resin is of highly hydrophobic nature and becomes more and more hydrophilic as the synthesis of the peptide evolves and the chain grows. Due to such phenomena certain peptide sequences are notoriously difficult to be synthesised on a standard *Merrifield* resin. In order to improve the synthesis of the decapeptide sequence 65-74 of the acylcarrier protein *Sheppard et al.*[117] altered the hydrophobic nature of the polystyrene polymer backbone by introducing a polyacrylamide polymer backbone which they felt is quite similar to a peptide.

The first polyacrylamide resins were synthesised by copolymerisation of *N,N*-dimethylacrylamide **1** (basic monomer), ethylenebisacrylamide **2** (crosslinking agent) and *N*-acryloyl-*N'*-(*tert*-butoxycarbonyl-β-alanyl)hexamethylene diamine **3** (functionalised monomer)[118] as shown in *Scheme 1.5.5.1.*

Scheme 1.5.5.1

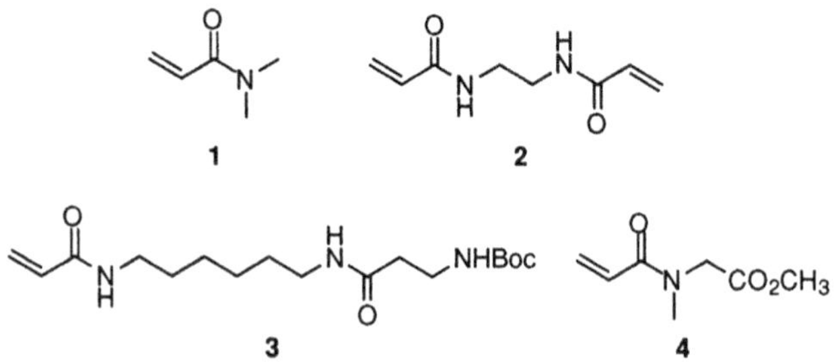

Monomer **3** essentially serves to functionalise the polyacrylamide polymer after liberation of the primary amino group of the β-alanine moiety. The hexamethylenediamine groups serve as spacers in order to spatially separate the reactive groups. Monomer **3** was later on replaced by acryloylsarcosin methylester **4** *(Scheme 1.5.5.1)*,[18] which is chemically closer to the basic monomer, in order to achieve a homogenous distribution of the functional groups.

The persulfate initiated copolymerisation of **1, 2** and **4** (performed in emulsion with 66% aqueous DMF, 1,2-dichloroethane and cellulose acetate/butyrate as emulgator) yields beads with a high degree of swelling (in H_2O, DMF, MeOH, pyridine or CH_2Cl_2 ten times the original volume, in dioxane about five times).[18,117] The aminolysis of the ester groups using ethylendiamine gives access to a resin containing free primary amino groups.

The polymerisation of polyacrylamide gels within the pores of Kieselguhr[119] or polyhipe-structures (PS-DVB-copolymer with 90% pore volume)[120] allows the synthesis of composite resins, which due to their low compressability are well suited for continous flow reactors. The Kieselguhr and polyhipe matrices serve thereby as skeletons whereas the polyacrylamide gels determine the chemical and swelling properties of the reactive matrix.

1.5.6. TentaGel® resins

TentaGel® polymers, originally developed by *Bayer* and *Rapp*,[52,74] consist of a polystyrene matrix covalently coated with polyethylene glycol (PEG) chains and have been developed primarily for solid-phase peptide synthesis similarly to the polyacrylamide resins. TentaGel® resins have become quite fashionable in the field of solid-supported synthesis of small molecules.[12,61] The PEG chains are introduced by anionic polymerisation of ethylene oxide with hydroxylated crosslinked polystyrene beads *(Scheme 1.5.6.1)*.

Scheme 1.5.6.1

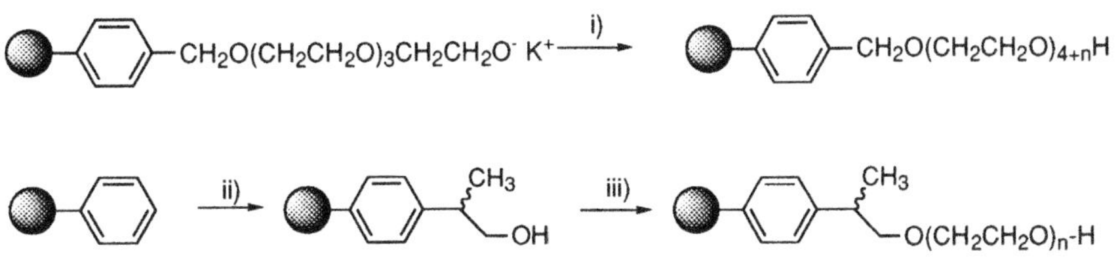

Reagents and conditions: i) ethylene oxide; ii) propylene oxide, $SnCl_4$, CH_2Cl_2;
iii) ethylene oxide, KOH, dioxane, 110°C

An alternative procedure uses the functionalisation of β-hydroxy polystyrenes which avoids the grafting as acid-labile benzylether groups to the polymer backbone.[103]

TentaGels are composed up to 60-80% w/w of PEG units determining largely their physical properties. They show good swelling properties in protic and polar solvents such as water, methanol, CH_2Cl_2, MeCN, THF or DMF whereas in apolar solvents such as ether they hardly swell at all.

The high flexibility of the PEG chains can be observed in NMR studies.[52,74]

The reactive centers are located at the end of the PEG spacers and are therefore easily accessible. Solvatised TentaGel®'s are pressure resistant and can be used in batch-procedures as well as in continuous flow reactors. Loadings of commercially available TentaGel®'s range between 0.15-0.3 mmol/g which is significantly lower than those of PS-resins.

1.5.7. Novel polymeric supports

PEGA is a copolymerisate of bis-(2-acrylamidoprop-1-yl)-PEG1900 (**1**), 2-acrylamidoprop-1-yl[2-aminoprop-1-yl]PEG300 (**2**) and *N,N*-dimethylacrylamide (**3**) as depicted in *Scheme 1.5.7.1*.[121]

Scheme 1.5.7.1

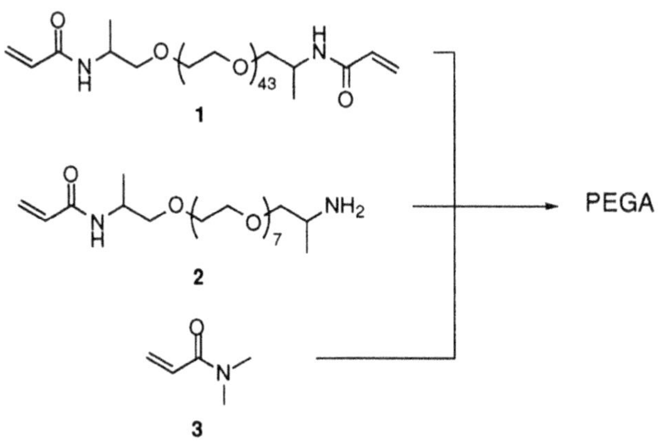

PEGA can be produced by bulk-polymerisation followed by granulation of the polymerisate or conversely by suspension polymerisation resulting in the formation of suitable beads for synthesis. PEGA beads were successfully used in solid phase peptide synthesis of the 65-74 fragment of the acylcarrier protein using the continous flow technique. The amino acid coupling reactions generally proceeded faster than the corresponding couplings on polyacrylamide gels on Kieselguhr.[121] The PEGA beads show excellent swelling properties in non-protic solvents such as CH_2Cl_2 and DMF as well as in protic solvents like alcohols, H_2O and aqueous buffers. PEGA resins are also compatible with enzyme catalysed reactions and have been employed for the enzymatic synthesis of a glycopeptide using a β-(1-4)-galactosyltransferase[122] as shown in *Scheme 1.5.7.2*.

Scheme 1.5.7.2

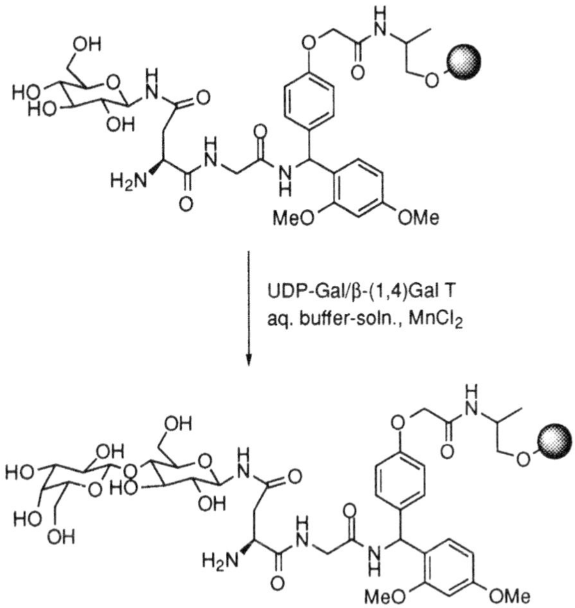

PEGA beads have been employed to study protein-ligand interactions *e.g.* by *Meldal et al.*[123] who generated a library of polymer-bound protease substrates, which were C-terminally linked to a fluorescence donor (2-aminobenzoic acid) and *N*-terminally fixed to a fluorescence acceptor (3-nitrotyrosine). The beads were subjected to the endoprotease *Subtilisin Carlsberg* whereby after proteolytic cleavage the fluorescence donor remains on the solid support. Fluorescent beads were subjected to sequence analysis, which not only revealed the sequence of the cleaved fragment but also the cleavage site.

The use of such beads as anchor for grafting and simultaneously test potentially interesting enzyme inhibitors was developed by coupling a fluorogenic protease substrate onto PEGA beads.[124] In a second reaction sequence a library of potential protease inhibitors was generated using the "portioning and mixing" strategy generating beads containing both a substrate and an inhibitor sequence. The beads were treated with *Subtilisin Carlsberg* resulting in fluorencent beads where proteolytic cleavage had occurred and non-fluorescent beads where cleavage was prevented. After sequencing, resynthesis and testing the peptides in solution, an inhibitory activity of 3 μM was obtained.[124]

Meldal et al.[125] presented recently two novel PEG-crosslinked resins derived from polyoxyethylene polystyrene (POEPS) and polyethylene polyoxypropylene (POEPOP) lacking amide derived polymer backbones resulting in an increased chemical stability. POEPS was obtained by "bulk polymerisation" of monomers **1** and **2** as depicted in *Scheme 1.5.7.3*.

Scheme 1.5.7.3

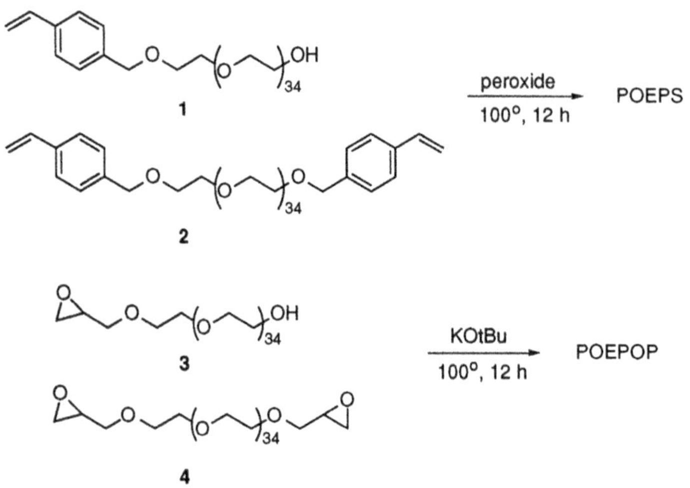

The POEPOP resins are obtained by anionic polymerisation of monomers **3** and **4**. The loading capacity can be controlled by the corresponding ratio of the monomers. Both POEPS and POEPOP resins show excellent swelling properties in solvents such as dichloromethane, DMF and water.

CLEAR resins:

CLEAR stands for **C**ross-**L**inked **E**thoxylate **A**crylate **R**esin. These resins belong to a group of highly crosslinked resins with good swelling properties, which are either obtained from "bulk polymerisation" or by suspension polymerisation[126] of monomers **1-6** as shown in *Figure 1.5.7.1*. CLEAR resins differ from the standard crosslinked polymers by the incorporation of branched crosslinking agents such as **1** and **6**. The amino groups present in monomers **2** and **3** constitute the attachment points for potential linker groups*. CLEAR resins show excellent

---

* *Su* and *Menger*[127] recently described tetrastyryl methane and proposed to use this compound as branched crosslinking agent to generate resins with novel properties.

swelling properties in protic and aprotic polar solvents such as DMF, TFA, H_2O, MeOH, CH_3CN, THF and CH_2Cl_2 whereas they swell gradually in more lipophilic solvents such as toluene, EtOAc and hexane.

Figure 1.5.7.1

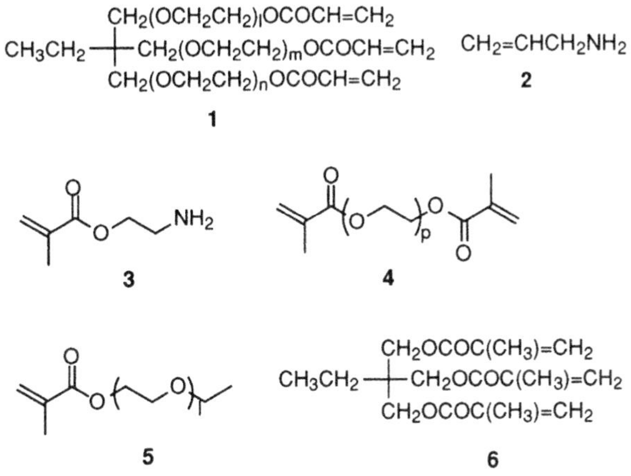

$$CH_3CH_2 \overbrace{\begin{array}{l} CH_2(OCH_2CH_2)_lOCOCH=CH_2 \\ CH_2(OCH_2CH_2)_mOCOCH=CH_2 \\ CH_2(OCH_2CH_2)_nOCOCH=CH_2 \end{array}} \qquad CH_2=CHCH_2NH_2$$

1   **2**

3   **4**

$$CH_3CH_2 \overbrace{\begin{array}{l} CH_2OCOC(CH_3)=CH_2 \\ CH_2OCOC(CH_3)=CH_2 \\ CH_2OCOC(CH_3)=CH_2 \end{array}}$$

5   **6**

	1	2	3	4	5	6	loading [mmol NH_2/g]
Clear I	1[a)	1					0.26
Clear II	3		3.6	7.0			0.3
Clear III	8		4.4		3.0		0.3
Clear IV	12	50				3.0	0.17
Clear V	13	50					0.13

a) Ratio of corresponding monomers employed in the polymerisation process; the relative rate of incorporation of the monomers does not correspond to the same ratio.

The high mechanical stability of CLEAR resins allow their use in continous flow reactors as used in solid phase peptide synthesis. The retroacylcarrier protein sequence 74-65 was synthesised successfully in the batch mode on these resins and their properties were compared to resins like polystyrene, polyhipe, PEGA and TentaGel®.

1.6. Solid-supported reagents

1.6.1. Introduction

Solid-supported reagents have been the subject of various books and review articles published since 1970.[2,5,66,128-136] Organic polymers such as polystyrene or poly(4-vinylpyridine) as well as many inorganic solid materials such as silica, alumina, graphite, clay minerals, or zeolites were used for the support of reagents and catalysts. In this chapter we will confine our discussion to reagents and catalysts supported by organic polymers and focus – after some general remarks – on their use in combinatorial chemistry.

Many applications of these reagents were presented in the literature over the last four decades but the importance of this work was only recently fully estimated in the context of combinatorial chemistry and parallel synthesis. Besides their most obvious use in transforming low molecular-weight substrates into products (for examples cf. *Tables 1.6.1-1.6.10*), they were applied as protective groups (*e.g.* monoprotection of symmetrically difunctionalised compounds[4]) or used to promote intramolecular reactions (*e.g.* macrocyclisations, see below). Polymer-bound analogues of toxic, explosive, or odorous reagents (cf. Ref.[131] for leading reference) are safer and more convenient to handle as the corresponding soluble compounds. Furthermore, polymeric reagents were applied for mechanistic studies,[137-139] as phase transfer catalysts (see 'triphase catalysis'[140]) and have been used in purification procedures involving the selective removal of products or contaminants from a solution.

Advantages and disadvantages of supported reagents as compared to their conventional analogues have been discussed by several authors.[66,128,131] The easy separation of consumed reagent by filtration, allowing the use of excess reagent in order to obtain complete conversion of soluble substrates without laborious purification is considered a major advantage compared to soluble reagents. The primary disadvantage of supported reagents constitute their higher costs resulting from additional synthetic steps required for their preparation. This drawback may be partially compensated if a particular reagent is recyclable.

Reagents (and catalysts) supported by organic polymers can be classified into two groups:
The first group consists of reagents with functional groups covalently attached to the polymeric matrix. Such functionalised polymers are obtained either by chemical modification of the basic polymer or by copolymerisation of functionalised monomers with the basic monomers and a crosslinking agents as outlined in the previous chapter. Functionalised divinylsubstituted monomers[141,142] or dendritic monomers with attached functional groups[143] were also applied as crosslinking agents.

The second group consists of reagents supported by ion exchange resins. Amberlyst® resins are often used to support anionic nucleophiles, oxidants or reducing agents. Many of these ion exchange resins are commercially available or easily prepared from the chloride form (*Scheme 1.6.1*).

Scheme 1.6.1

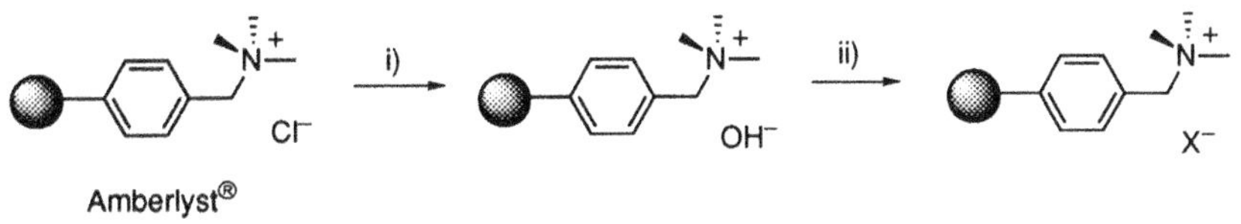

Amberlyst®

Reagents and conditions: i) NaOH; ii) X⁻

In general, the analysis of polymeric species is rather difficult and therefore, their exact structure is not always known. Polymeric trityl lithium[144] for example was prepared by reacting polystyrene with benzhydryl chloride in a *Friedel-Crafts* alkylation, affording polymeric triphenyl methane **1**, which was subsequently deprotonated producing polymer **2** (*Scheme 1.6.2*). Benzhydryl substitution probably also occurs on the initially generated trityl groups, resulting in grafting of the polystyrene.[144] The polymer may therefore possess functional groups in different environments, with different accessibilities.

Scheme 1.6.2

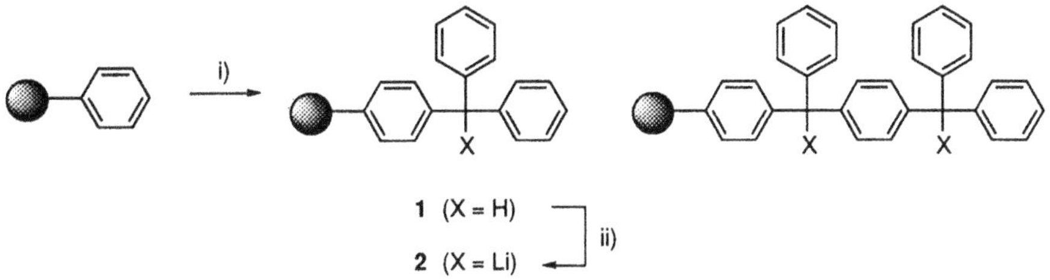

1 (X = H)
2 (X = Li)

ii)

Reagents and conditions: i) benzhydrol or benzhydryl chloride, AlCl₃, nitrobenzene; ii) BuLi, THF

Chemical transformations of polymers always include the possibility of undesired modifications which are difficult to recognise but might be responsible for low yields or purity of products if such polymers are used as reagents. Examples of intrapolymer reactions leading to additional crosslinking have been discussed in the previous chapter.

The reactivity of functionalised crosslinked polymers may significantly differ from that of analogous soluble reagents. Polymer-bound benzyne, generated by Pb(OAc)$_4$-oxidation of resin-bound 1-aminobenzotriazole **3**, for example was converted into aryl acetates **4**. This reaction was not observed if an analogous 1-aminobenzotriazole derivative **6** was oxidised in solution, where the formation of dimers (**7** and **8**) was predominant[145] (*Scheme 1.6.3*).

Scheme 1.6.3

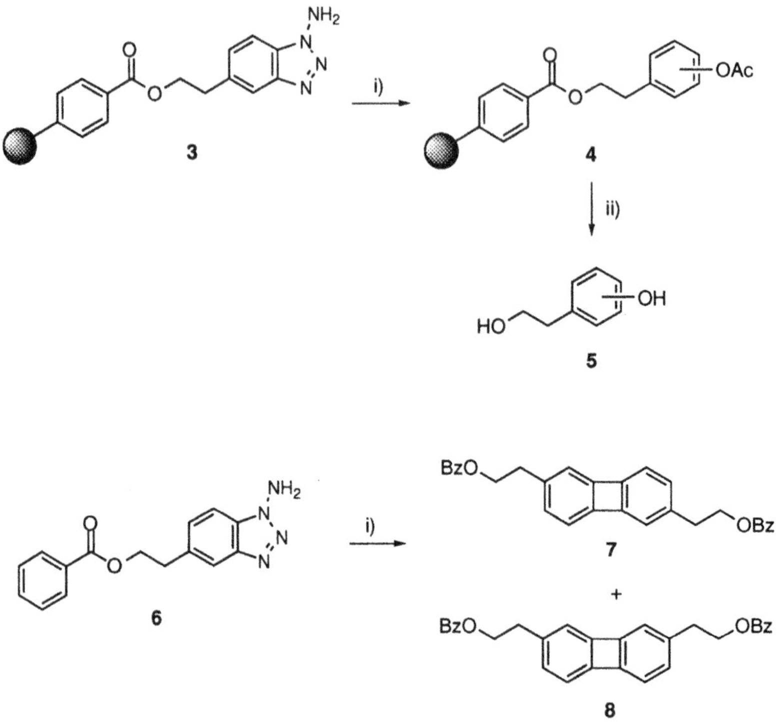

Reagents and conditions: i) Pb(OAc)$_4$; ii) MeMgBr

Immobilisation prevents the benzyne intermediates from dimerisation and allows the alternative transformation which in solution is too slow to compete with dimerisation. Applications of functionalised polymers for monoprotection (many examples were contributed by *Leznoff*,[4,116,146-150] further examples see Refs.[151,152]) or monoactivation of symmetrically difunctionalised molecules as well as their use to promote the formation of large rings are based on this site isolation or "pseudodilution effect".[145]

Crosby et al.[114] investigated the monooxidation of heptanediol **10** using macroreticular poly(4-methylmercaptostyrene) **9** and chlorine (*Scheme 1.6.4*). The product distribution was found to depend on the substitution degree of the polymer. Higher degree of functionalisation afforded increasing conversion of diol **10** but went along with increasing formation of dialdehyde **12** resulting from intrapolymer reaction.

Scheme 1.6.4

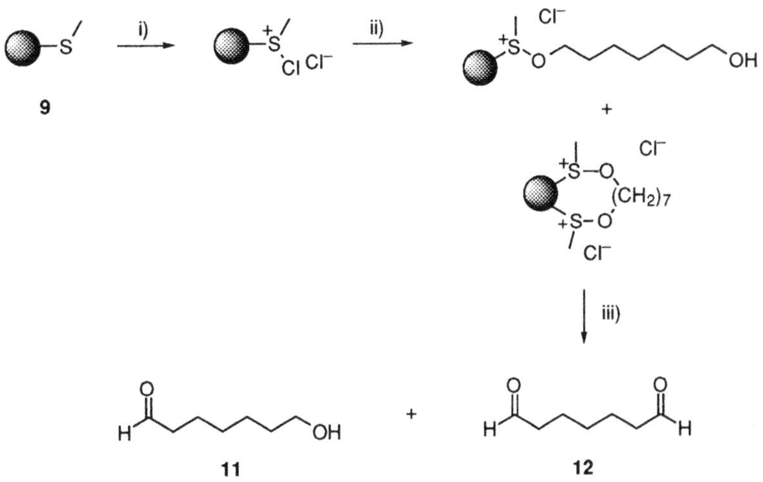

Reagents and conditions: i) Cl₂, CH₂Cl₂; ii) HO(CH₂)₇OH (**10**); iii) NEt₃

Loading (mmol/g)	Product distribution		starting material
	11	**12**	**10**
0.66	50.2 %	2.2 %	47.6 %
1.13	46.8 %	7.5 %	45.7 %
3.56	44.3 %	40.3 %	15.4 %

Appart from the degree of polymer loading, many other factors such as distribution of the functional groups in the polymeric matrix, conformational flexibility of the macromolecule, which depends largely on the degree of crosslinking, or the reaction rate determine the degree of site isolation. Apparent site isolation may result from kinetic factors if dissolved and attached substrate molecules compete for the remaining reactive groups of a polymer.

Intramolecular formation of large rings can be carried out efficiently on solid support, which may help to suppress competitive intermolecular reactions, as illustrated by some recent examples: Macrocyclisation of the acrylamido peptide-3-iodobenzylamide **13** via *Heck* reaction on solid support proceeds more rapidly than the corresponding solution phase reaction[22] (*Scheme 1.6.5*).

Scheme 1.6.5

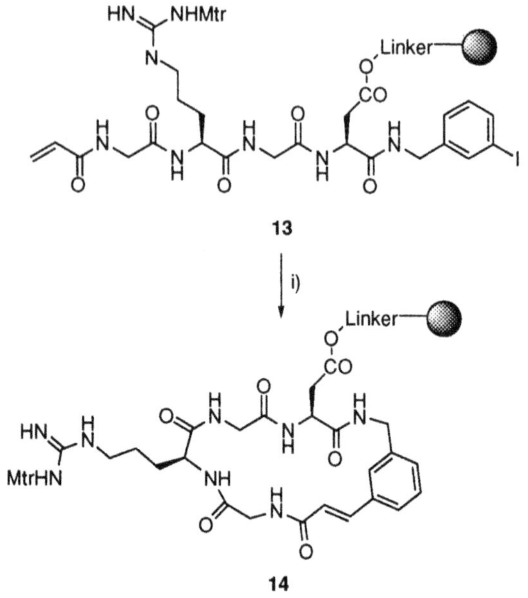

13

i)

14

Reagents and conditions: i) Pd(OAc)$_2$, PPh$_3$, Bu$_4$N$^+$Cl$^-$, DMF/H$_2$O/NEt$_3$, 37°C, 4h.

Treatment of the hydroxy acid **15** with methyl azodicarboxylate resin **17** yielded lactone **16** in 42% yield.[21] The comparable reaction with soluble azodicarboxylate gave **16** in 8% only (*Scheme 1.6.6*).

Scheme 1.6.6

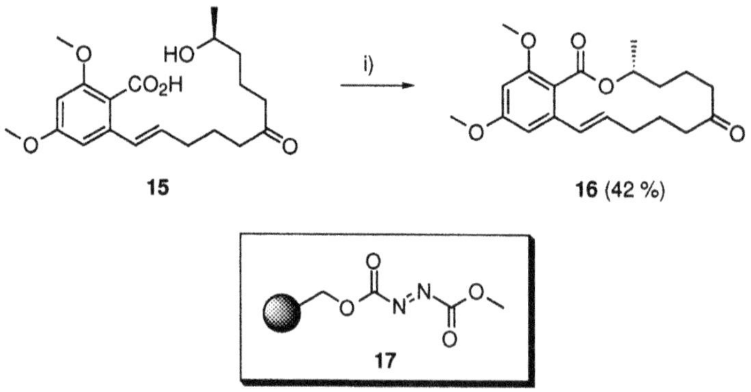

15   **16** (42 %)

17

Reagents and conditions: i) **17**, PPh$_3$, r.t., 48h

On the other hand, *Story et al.*[153] reported an unsuccessful attempt to prepare ten membered lactam ring **19** on solid support (*Scheme 1.6.7*); the HBTU mediated coupling produced linear products such as dimer (**21**) instead.

Scheme 1.6.7

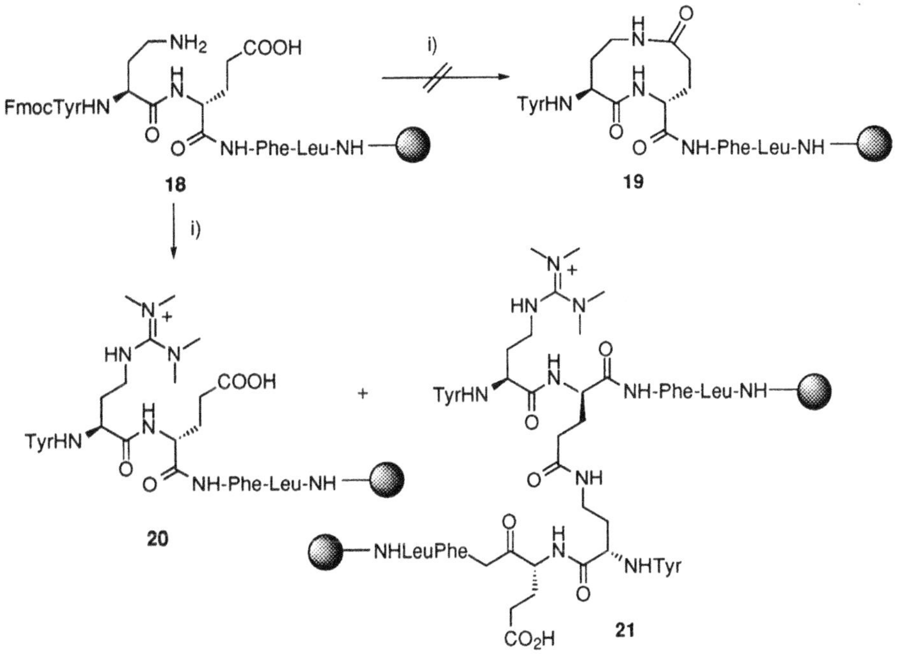

Reagents and conditions: i) a) HBTU, HOBt, DMF/DMA (1:1), DIEA, 2h, b) piperidine

1.6.2. Multipolymer systems

1.6.2.1. Insoluble polymeric reagents

An important property of polymeric reagents was described by *Rebek et al.*[137] and *Cohen et al.*,[154] namely that functional groups bound to different insoluble polymers cannot react with one another, as only a negligible number of reactive sites is located on the outer surface of the beads. Thus, two or more otherwise incompatible reagents such as acid / base[155] or oxidant / reducing agent[156] immobilised on a polymeric support, may be simultaneously used in the same reaction compartment.

For mechanistic investigations, *Rebek* introduced the three-phase method,[137,157-159] which is based on the site isolation of reactive groups. The test system consists of two polymers which can – if necessary – be separated by a frit or distinguished by particle size. An intermediate is released from the first polymer (polymeric substrate) by action of a soluble reagent. The second polymer suspended in the same solution is used to trap the intermediate. Any observed reaction between the two solid phases requires free intermediates in solution and the method therefore supplies evidence for their existence (*Scheme 1.6.8*). The structure of the intermediate may be revealed indirectly by cleavage of the product from the second polymer followed by spectroscopic investigation. Such polymer systems are particularly well suited for elucidation, whether a given reaction is proceeding via an intramolecular or through a dissociative mechanism involving free intermediates.[160]

Scheme 1.6.8

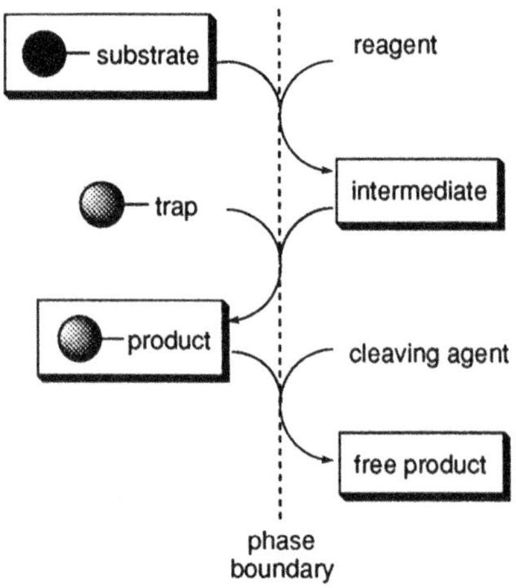

Cohen et al.[144,154] provided the first synthetic application for the simultaneous use of antagonistic reagents. A two polymer system consisting of a strong base and an active ester was used to acylate enolates derived from ketones, nitriles, secondary amides or esters. The benzoylation of acetophenone **22** for example (*Scheme 1.6.9*) produced the β-diketone **23** in 96% yield, whereas the control reaction with soluble reagents afforded only 48% of **23**. Since the diketone **23** is more acidic than the starting ketone, the latter is reformed by proton exchange with the product and the yield of the reaction in solution does therefore not exceed 50%.[137] Unlike their soluble anlogues, polymeric trityl lithium and polymeric active ester do not interact in suspension and excess of the base can thus be used to obtain complete conversion.

Scheme 1.6.9

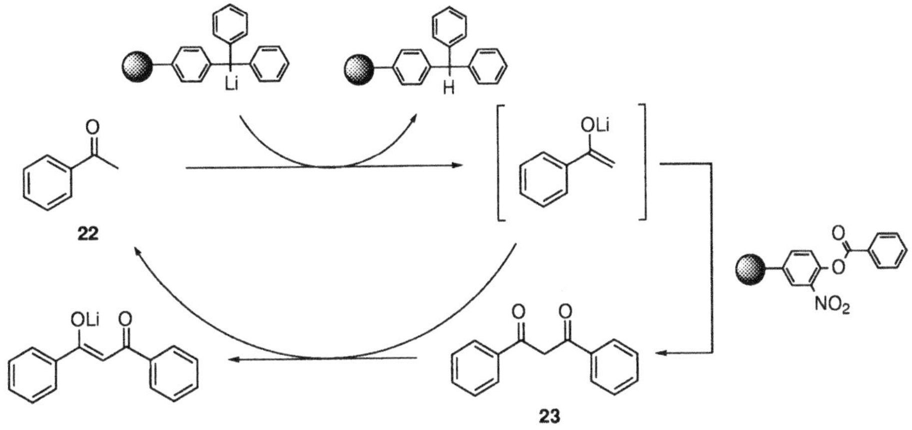

22

23

Dioxolan protected ketones **24** and aromatic or α,β-unsaturated aldehydes were hydrolysed and directly transformed into olefins using an acidic cation exchange resin and a phosphonate resin in aqueous THF[155] (*Scheme 1.6.10*). Again, such a combination of reagents would be impossible in solution as the ylide would neutralise the acid catalyst.

Scheme 1.6.10

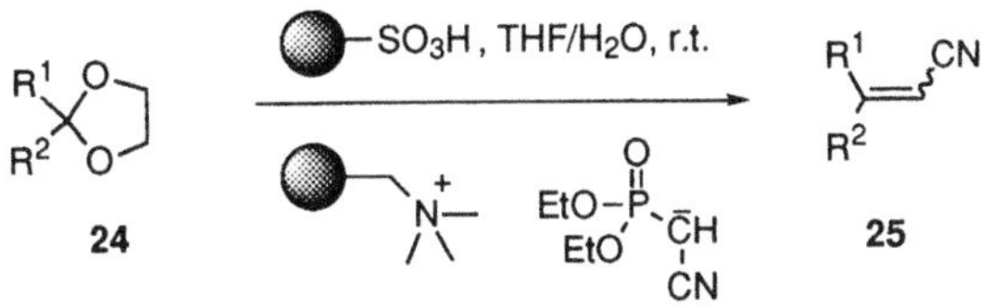

Until now, only a few synthetic examples have been described using a combination of more than two polymeric reagents either simultaneously or sequentially. Such methods avoid the isolation of intermediates and thus improve the efficiency of a particular reaction sequence.

An early example was based on *Cohens* β-diketone synthesis (*Scheme 1.6.11*) and involved treatment of product **23** with a hydrazinium cation exchange resin to generate pyrazole **26**.[154]

Scheme 1.6.11

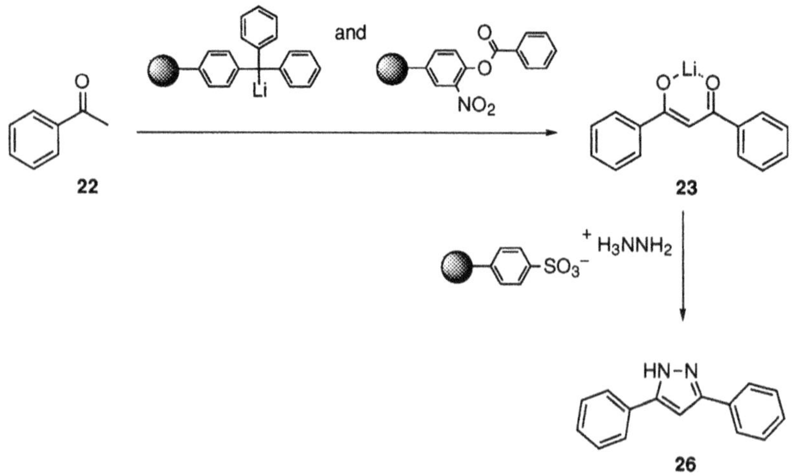

22   **23**

26

Recently, *Parlow*[161] presented a synthesis of α-substituted acetophenones using three polymeric reagents sequentially or simultaneously (cf. *Chapter 3.5.5*)

1.6.2.2. Soluble and insoluble polymeric reagents

Unlike reagents bound to crosslinked polymeric supports, soluble macromolecules are able to interact with reactive groups attached to insoluble polymers. This fact was demonstrated by *Frank* and *Hagenmaier*,[162] who developed an alternating liquid-solid phase peptide synthesis procedure. Amino acids attached to crosslinked polystyrene via a carbamate linker were condensed with a peptide ester of polyethyleneglycol monostearylether (*Scheme 1.6.12*).

Scheme 1.6.12

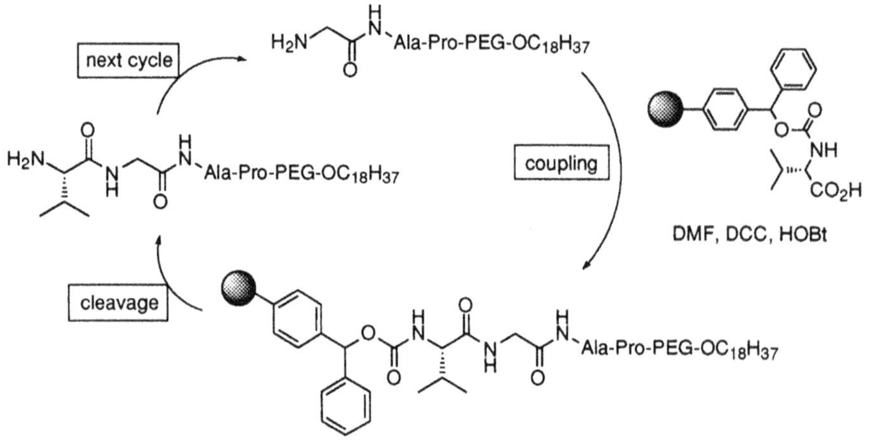

A related example was provided by *Jung* and coworkers,[163] who reacted the insoluble polymeric active ester **27** (*Scheme 1.6.13*) with the free amino group of a soluble aminoacid- or peptide polyethyleneglycol ester **28**.

Scheme 1.6.13

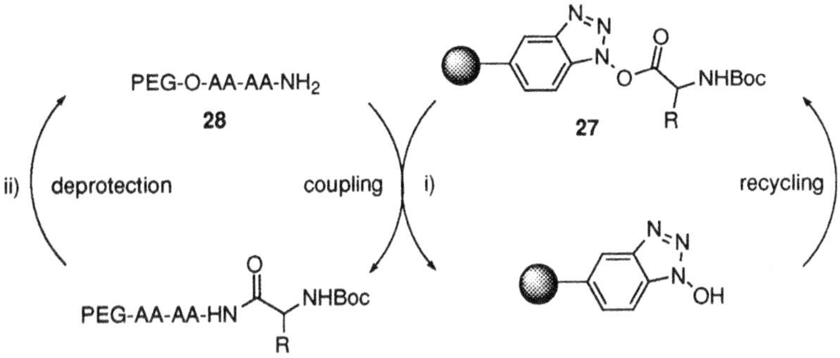

Reagents and conditions: i) N-methylmorpholine, CH$_2$Cl$_2$; ii) HCl / AcOH / H$_2$O

Han and *Janda*[164] prepared the soluble polymeric monomethyl-polyethyleneglycol derived catalyst **29**, which was used in asymmetric dihydroxylation of the TentaGel® attached cinnamic acid ester **30** to produce the diol **31** with high e.e. (*Scheme 1.6.14*).

Scheme 1.6.14

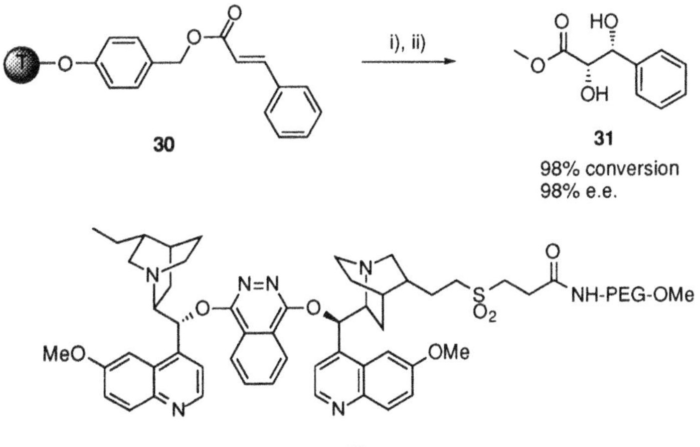

31
98% conversion
98% e.e.

29

Reagents and conditions: i) **29**, OsO$_4$, K$_3$Fe(CN)$_6$, K$_2$CO$_3$, CH$_3$SO$_2$NH$_2$, tBuOH / H$_2$O (1:1), r.t., 24h; ii) NaOMe, MeOH

Sterically more demanding soluble polymeric species, however, might be kinetically isolated from the reactive sites of crosslinked macromolecules. This was demonstrated by *Bergbreiter* and *Chandran*,[165] who described a concurrent alkene reduction and alcohol oxidation, using the soluble polymeric Rh(I) hydrogenation catalyst **32** and poly(4-vinylpyridinium chlorochromate) (PVPCC) as a stoichiometric oxidant (*Scheme 1.6.15*). A control experiment proved, that *Wilkinson's* catalyst (ClRh(PPh₃)₃) could not replace the polymeric catalyst as the Rh(I) as well as the triphenyl phosphine ligands were rapidly oxidised. The polymeric catalyst was inactive only after being exposed to PVPCC over 40 hours. The size of the bulky macromolecular catalyst seems to be sufficient to allow only very slow diffusion of the catalyst into the matrix of the insoluble polymeric oxidant and the much faster proceeding hydrogenation is therefore completed before desactivation of the catalyst.

Scheme 1.6.15

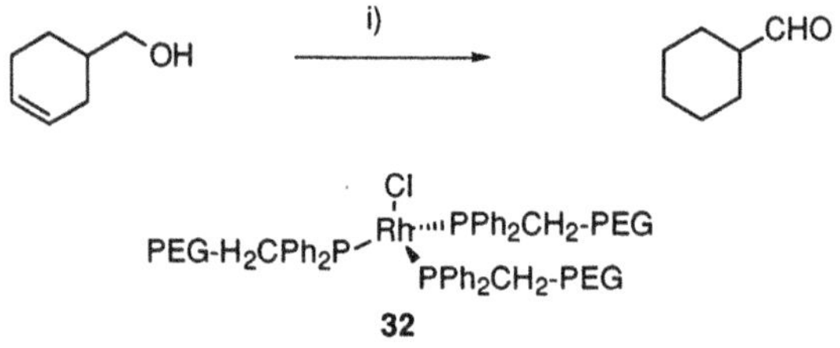

Reagents and conditions: i) **32**, PVPCC, H₂, xylene, 100°C, 12h, 80%

1.6.3. Application of functionalised polymers to purification

Polymeric reagents can be used to selectively remove a component from a mixture. This is effected by attachment of the compound in question to a solid support, subsequent filtration, and – if necessary – release from the support.

The most classical examples of this application of crosslinked polymeric reagents were provided by *Fréchet* and comprise the separation of cis and trans 1,2-diols (cf. *Chapter 3*) and the isolation of cis 1,3-diols from crude reaction mixtures using polystyrylboronic acid resin[166] (*Scheme 1.6.16*).

Scheme 1.6.16

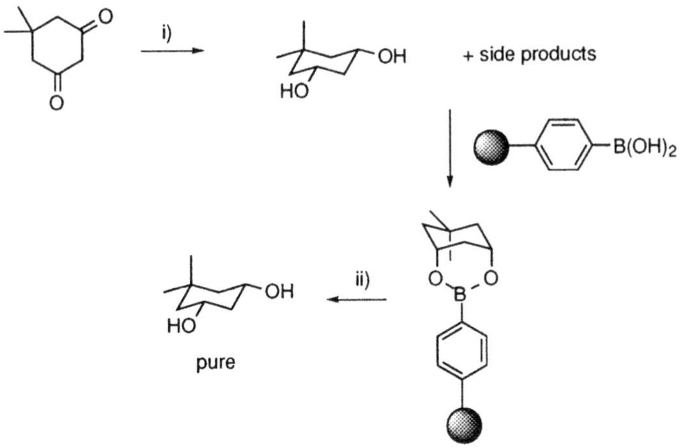

Reagents and conditions: i) Na / EtOH; ii) H₂O

Carpino et al.[167,168] presented polymer-bound piperazines **33** which caused deblocking of Fmoc amino protective group with subsequent scavenging of the released dibenzofulvene to give **35**, affording essentially pure amine **34** (*Scheme 1.6.17*).

Scheme 1.6.17

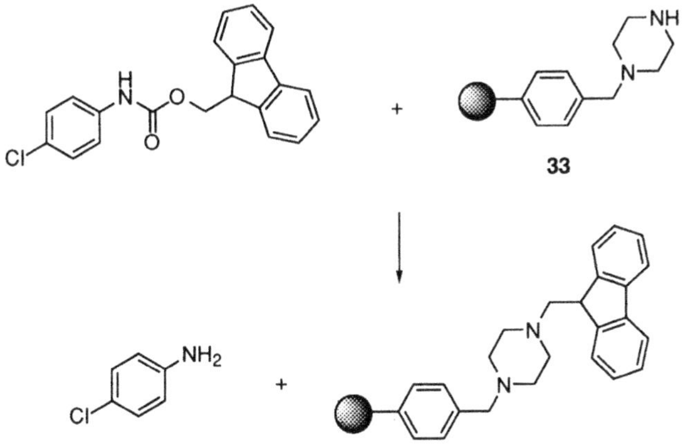

Functionalised polymers may simplify the isolation of an isomer in case of non-specific transformations.

Crowley and *Rapoport*[169] described the *Dieckmann* condensation of solid-supported radiolabeled mixed diesters **36** (*Scheme 1.6.18*) which afforded two β-ketoesters, one being released into solution and the other staying grafted onto the polymer.

Scheme 1.6.18

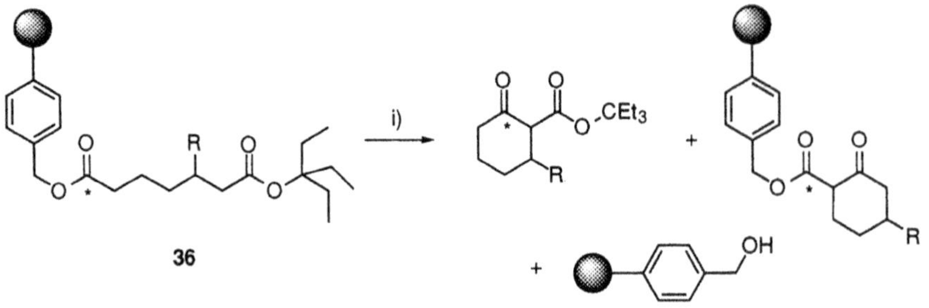

Reagents and conditions: i) (Et)₃COK, toluene, 110°C; AcOH.

Leznoff and *Svirskaya*[170] achieved the synthesis of nonsymmetric tetraaryl porphyrine **39** (*Scheme 1.6.19*) on solid support. The polymer-bound product **37** was separated from the soluble symmetric porphyrin byproduct **38** by continuous extraction and **39** was finally released by methanolysis of the ester linkage.

Scheme 1.6.19

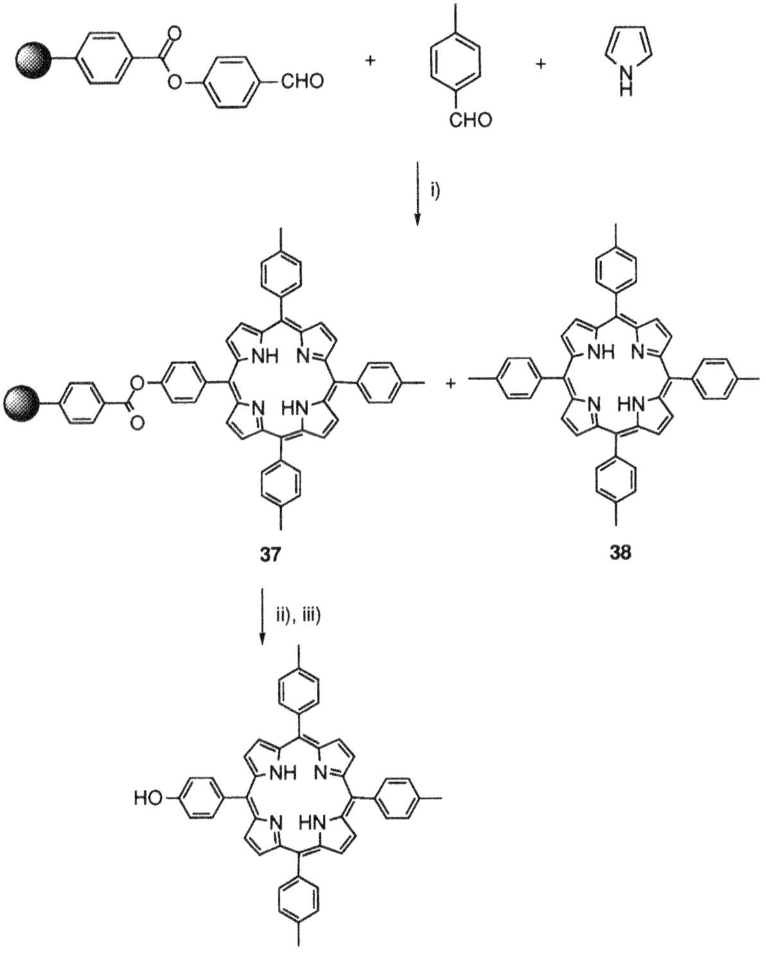

37　　　　　　　　　　　**38**

39

Reagents and conditions: i) propionic acid, Δ; ii) continuous extraction; iii) K₂CO₃, MeOH.

During the last two years the interest in separation techniques based on reactive polymers had a renaissance and many applications of supported quenching reagents in solution phase parallel syntheses were described. Ion exchange resins[171-174] as well as polymers with covalently attached reactive groups[175-178] were used to selectively trap either reaction products or contaminants. Excess of a reagent is usually removed by polymers bearing functional groups which mimic the reactivity of the starting material (cf. *Scheme 1.6.20* for general representation; for detailed discussion of specific examples see *Chapter 3*).

Scheme 1.6.20

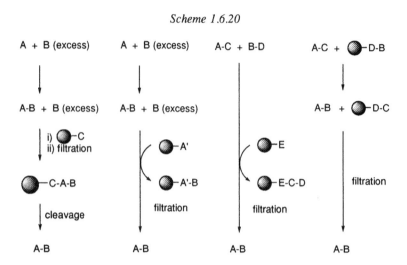

By simplifying workup and purification procedures, polymeric reagents may help to circumvent the major obstacle to solution phase parallel syntheses. Separation by supported reagents is based on the chemical reactivities of the components of a mixture rather than on their physical properties. Therefore, this method can be used to purify library mixtures where conventional techniques such as chromatography, crystallisation or precipitation are no longer applicable.

Mixtures of ureas were generated by reacting mixtures of amines with excess of an isocyanate and subsequent removal of remaining reagent using aminomethyl polystyrene[179] (*Scheme 1.6.21*).

Scheme 1.6.21

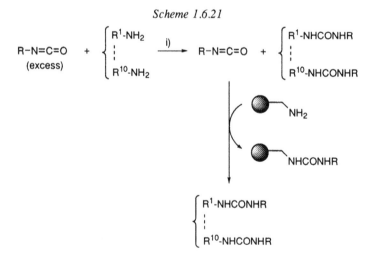

Reagents and conditions: i) CHCl$_3$, r.t.

Supported reagents were further applied to achieve post-cleavage purification of products obtained by solid-phase synthesis. This is illustrated by an example from *Ellman's* laboratory[180] (*Scheme 1.6.22*). The β-turn mimetic precursor **43** was released from solid support by TCEP (**41**) mediated cleavage of a disulfid linkage. The polymeric guanidine **44** was used to remove excess of **41** and of the phosphinoxide **42**, to promote formation of the cyclic sulfide **45**, and to trap the byproduct HBr from solution.

Scheme 1.6.22

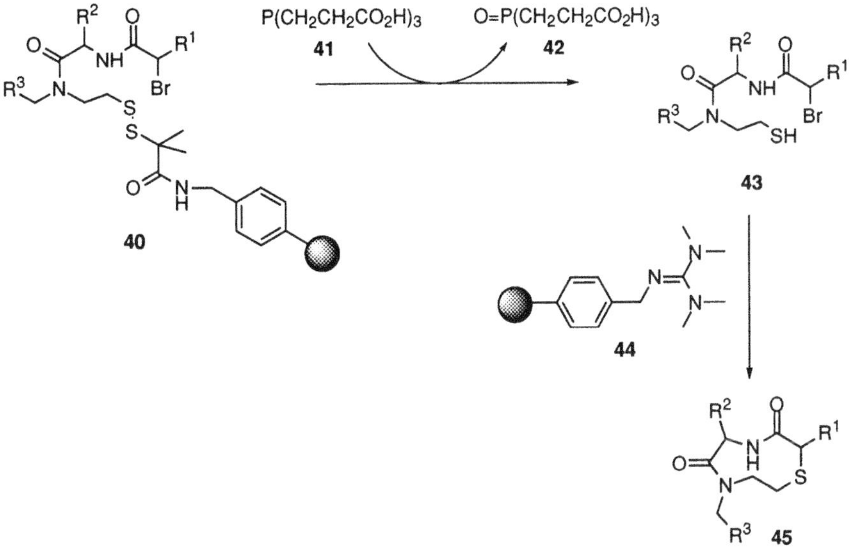

1.6.4. Polymer-supported reagents and catalysts

In general, polymer-supported reagents are used in excess. On one hand, it should be considered that not all functional groups are solvent-accessible which is important if the loading of a polymer was determined by elemental analysis. On the other hand, however, polymers may adsorb products. Hence, there is some optimal amount of reagent to be applied. Polymeric reagents are not always used in the same solvents as their low-molecular weight analogues. Chromate supported on ion exchange resin for example gave best results when used in cyclohexane rather than in CH_2Cl_2,[181] and polymeric perbenzoic acid performed better in THF than in CH_2Cl_2.[182] Macroreticular ion exchange resins are often used in apolar solvents (benzene, toluene, hexane, pentane) to which ionic groups are well exposed in the permanent pores of the material. Such resins are offering an alternative to classical phase transfer conditions.[183,184] Most of these ion exchange resins are commercially available.

Some of the most widely used solid-supported reagents and catalysts are summarised in the following tables:

Table 1.6.1. Solid-supported nucleophiles

Table 1.6.2. Solid-supported electrophiles

Table 1.6.3. Solid-supported oxidants

Table 1.6.4. Polymer-supported reducing agents

Table 1.6.5. Polymer-supported coupling and dehydrating agents

Table 1.6.6. Solid-supported halogen-carriers

Table 1.6.7. Polymeric bases

Table 1.6.8. Reagents derived from polystyryl diphenylphosphine

Table 1.6.9. Chiral polymer-bound catalysts

Table 1.6.10. Miscellaneous reagents

Table 1.6.1. Solid-supported nucleophiles

structure	preparation	application
$X^- = F^-$	From Amberlyst® A-26 (Cl⁻) or Amberlite® IRA 900 (Cl⁻) by treatment with NaOH and HF[185,186]	Nucleophilic displacement of halides or sulfonates in primary or secondary substrates affording fluorides. Olefins are formed as by-products.[185,186] Halide displacement in α-haloesters and in α-halo methylketones.[185] Cleavage of diphenyl-*tert*-butylsilyl ethers.[184] For other applications cf. *Table 1.6.7*
$X^- = CN^-$	From Amberlyst® A-26 (Cl⁻) by treatment with aq. KCN[187]	Synthesis of nitriles from alkyl and acyl halides[187]
$X^- = SCN^-$	From Amberlyst® A-26 (Cl⁻) by treatment with aq. KSCN[187,188]	Thiocyanates or isothiocyanates were obtained from primary or secondary alkyl halides. [187] Acyl isothiocyanates or sulfonyl isothiocyanates were prepared from the corresponding acid chlorides. [187]
$X^- = NCO^-$	From Amberlyst® A-26 (Cl⁻) by treatment with aq. KOCN[188]	Symmetrical *N,N'*-dialkyl ureas from primary and secondary as well as from benzylic and allylic halides if the reaction was performed with wet resin in apolar solvents.[188] *N*-Substituted ethyl urethanes were obtained from alkyl and benzyl halides if the reaction was carried out in EtOH.
$X^- = NO_2^-$	From Amberlyst® A-26 or Amberlite® IRA 400 From Amberlite® IRA 900 (Cl⁻) by treatment with NaNO₂ soln.[183]	Synthesis of nitroalkanes and alkyl nitrites from alkyl halides.[189,190] α-Nitro carboxylic esters from α-bromo esters; alkyl-nitro compounds from primary, secondary and benzylic bromides (formation of nitrites was not detected).[183]
$X^- = Ph\text{-}Se^-$	From Amberlyst® A-26 (BH₄⁻) by treatment with (PhSe)₂[191]	Organyl phenyl selenides were prepared from alkyl iodides, allyl bromides, α-bromo esters or benzyl halides.[191]

61

Table 1.6.1. Solid-supported nucleophiles (continued)

structure	preparation	application
X⁻ = Ar-O⁻	From Amberlite® IRA 900 (Cl⁻) by treatment with NaOH and ArOH in EtOH.[171]	Aryl alkyl ethers and heteroaryl alkyl ethers from alkylhalides.[171,183]
	From Amberlite® IRA 400 (Cl⁻) by treatment with ArONa soln.[192]	Diaryloxymethanes from dichloromethane.[192]
	From Amberlyst® A-26	Alkyl aryl ethers from alkyl sulfonates[193]
X⁻ = AcO⁻	Amberlyst® A-26	Acetates from prim. iodides;[193] acetamido diols from iodo amino alcohol hydro-chlorides (via aziridines).[194,195]
X⁻ = Ar-CO₂⁻ X⁻ = Ph-CH₂-CO₂⁻ X⁻=PhCH=CHCO₂⁻	From Amberlite® IRA 400 (Cl⁻)[196]	Synthesis of acetonylesters from chloroacetone
X⁻ = CO₃²⁻Na⁺	From Amberlyst® A-26 (Cl⁻) by treatment with Na₂CO₃ soln.[197]	Synthesis of alcohols from primary, benzylic and allylic halides.[197]
X⁻ = Z= CN, CO₂Me	From Amberlyst® A-26 (Cl⁻)[155]	Olefins from aldehydes and ketones by *Horner-Wittig* reaction; *E*-isomer or mixtures of *E/Z*-isomers were obtained.[155]
	From PTBD-resin (cf. Table 1.6.7, polymeric bases) by treatment with ArOH in MeCN.[198]	Alkyl aryl ethers from primary, allylic and benzylic bromides as well as from α-bromo ketones, esters and amides.[198] Bis-aryl ethers from aryl fluorides.[198]

Table 1.6.1. Solid-supported nucleophiles (continued)

structure	preparation	application
	see *Chapter 1.5*	
	From chloromethyl polystyrene by treatment with tris(2-aminoethyl)amine in DMF.[177]	Scavengers; to remove excesses of isocyanates, carboxylic acid chlorides or sulfonyl chlorides from reaction mixtures.[176,177,179]

Table 1.6.2. Solid-supported electrophiles

structure	preparation	application
	see *Chapter 1.5*	Scavenger; to remove excess of primary amine from reaction mixture (reductive amination).[176]
	From aminomethyl polystryene by treatment with triphosgene.[177]	Scavenger; to separate secondary amines from tertiary amines.[177]
	From carboxylated polystyrene by treatment with SOCl$_2$.[199]	Scavenger; to separate secondary amines from tertiary amines.[176]
	From polymer-supported amines by treatment with CS$_2$/TsCl or with CSCl$_2$.[200]	

Table 1.6.3. Solid-supported oxidants

structure	preparation	application
N⁺ CrO₄H⁻	From Amberlyst® A-26 (Cl⁻) by treatment with CrO₃/H₂O.[201,202]	Aldehydes or ketones from primary or secondary alcohols, respectively.[201,202] Aldehydes and ketones from allylic and benzylic halides.[203]
NH ClCrO₃⁻	From crosslinked poly(vinylpyridine) by treatment with CrO₃, aq. HCl. [181,204]	Oxidation of primary, secondary, benzylic and allylic alcohols.[181,204]
NH Cr₂O₇⁻	From crosslinked poly(vinylpyridine) by treatment with CrO₃/H₂O.[181,205]	Oxidation of primary, secondary, benzylic and allylic alcohols to aldehydes and ketones.[181,205]
N⁺ Cl(CrO₃)ₙ⁻	From Amberlyst® A-26 (Cl⁻) (*Scheme 1.6.23*)[181]	Oxidation of primary and secondary alcohols.[181]
Cl S⁺ Cl⁻	From poly(p-bromostyrene)[114]	Oxidation of primary and secondary alcohols.[114] Monooxidation of diols.
N OsO₄	Treatment of poly (vinylpyridine) with OsO₄.[206]	Bis-hydroxylation of alkyl- and arylsubstituted olefins, as well as of α,β-unsat. esters, ketones or allylic alcohols.[206]
N⁺-N Cl⁻ OsO₄	Reaction of chloromethyl polystyrene with DABCO and OsO₄.[206]	In the presence of a sec. oxidant catalytic amounts of polymer-bound OsO₄ are needed.[206] Aldehydes from olefins by oxidative cleavage using simultaneously polymer-bound OsO₄ and NaIO₄.[207]
N⁺ IO₄⁻	From Amberlyst® A-26 OH⁻ form by treatment with aq. HIO₄ soln.[208] From Amberlite® IRA 904 Cl⁻ form by treatment with aq. NaIO₄ soln.[208]	Oxidative cleavage of 1,2-diols.[156,208] Quinones by oxidation of quinols.[208] Sulfoxides from sulfides (in MeOH).[208] Other transformations: azobenzene from hydrazobenzene, nitrosocarbonylbenzene from benzohydroxamic acid.[208]

Table 1.6.3. Solid-supported oxidants (continued)

structure	preparation	application
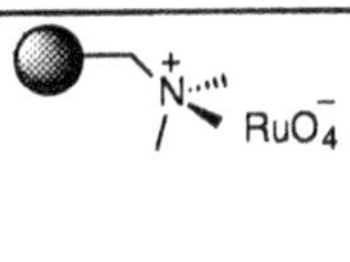 ⬤—N⁺—RuO₄⁻ (PSP)	From Amberlyst® IR 27 by treatment with aq. KRuO₄[209,210]	Oxidation of primary and secondary alcohols.[210] *In situ* oxidation of hydroxylamines to nitrones and subsequent [3+2] cycloaddition reactions.[209]
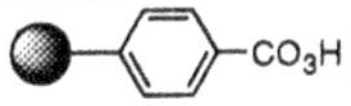 ⬤—⬡—CO₃H	From carboxyl substituted polystyrene (cf. *Schemes 1.5.4.5* and *1.5.4.10*) by treatment with 85% H₂O₂ in methansulfonic acid.[182]	Epoxides from di- and tri-substituted olefins.[182,211] Sulfoxides and sulfones from sulfides; some sulfides (penicillins, deacetoxycephalosporins) could be oxidised selectively to give sulfoxides.[212] Stereoselectivities of epoxide or sulfoxide formation compare well with the results observed for monomeric *m*-CPBA.[211,212]

Scheme 1.6.23

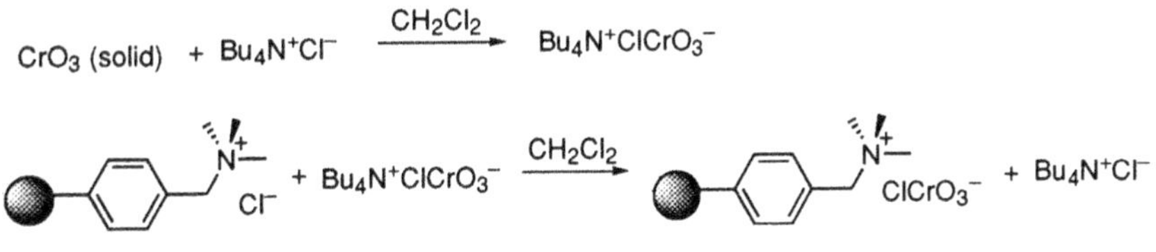

$$CrO_3 \text{ (solid)} + Bu_4N^+Cl^- \xrightarrow{CH_2Cl_2} Bu_4N^+ClCrO_3^-$$

⬤—⬡—CH₂—N⁺—Cl⁻ + Bu₄N⁺ClCrO₃⁻ $\xrightarrow{CH_2Cl_2}$ ⬤—⬡—CH₂—N⁺—ClCrO₃⁻ + Bu₄N⁺Cl⁻

Treatment of quarternary ammonium chloride resins in the presence of a phase transfer catalyst in a solid-liquid-solid system (one pot procedure)

Table 1.6.4. Polymer-supported reducing agents

structure	preparation	application
	From Cl⁻ form of exchange resin by treatment with aq. NaBH$_4$ soln.[213-215] From Amberlyst® A-26 or Amberlite® IRA 400[213-216]	Reduction of benzaldehyde.[216] Reduction of ketones (supported BH$_4^-$ afforded reverse diastereoselectivity as compared with NaBH$_4$).[217] Allylic alcohols from α,β-unsaturated aldehydes and ketones.[213] Reduction of α,β-unsaturated nitroalkenes to nitroalkanes.[214] Anilines from aryl azides; aryl sulfonamides from aryl sulfonyl azides.[215]
	From Seralite SRA 400[218]	Sat. alcohols from α,β-unsat. aldehydes; sat. ketones from aliph. α,β-unsat. ketones; sat. esters from α,β-unsat. esters (including malonates and cyanoacetates).[218]
	From etheral solution of Zn(BH$_4$)$_2$ and pyrazine[219]	Reduction of aldehydes and ketones; allylic alcohols from α,β-unsaturated aldehydes and ketones; alcohols from carboxylic acid chlorides; amines from aliphatic azides.[219]
	From Amberlyst A-26 (Cl⁻)by treatment with NaOH and ascorbic acid[178]	Reduction of DDQ.[178]
	From chloromethyl polystyrene by treatment with KSCH$_3$[220]	Reduction of acetophenone.[220]
	From macroreticular bromo polystyrene[113]	Reduction of aldehydes and ketones. In case of terephthaldehyde monoreduction (86:14) predominates even if excess of red. agent is used.[113] Chemoselective reduction of alkyl halides in the presence of ketones.[113]

Table 1.6.4. Polymer-supported reducing agents (continued)

structure	preparation	application
	From chloromethyl polystyrene by treatment with L(−)-ephedrine.[221]	Study of the enantioselective reduction of acetophenone.[221]
  and [Ir(COD)Cl]$_2$ or [Ru(C$_6$H$_6$)Cl$_2$]$_2$	Obtained by copolymerisation of the vinylsubstituted monomer with styrene and DVB.[222]	Ir(I) and Ru(II) catalysed hydride transfer reduction of acetophenone.[222] Ir(I): 96% conversion, 94% ee. Ru(II): 96% conversion, 31% ee.

Table 1.6.5. Polymer-supported coupling and dehydrating agents

structure	preparation	application
	From aminomethyl polystyrene (*Scheme 1.6.24*)[91,92]	Symmetric anhydrides from carboxylic acids.[91] *Moffatt* oxidation of prim. and sec. alcohols.[92,223]
	From aminomethyl polystyrene (*Scheme 1.6.24*)[93]	Dipeptides from appropriately protected amino acids.[93]
	From chloromethyl polystyrene (*Scheme 1.6.24*)[224]	Amides from amines and carboxylic acids.[224] Thiolesters from thiols and carboxylic acids.[225]
		Suggested to be applied in DMSO oxidation of alcohols or in sulfatations.[226]

Scheme 1.6.24

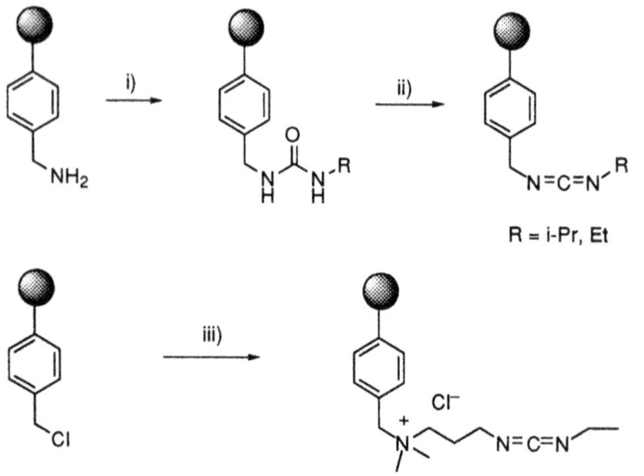

R = i-Pr, Et

Reagents and conditions: i) R-N=C=O, THF or CH_2Cl_2 / Et_2O; ii) TsCl, NEt_3, CH_2Cl_2, 40°C or TsCl, pyridine, 70°C; iii) EDC, DMF, 100°C, 15h.

Table 1.6.6. Solid-supported halogen-carriers

structure	preparation	application
ICl$_2^-$	From Amberlyst® A-26 I⁻ form by treatment with Cl$_2$[227]	α-Chlorination of aldehydes and ketones.[227] vic. dichloroalkanes from alkenes.[227]
Br$_3^-$	From Amberlyst® A-26 Br⁻ form by treatment with Br$_2$[228,229]	α-Bromination of aldehydes,[228] aliphatic ketones,[228,229] acetophenones,[228-230] (2-furyl)alkylketones,[230] acetates,[228] malonic esters.[228] Unsymmetrical ketones are selectively brominated at the more highly substituted position.[228] Vicinal dibromides from olefins, acetylenes and α,β-unsaturated ketones. [228] Selective *para*-bromination of phenols.[231]
BrCl$_2^-$	From Amberlyst® A-26 Br⁻ form by treatment with Cl$_2$[228]	Chlorobromination of olefins; unsymmetrical olefins produce mixtures of isomers. [228] Chlorobromination of terminal acetylenes (affording 1-bromo-2-chloro-alkenes) [228]
	Radical polymerisation of maleimide and subsequent treatment with Br$_2$/NaOH[232]	Radical bromination of cumene **46** produced the tribromide **47** (85%); **47** is not obtained in the solution phase reaction with NBS.[232]
CO$_3^{2-}$ ·I$_2$	From Amberlyst® A-26 CO$_3^{2-}$ form by treatment with I$_2$.[193]	(Iodomethyl)oxazolidinones in high yields (≥90%) with moderate to high diastereoselectivities from allylic amines.[193] (R = CH$_2$OBn; R' = Bn: yield 90%, cis / trans 1:99)

Table 1.6.7. Polymeric bases

structure	preparation	application
	From chloromethyl polystyrene by treatment with piperidine.[233]	As auxiliary base in: -acylation and sulfonylation of amines[177] -synthesis of mixed anhydrides from carboxylic acids[177]
	From chloromethyl polystyrene by treatment with morpholine[177,233]	
		Auxiliary base in oxidations of amines with benzoyl peroxides.[234]
	Amberlyst® A 21	Nitroaldol reaction[235] 1,4-Addition of primary nitroalkanes to methyl acrylate[236]
	PTBD-resin[198]	As auxiliary base in: *O*-alkylation of phenols with alkyl halides;[198] substitution of aryl fluorides with phenolates (the polymer also acts as a scavenger for excess phenol);[198] *N*-alkylation of sulfimides and imides.[198]
	from chloromethyl polystyrene by treatment with the corresponding triamino(imino)-phosphorane.[237]	
	Amberlyst® A-26 (cf. *Table 1.6.1*)	β-Elimination of HBr from β-bromoaldehyde;[238] C-methylation of malonic esters;[239] Synthesis of α-hydroxy ketones from iodo carbonates via dehydroiodination and successive hydrolysis (*Scheme 1.6.25*);[240] Synthesis of ketoesters from iodolactones (*Scheme 1.6.25*).[240]

Table 1.6.7. Polymeric bases (continued)

structure	preparation	application
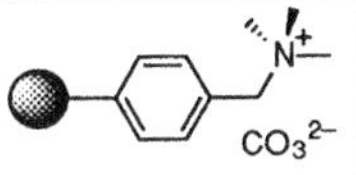	Amberlyst® A-26 (cf. *Table 1.6.1*)	Azirines from 1-iodo-2-aminoalcohol hydrochloride.[194,195] 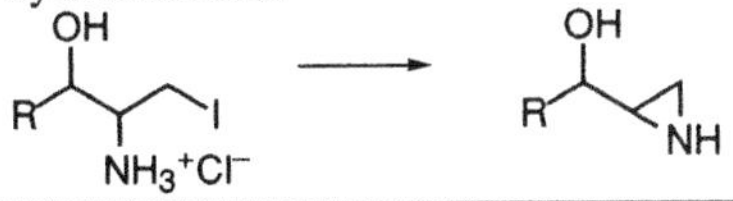
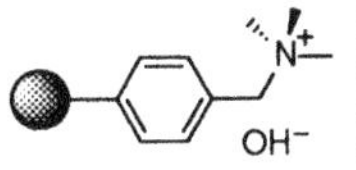	Amberlyst® A-26	The polymeric base is needed to promote the *Dieckmann* condensation of amid ester **I** affording pyrrolidinone **II** and for the purification of the condensation product as **II** remains attached to the resin (cf. *Scheme 1.6.26*)[241]

Scheme 1.6.25

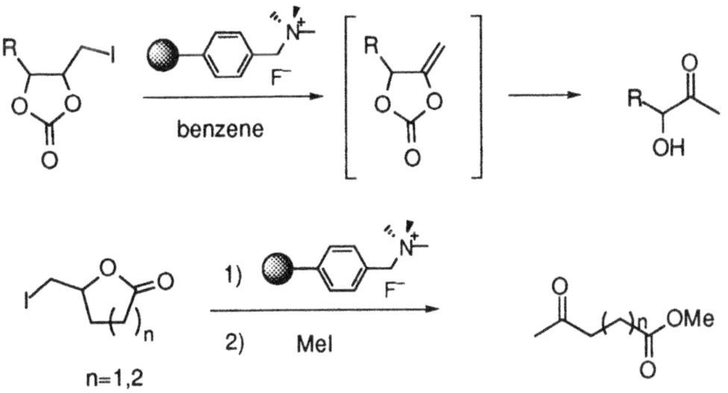
benzene
1)
2) MeI
n=1,2

Scheme 1.6.26

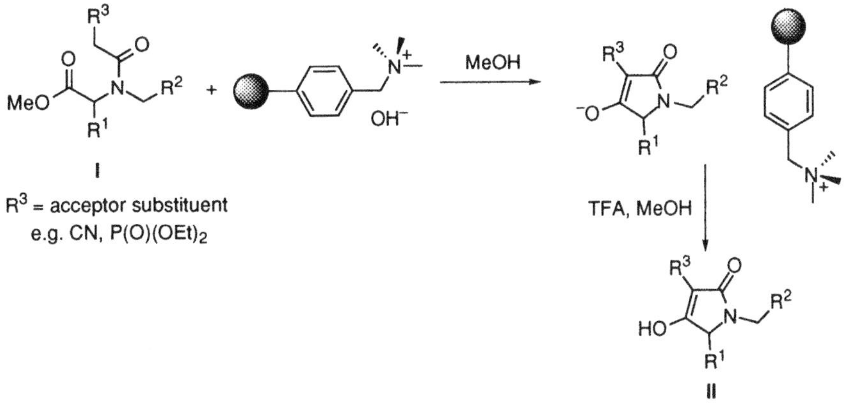
MeOH
TFA, MeOH
I
II
R³ = acceptor substituent
e.g. CN, P(O)(OEt)₂

Table 1.6.8. Reagents derived from polystyryl diphenylphosphine

structure	preparation	application
⬤—◯—P̶(Ph)₂CH₂R X⁻	Treatment of polystyryl diphenyl phosphine[a)] with alkylhalides	*Wittig* reaction
R = Ph; X = Cl R = H, CH₃, Ph; X = I, Br R = Ph; X = Cl, Br R = CO₂Me; X = Br		*Cis / trans* stilbene from benzaldehyde.[242] Olefins from aromatic and aliphatic aldehydes and ketones.[243] Mono-olefins from terephthaldehyde and iso-phthaldehyde (ylide generation by treatment of benzyl diphenyl polystyrylphosphonium bromide with ethylene oxide afforded mono ethylene acetals as by-products).[244]
R= Me Me X= Br Me	From polystyryl diphenylphosphine and β-cyclogeranyl bromide of β-geraniol (cf. *Scheme 1.6.27*)[245]	Synthesis of ethylretinoate (**48**) from aldehyde **49** (*Scheme 1.6.27*)[245]
R = H, Ph	From polystyryl diphenylphosphine crosslinked with 2,8 and 20% DVB.[115]	Comparison of olefin formation from aldehydes and ketones using polymeric phosphonium halides with different degree of crosslinking.[115]
⬤—◯—PPh₂ CCl₄		Amides from primary amines and Cbz-, Boc- or Fmoc-protected α-amino acids. No epimerisation was detected (detection limit 2%)[246] Alkyl chlorides from primary, secondary and benzylic alcohols.[247]
⬤—◯—PPh₂ X₂ X = Cl, Br, I	Treatment of poly-styryl diphenyl-phosphine with Cl₂, Br₂ or I₂ in CH₂Cl₂.[248]	Halohydrins from epoxides; yields and selectivities compare well with those obtained with monomeric triphenylphosphine-halogen complexes.[248] Open and cyclic acetals and thioacetals by treatment of aldehydes and ketones with alcohols or thiols in the presence of polystyryl diphenylphosphine iodine complex.[249]

a) Polystyryl diphenylphosphine was prepared from bromo-polystyrene[115,243,244] as outlined in *Scheme 1.5.4.10* or by copolymerisation of styrene, p-styryl diphenylphosphine and DVB.[243]

Scheme 1.6.27

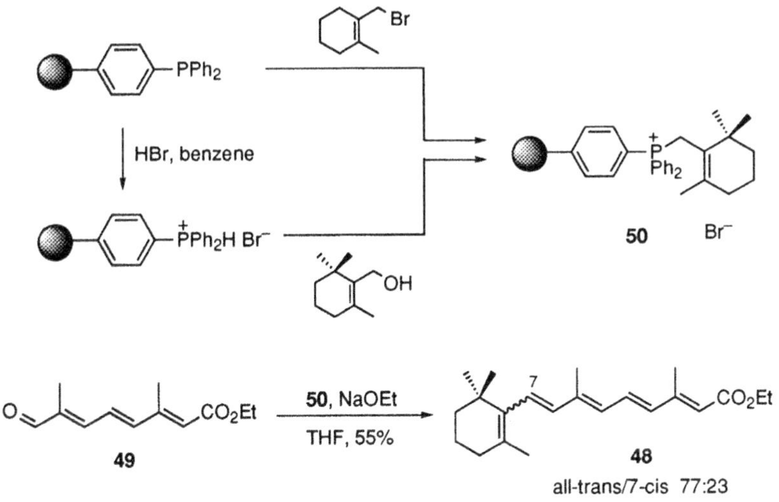

Table 1.6.9. Chiral polymer-bound catalysts (Structures, see *Figure 1.6.1*)

structure	preparation	application
52 / triallylborane	Copolymerisation of monomeric precursor 51, styrene and DVB (1:8:1) and treatment of the polymer with triallyl borane prior to use (*Scheme 1.6.28*)[250]	Enantioselective synthesis of homoallyl amines by allylation of N-trimethylsilyl benzaldehyde imine (*Scheme 1.6.28*).[250]
53 / Et$_2$Zn	From chloromethyl polystyrene and (–)-ephedrine.[221]	Enantioselective addition of Et$_2$Zn to aromatic and aliphatic aldehydes.[251] enantioselective alkylation of N-(diphenyl)phosphinoyl imines[252] (*Scheme 1.6.29*)
54 / Et$_2$Zn	Treatment of 6-iodohexyl-polystyryl ether with (1S,2R)-N-butyl norephedrine[253]	Enantioselective addition of Et$_2$Zn to benzaldehyde (cf. *Chapter 3.6.3*)

Table 1.6.9. Chiral polymer-bound catalysts (continued)

structure	preparation	application
55 / Ti(OiPr)$_4$ / Et$_2$Zn	From chloromethyl polystryene or by co-polymerisation of the chiral monomer with styrene or DVB. Subsequent transformation into solid-supported Ti-TADDOLates.[254]	Enantioselective addition of Et$_2$Zn to benzaldehyde (cf. *Chapter 3.6.3*). Enantioselective *Diels-Alder* reactions (cf. *Chapter 3.6.3*).[254]
56	(cf. *Chapter 3.6.4*)	Catalyst in three component coupling of aldehydes, aromatic amines and olefines affording quinolines (cf. *Chapter 3.6.4*).[255] Chemoselective addition of silyl enol ethers to imines in the presence of aldehydes.[256]
57 / BH$_3$	Copolymerisation of the vinyl-substituted monomer with styrene and DVB.[257,258]	Promotor in *Mukaiyama* aldol reactions (*Scheme 1.6.30*). Yields are comparable but enantiomeric excesses are lower than those obtained with the soluble Ts-Val-OH.[257] Catalyst for *Diels-Alder* reaction of cyclopentadiene and methacrolein (*Scheme 1.6.30*).[257]
58	Copolymerisation of the monomeric divinyl substituted Mn complex with styrene and DVB.[141]	Enantioselective epoxidation of styrene and *cis*-β-styrene using various oxidants.[141] Cis / trans ratio of epoxides range from 73:27 to 93:7; enantiomeric excesses from 4 to 41%.
59 / OsO$_4$	Copolymerisation of 1,4-bis(9-quinyl)-phthalazine with methyl methacrylate.[142]	Asymmetric dihydroxylation (cf. *Chapter 3.6.1*). other polymer-supported catalysts for asymmetric dihydroxylation have been presented in the recent literature.[259-266]

Figure 1.6.1

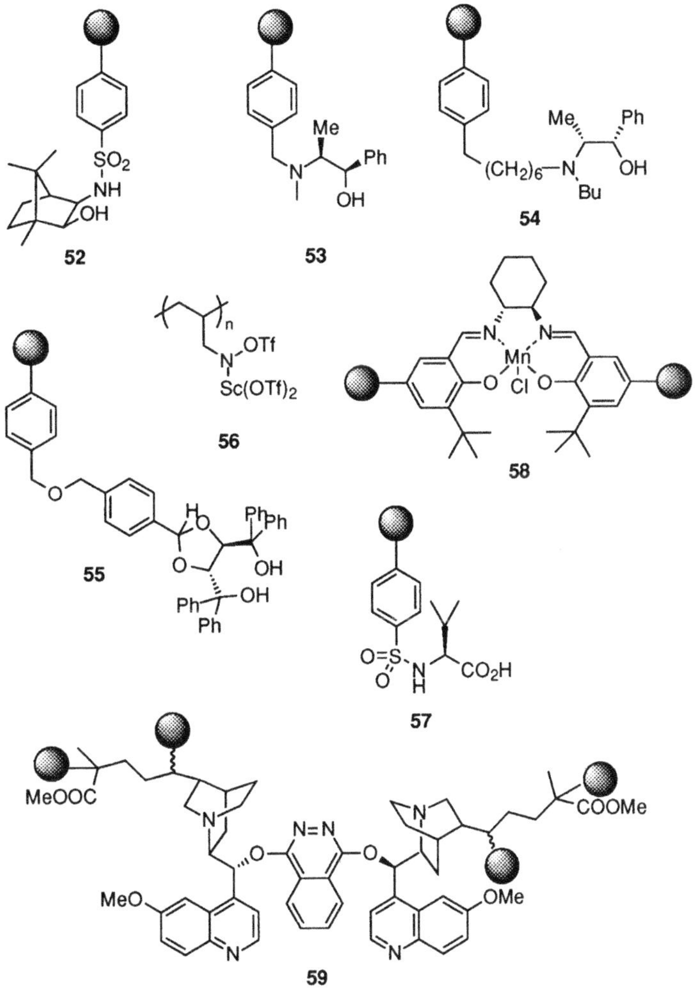
SO2
NH
OH
52
Me
Ph
N
OH
53
Me
Ph
(CH2)6
N
OH
Bu
54
n
OTf
N
Sc(OTf)2
56
N
N
Mn
O
Cl
O
58
O
H
O
O
Ph
Ph
OH
Ph
OH
Ph
55
O=S
O
N
H
CO2H
57
MeOOC
N
N=N
O
O
N
COOMe
MeO
OMe
N
N
59

Scheme 1.6.28

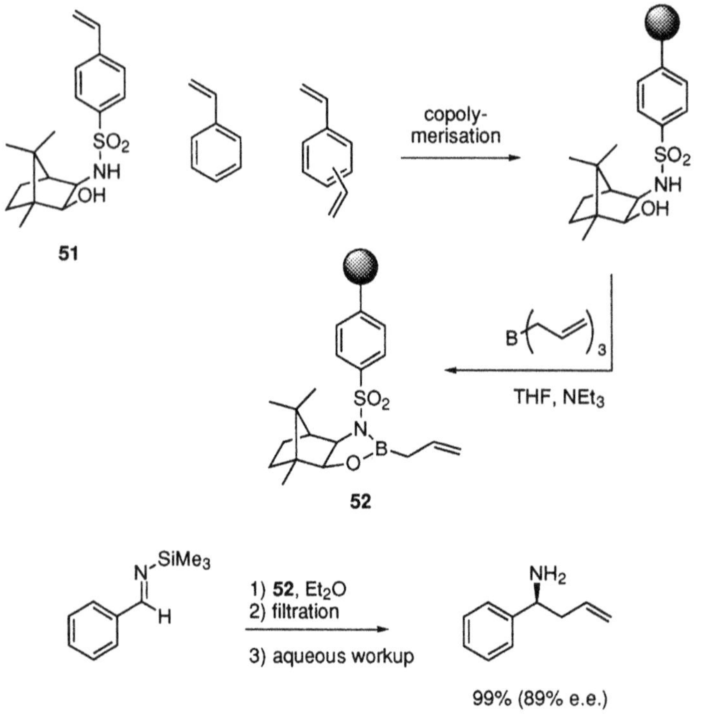

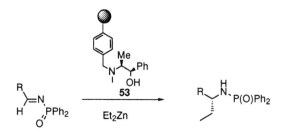

Scheme 1.6.29

R = phenyl: 80% yield, 80% e.e.
R = naphthyl: 61% yield, 85% e.e.

Scheme 1.6.30

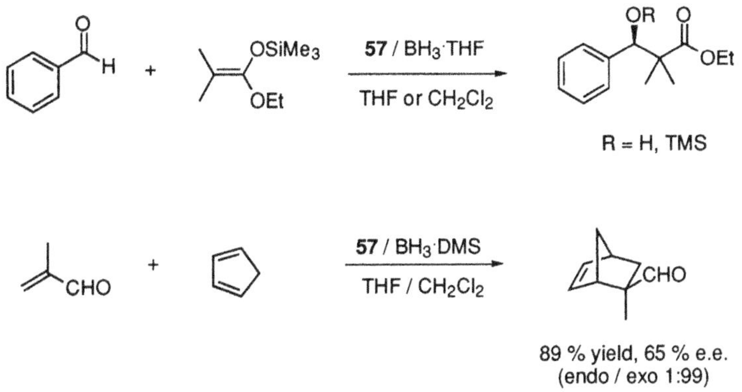

R = H, TMS

89 % yield, 65 % e.e.
(endo / exo 1:99)

Table 1.6.10. Miscellaneous reagents

structure	preparation	application
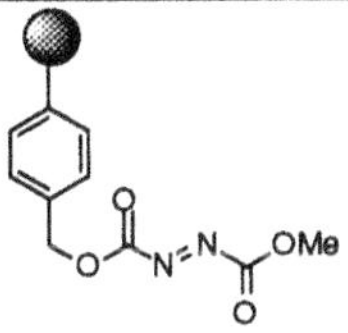	From hydroxymethyl polystyrene by treatment with $COCl_2$, H_2NNHCO_2Me and NBS or Cl_2.[21]	*Mitsunobu* reactions: esters from carboxylic acids and alcohols; lactones from hydroxy acids; *N*-alkylation of phthalimides, α-alkylation of cyanoacetate; carbodiimides from thioureas.[21]
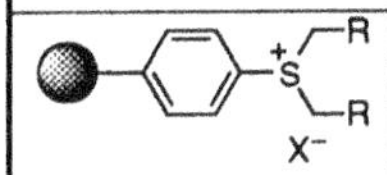 R=H, Me; X=FSO$_3^-$	Lithiation of bromo polystyrene and subsequent treatment with alkyl disulfide and alkylation.[267]	Epoxides from aldehydes and ketones under phase transfer conditions.[267]
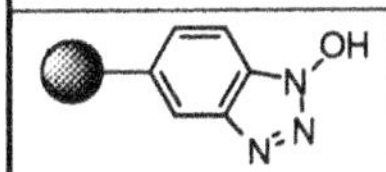	From polystyrene in 3 steps (*Scheme 1.6.31*).[268]	Preparation of polymeric activesters by reaction with *N*-protected amino acids and DCC.[268] Synthesis of medium ring lactams by coupling ω-amino acids to the HOBt resin and subsequent cyclisation.[269]
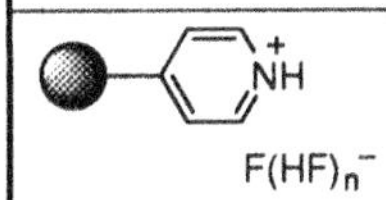 F(HF)$_n^-$	Treatment of crosslinked poly-(4-vinylpyridine) with anhydrous hydrogen fluoride.[270]	Hydrofluorination of alkenes (→fluorides) and alkynes (→gem. difluorides).[270] Alkylfluorides from sec. and tert. alcohols.[270] Treatment of alkenes with the polymeric HF-equivalent in the presence of NBS afforded vic. bromofluorides.[270]

Scheme 1.6.31

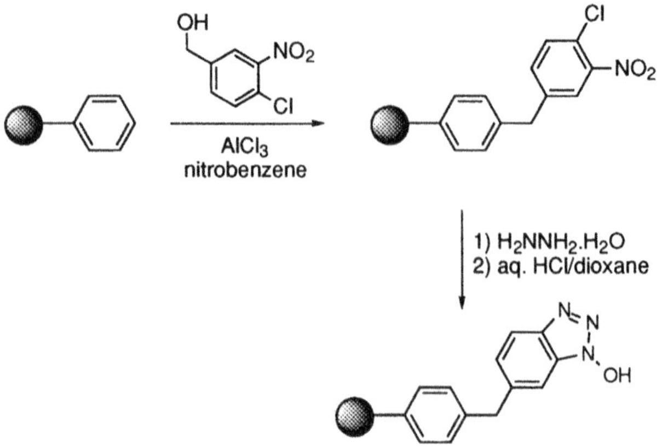

1.6.5. Polymer-supported chiral auxiliaries

In the context of library generation the ever-growing interest in asymmetric synthesis on solid phase is documented by various recent publications.[271-276]

In contrast to polymeric chiral catalysts and reagents presented in *Tables 1.6.1-1.6.10* interacting with soluble substrates, supported chiral auxiliaries are used as linkers to attach substrates and products to the polymer and act as source for optical induction. Chiral auxiliaries are regenerated at the end of a reaction sequence. It is therefore particularly attractive to attach these expensive reagents to crosslinked polymers, thus facilitating their recovery as well as the isolation of reaction products. Asymmetric syntheses using polymeric auxiliaries allowed further to investigate the steric influence of the polymer backbone on the diastereoselectivity of a given transformation.

Scheme 1.6.32

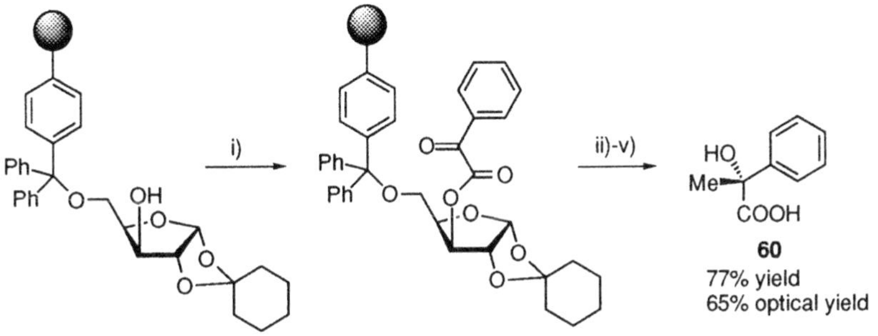

Reagents and conditions: i) phenylglyoxylic acid, py, benzene; ii) MeMgCl, Et$_2$O, benzene; iii) KOH, MeOH/H$_2$O; iv) filtrate; v) H$_3$O$^+$

Early examples of such studies date back to the seventies. *M. Kawana* and *S. Emoto*[277,278] achieved a synthesis of atrolactinic acid **60** using a polymer-bound D-xylose derivative as auxiliary (*Scheme 1.6.32*) to which phenylglyoxylic acid was attached via an ester linkage. Addition of MeMgBr to the chiral ester was the key step of the sequence. *Leznoff* and coworkers[279,280] immobilised cyclohexanone via imine formation with a polymeric chiral amine and investigated the α-alkylation (*Scheme 1.6.33*). Both groups reported optical yields at least as high as in the comparable solution phase reactions and found that their polymeric auxiliaries could be re-used successfully after recovery.

Scheme 1.6.33

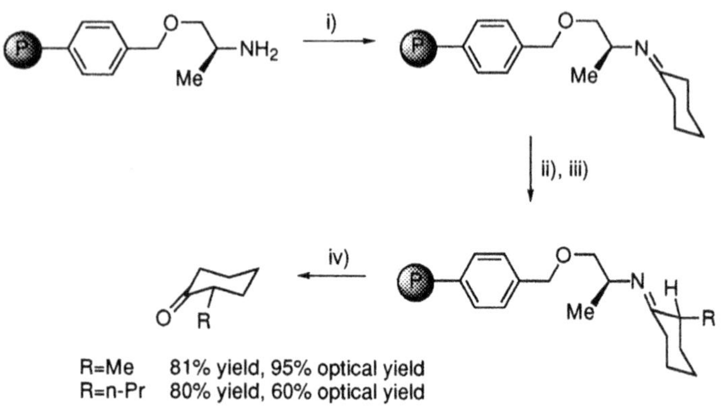

R=Me   81% yield, 95% optical yield
R=n-Pr 80% yield, 60% optical yield

Reagents and conditions: i) cyclohexanone, benzene, molecular sieves;
ii) LDA, THF, 0°C; iii) R-I, r.t.; iv) H₃O⁺

Colwell et al.[281] studied the α-benzylation of polymeric oxazoline **61** and obtained optically active ester **63** after ethanolysis (*Scheme 1.6.34*). The chemical yield of **63** was poor due to incomplete cleavage. The recovered polymer consisted of amino alkohol **64**, amino ester **65** and oxazoline **62**. Recycling of the chiral polymer-bound auxiliary was therefore difficult.

Scheme 1.6.34

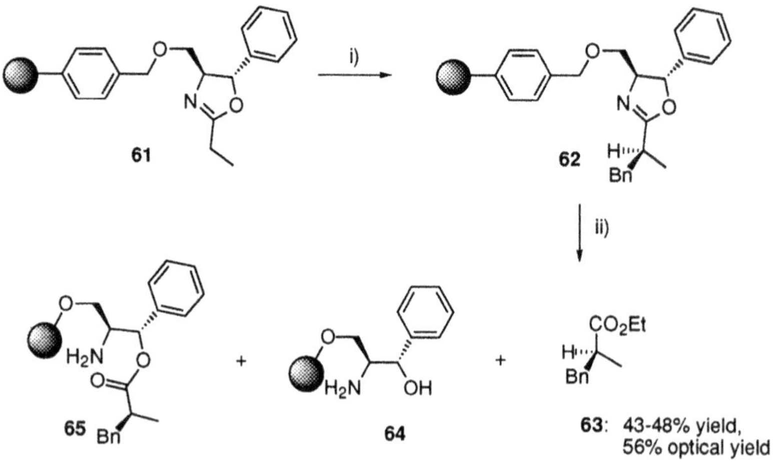

63: 43-48% yield,
56% optical yield

Reagents and conditions: i) nBuLi, BnCl, THF; ii) H₂SO₄, EtOH

The synthesis of γ-butyorolactones was achieved by *Kurth et al.*[271,272] by α-alkylation and subsequent iodo-lactonisation of *N*-acylated chiral amines (*Scheme 1.6.35*). Pyrrolidine **67**[272] was more selective in both alkylation and cyclisation as supported L-prolinol **66**.[271] Both chiral polymers were re-usable.

Scheme 1.6.35

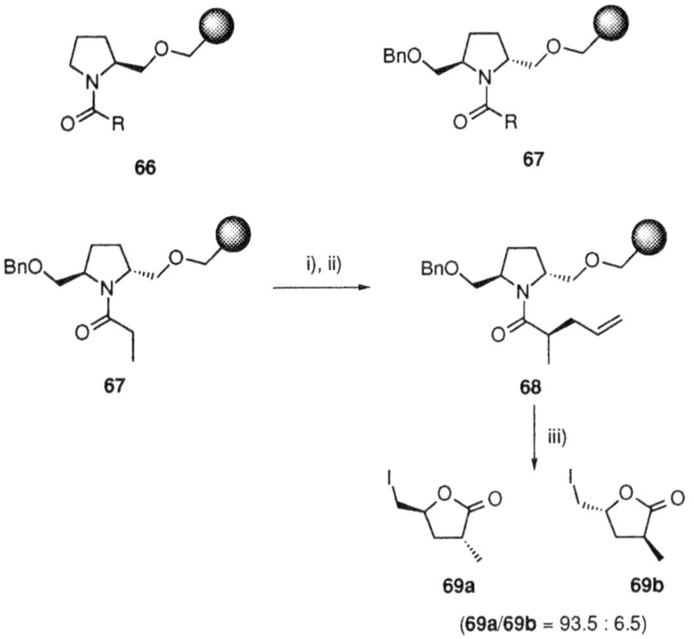

(69a/69b = 93.5 : 6.5)

Reagents and conditions: i) LDA, THF, 0°C, 30 min; ii) allyl iodide; iii) I₂, THF, H₂O

Polymer-bound *Evans* oxazolidinones were recently studied by several groups.[274-276] *Burgess* and *Lim*[275] compared the benzylation of propionylated auxiliary **71** attached to *Merrifield* resin **70a**, *Wang* resin **70b** and TentaGel® R PHB **70c**, respectively (*Scheme 1.6.36*). Products **73** were released by reductive cleavage. In terms of yields and enantiomeric excesses, best results were obtained with the *Wang* resin supported auxiliary. Furthermore, prolonged reaction times resulted in decreased yields. Other electrophiles such as isopropyl bromide and benzyloxymethyl chloride gave lower chemical yields. Comparably low diastereoselectivity was observed when the benzylation was performed with the auxiliary directly attached to *Merrifield* resin. This is in contrast to the results obtained by *Shuttleworth* and *Allin*[276] who reported high diastereoselectivity for the benzylation of supported serine-derived *N*-acyl oxazolidinone **74**.

Scheme 1.6.36

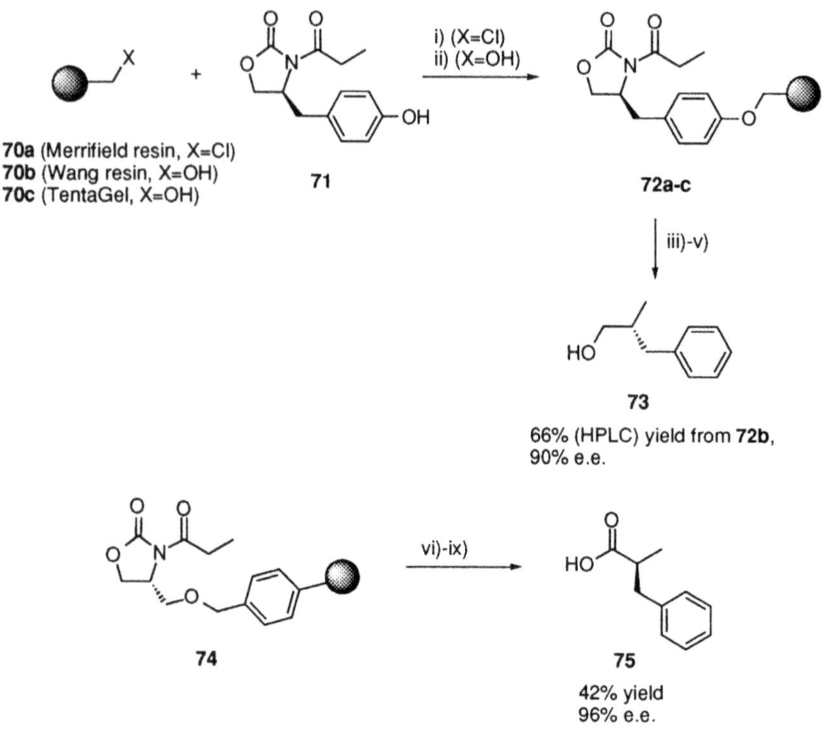
X
+
O
O
O
N
OH
i) (X=Cl)
ii) (X=OH)
70a (Merrifield resin, X=Cl)
70b (Wang resin, X=OH)
70c (TentaGel, X=OH)
71
72a-c
iii)-v)
HO
73
66% (HPLC) yield from 72b,
90% e.e.
vi)-ix)
74
75
42% yield
96% e.e.

Reagents and conditions: i) KOtBu, 18-Crown-6, $Bu_4N^+I^-$, DMF; ii) PPh_3, DEAD; iii) LDA (3 eq), 0°C; iv) BnBr (5 eq), 0-25°C; v) $LiBH_4$; vi) LDA (2 eq), THF, 0°C; vii) BnBr; viii) aq. NH_4Cl; ix) $LiOH{\cdot}H_2O$, THF, H_2O

Polymeric *Evans* auxiliary **76** was also applied to aldol reactions[274] (*Scheme 1.6.37*). In a model reaction, immobilised dihydrocinnamic acid was converted into the boron enolate and reacted with isovaleraldehyde. Products were released by treatment of the resin with NaOMe. The β-hydroxy ester **79** was obtained in a 20:1 diasteromeric ratio, along with some ester **80** derived from unreacted starting material.

Scheme 1.6.37

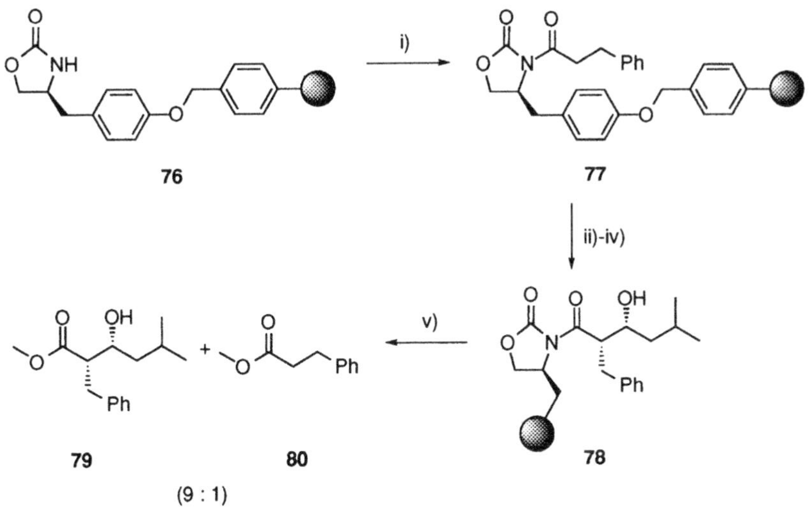

Reagents and conditions:i) LiHMDS, THF, Ph(CH$_2$)$_2$COCl; ii) nBu$_2$BOTf, DIEA, CH$_2$Cl$_2$, -20°C; iii) isovaleraldehyde, -20°C; iv) H$_2$O$_2$, DMF; v) NaOMe, THF, -20°C.

Reggelin and *Brenig*[273] reported a different approach for the asymmetric synthesis on solid support (*Scheme 1.6.38*). An acylated *Evans* auxiliary was used as a soluble reagent for the transformation of polymer-supported aldehyde **81** into imide **82**. The latter was converted into the *Weinreb* amide **83** which was – after protection of the hydroxyl group – submitted to DIBAH reduction to generate aldehyde **84.**

The configurational course of the reaction was studied by liberation of diastereomers **85** from the polymer **82.**

Scheme 1.6.38

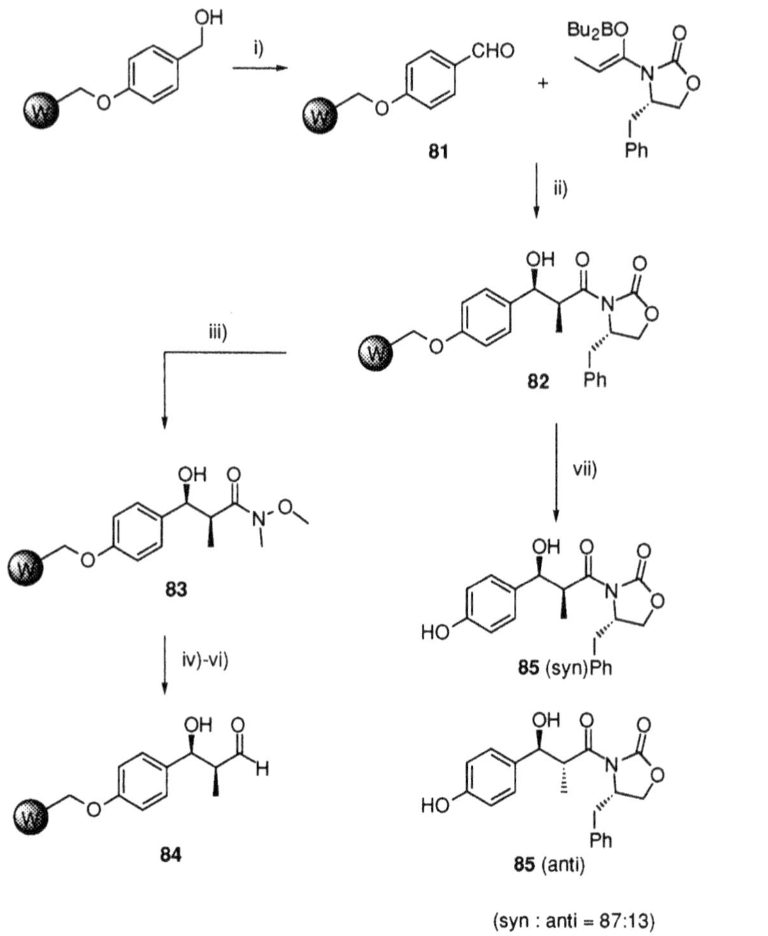

Reagents and conditions: i) py SO$_3$, DMSO, NEt$_3$; ii) H$_2$O$_2$, MeOH; iii) Me$_3$Al, MeNH(OMe) HCl; iv) TIPSOTf, 2,6-lutidine; v) DIBAH; vi) HCl / EtOH; vii) BCl$_3$, CH$_2$Cl$_2$; -78°C

1.7. Linker molecules and cleavage strategies

1.7.1. Introduction

Linker molecules play a key role in any successful synthetic strategy on solid phase as they covalently link the polymeric support and the molecules that are synthesised. Linkers are usually bifunctional spacer molecules which contain on one end an anchoring group for attachment to the solid support and on the other end a selectively cleavable functional group used for the subsequent chemical transformations and cleavage procedures. Whereas linkers were traditionally designed to release one specific functional group, *e.g.* carboxylic acids and amides in peptide synthesis, the synthesis of small-molecular-weight compound libraries has required additional linker- and cleavage strategies. Among those, the use of traceless linkers and linkers that allow cyclisation-assisted and multidirectional cleavage have emerged as powerful tools in solid-phase organic synthesis.

Since a given linker molecule has to be chemically inert during the construction of the target molecules and thus be resistant to a wide variety of reaction conditions, the so-called "safety-catch" linker principle was rediscovered for combinatorial synthesis. A "safety-catch" linker is usually converted from a chemically inert entity into a reactive species in the very last step before cleavage. This rather intriguing strategy allows most of the time multidirectional cleavage procedures, which will be discussed in *Chapter 1.7.4* and constitutes one of the most useful tools in COS. Consequently, we have grouped linker molecules and strategies into the following sections:

- Linker molecules releasing one specific functional group ("monofunctional cleavage")

- Cyclisation-assisted cleavage strategies

- Multidirectional cleavage strategies
    - by direct nucleophilic or electrophilic substitution
    - using "traceless" linkers
    - by activation of the linker molecules ("saftey-catch principle")

1.7.2. Linker molecules releasing one specific functional group

Monofunctional resin-cleavage has found high attention in the field of organic synthesis on solid support and many excellent compilations have emerged in the recent literature.[282-284] Monofunctional resin-cleavage strategies are well suited for the construction of focused combinatorial libraries, where a given important pharmacophoric group which remains constant is released in the very last step. Hence, the linkage to the resin serves as a protective group throughout the synthesis. We have classified the linker groups according to the functional groups that are released from the resin in the following order:

- carboxylic acids (*Table 1.7.1*)
- amides (*Table 1.7.2*)
- sulfonamides (*Table 1.7.3*)
- hydroxamic acids (*Table 1.7.4*)
- amines (*Table 1.7.5*)
- alcohols and phenols (*Table 1.7.6*)
- miscellaneous

Carboxylic acids

The largest number of linkers have been certainly described for the release of carboxylic acids and amides as they are closely related to peptide synthesis. Nevertheless, carboxylic acids play an important role in drug discovery and many prominent classes of pharmacologically active compounds such as aryl acetic acids (*e.g.* present in Voltaren®), 2-aryl propionic acids (*e.g.* present in Naproxen®) used as non-steroidal antiinflammatory agents, angiotensin converting enzyme (ACE) inhibitors such as Vasotech® or antibiotics such as Ciproxin® and Ceclor® contain this moiety (*Figure 1.7.1*).

Figure 1.7.1

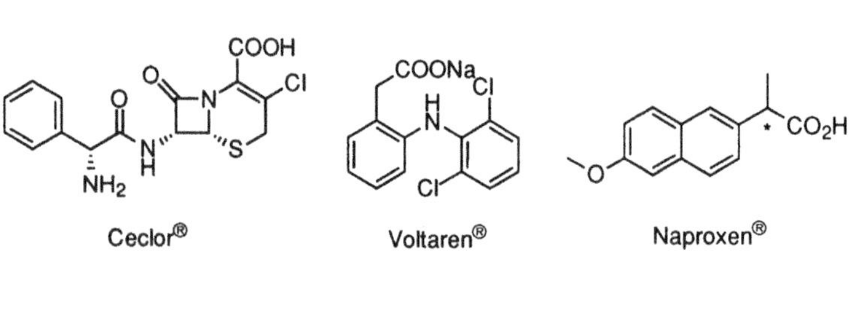

Vasotec®

Ciproxin®

Ceclor®

Voltaren®

Naproxen®

In *Table 1.7.1* some of the most widely used linkers to release carboxylic acids are grouped. The first family of linker groups (entries 1-8) comprise groups that can be cleaved under acidic reaction conditions. Entries 9-15 show examples of linkers that are cleaved under basic, entries 13 and 16 under neutral and palladium-catalysed conditions. Finally, entries 17-20 depict the class of photocleavable linker groups playing an increasingly important role in COS.

Table 1.7.1: Release of carboxylic acids

entry	structure	abbreviation	cleavage conditions	references, comments
1		*Merrifield* resin	HF, CF$_3$SO$_3$H	285
2		Hydroxymethyl PS	HF, CF$_3$SO$_3$H	
3		PAM	HF, TFA	88
4		*Wang*-resin	95% TFA	287
5		*Sasrin*-resin	1% TFA	288
6		*Rink*-resin	1% TFA	289
7	(R = H, Cl)	Trityl resin	1% TFA / AcOH HFIF	R = H[290] R = Cl[291]
8		Acetal resin	80% TFA / CH$_2$Cl$_2$	292
9		a)	NaOH	293
10			NH$_3$ / TFE	294

Table 1.7.1: Release of carboxylic acids (continued)

entry	structure	abbreviation	cleavage conditions	references, comments
11			DBU, piperidine	295
12			Bu$_4$NF	296
13			Bu$_4$NF, CsF	297
14			Bu$_4$NF	298
15		*Kaiser*-oxime resin	H$_2$NNH$_2$ H$_2$O	106
16		Hycram	Pd(0)	299
17			hv / 350 nm	300
18		brominated *Wang*-resin	hv / 350 nm	301
19			hv / 350 nm	73
20			hv / 350 nm	302

a) X = O, NH

Amides

Similarly to the release of carboxylic acids, amides have been grouped according to acidic and photochemically cleavable linkers (*Table 1.7.2*)

Table 1.7.2: Release of amides

entry	structure	abbreviation	cleavage conditions	references, comments
1		BHA	HF, CF₃SO₃H	303,304
2		MBHA	HF, CF₃SO₃H	305
3		*Rink*-amide [a)] (RAM)	TFA	commonly used resin for amide liberation in combinatorial chemistry[289]
4		PAL	TFA	306
5			TFA	307
6		SCAL (safety catch amide linkage)	TFA / (EtO)₂(PS)SH	308
7		*Sieber*	1% TFA	309
8		2-XAL 3-XAL	1% TFA	310
9		HMB (*Sheppard*)	NH₃	286

Table 1.7.2: Release of amides (continued)

entry	structure	abbreviation	cleavage conditions	references, comments
10	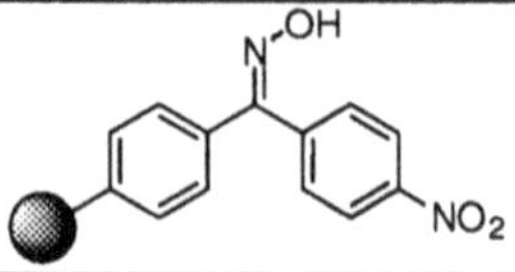	*Kaiser* resin	NH_3 $NH_2NH_2 \cdot H_2O$	106
11	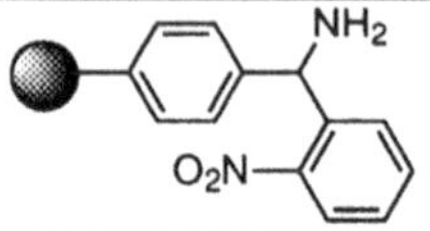		hv / 350 nm	311
12	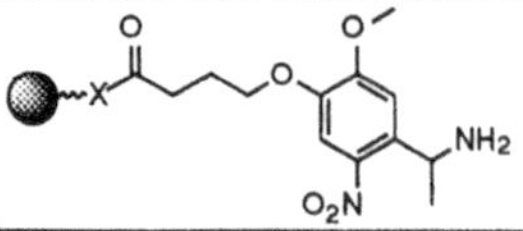		hv / 350 nm	312

a) X = O, NH

Sulfonamides

Sulfonamides are a pharmacologically important class of molecules. Indeed, many prominent drugs[313] contain this functional group as depicted in *Figure 1.7.2*.

Figure 1.7.2

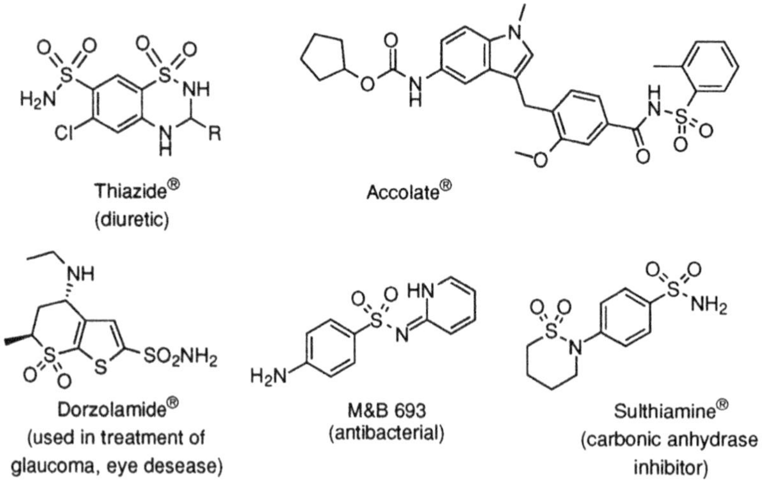

Thiazide®
 (diuretic)

Accolate®

Dorzolamide®
 (used in treatment of <br> glaucoma, eye desease)

M&B 693
 (antibacterial)

Sulthiamine®
 (carbonic anhydrase <br> inhibitor)

Some of the most useful linker molecules for the release of sulfonamides are depicted in *Table 1.7.3*.

Table 1.7.3: Release of sulfonamides

entry	structure	cleavage conditions	references, comments
1	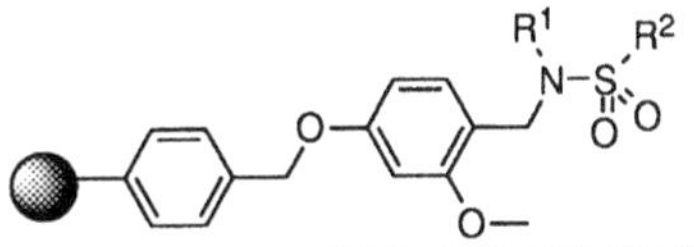	TFA	314
2	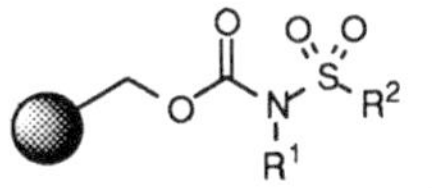	LiOH or NaOMe	315

Further sulfonamide linkers can be found in the literature.[316,317]

Hydroxamic acids

Hydroxamic acids have emerged as an important class of inhibitors of matrix metalloproteinases (MMP's) such as collagenase, gelatinase and others.[28] MMP's are becoming prime targets in many therapeutic areas (*e.g.* antiinflammatory and anticancer agents). In *Figure 1.7.3* are depicted several drugs or development candidates containining a hydroxamic acid moiety.

Figure 1.7.3

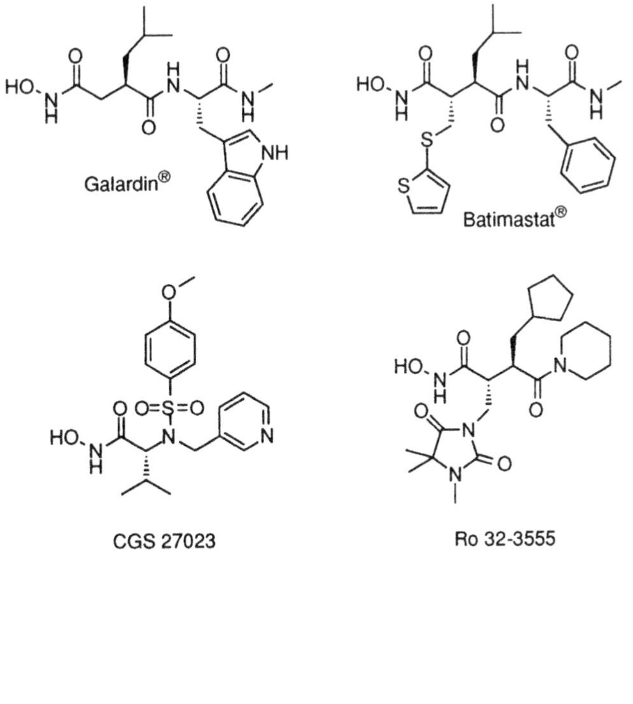

Galardin®

Batimastat®

CGS 27023

Ro 32-3555

Therefore, linker molecules releasing the hydroxamic acid moiety are most valuable tools for the construction of targeted libraries for interactions with MMP's *(Table 1.7.4)*

Table 1.7.4: Release of hydroxamic acids

entry	structure	cleavage conditions	references, comments
1		50% TFA, 5% iPr$_3$SiH in CH$_2$Cl$_2$	318
2	R=H, Cl	R=H: TFA/CH$_2$Cl$_2$ R=Cl: 2% TFA, Et$_3$SiH in CH$_2$Cl$_2$	319
3		PPTS, EtOH	320
4		1% TFA, iPr$_3$SiH in CH$_2$Cl$_2$	321

Amines

The linker groups releasing amines are depicted in *Table 1.7.5* and grouped according to their releasing reaction conditions.

Table 1.7.5: Release of amines

entry	structure	abbreviation	cleavage conditions	references, comments
1		*Rink*-chloride X=Cl →X=NR$_1$R$_2$	TFA / CH$_2$Cl$_2$	322
2		Chloro-trityl resin X=Cl →X=NR$_1$R$_2$	TFA	291
3	NR^{1}R^2	BAL	TFA / Et$_3$SiH	323
4	NR^{1}R^2		Cl$\diagdown$O$\diagup$Cl (O)	324
5	NR^{1}R^2		LAH / THF; TFA / CH$_2$Cl$_2$	325
6	NR^{1}R^2 SO$_2$ NO$_2$		PhSH, K$_2$CO$_3$, MeCN	326
7	NR^{1}R^{2}R^3 X$^-$	REM	DIEA	327,328
8	NR^{1}R^2		Pd(0)	329
9	NR^{1}R^2 SiMe$_3$		Bu$_4$NF, CsF	297
10	SiMe$_3$ NR^{1}R^2		Bu$_4$NF, CsF	297
11	X	X=OH →X=NHR	2% H$_2$NNH$_2$/DMF	330

93

Alcohols, diols and phenols

Various linker strategies for the attachment of alcohols and phenols have been devised and the most recent examples are listed in *Table 1.7.6*.

Table 1.7.6: Release of alcohols, diols and phenols

entry	structure	abbreviation	cleavage conditions	references, comments
1		Chloro-trityl resin X=Cl →X=OR	TFA / CH$_2$Cl$_2$	291
2			TFA / CH$_2$Cl$_2$ PPTS / EtOH	331
3		*Rink*-chloride X=Cl →X=OR	TFA / CH$_2$Cl$_2$	322
4		*Wang* (R^1=H) *Sasrin* (R^1=OMe)	TFA / CH$_2$Cl$_2$	287
5		X=Cl →X=OR	TFA / CH$_2$Cl$_2$	
6			H$_3$O$^+$	5
7			H$_3$O$^+$	5
8		X = O, NH	HF, anisole; NH$_3$ or NH$_2$NH$_2$	332, 333, 334
9		→ RCH$_2$OH	LiBH$_4$	335
10			Bu$_4$NF, CsF	297

Table 1.7.6: Release of alcohols, diols and phenols (continued)

entry	structure	cleavage conditions	references, comments
11		Bu₄NF, CsF	297
12		HF, anisole	336
13		hv, 350 nm	336

Miscellaneous

Release of guanidines

Refs. 337,338

Release of aldehydes

Ref. 335

Ref. 5

Release of phthalazinones

Ref. 339

1.7.3. Cyclisation-assisted cleavage

Among the various cleavage strategies described in the literature there is an increasingly important number dealing with cyclisation-assisted cleavage procedures. One of the first examples was the pioneering benzodiazepine synthesis of *Camps* and coworkers[340] in 1974, where the final benzodiazepine ring formation occurred by simultaneous cleavage from the resin. Many applications of this principle have appeared and some of the most recent examples will be discussed in Chapter 4.

Cyclisation-assisted cleavage offers the following advantages:

• Only molecules that have gone through the whole reaction sequence necessary for the cyclisation reaction will be cleaved.

• Even if the single reaction steps do not proceed quantitatively, the cyclisation will nevertheless lead to pure products.

As expected, cyclisation-assisted cleavage is largely independent of the nature of the linker molecule but rather depending on the synthesis that is necessary to create the precursor molecules. Therefore, it is not surprising that many linkers are compatible with cyclisation-assisted cleavage procedures. Nevertheless, *Figure 1.7.4* shows some of the most prominent examples of this methodology.

Figure 1.7.4

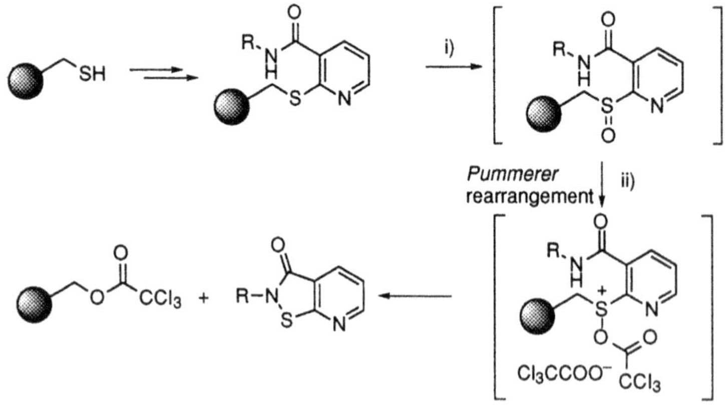

b:

c:

d:

e:

f:

Reagents and conditions: i) *Davies* reagent; ii) trichloroacetic anhydride;[341] iii) DMAP, toluene;[342] iv) acryloyl chloride, NEt$_3$, CH$_2$Cl$_2$; v) R$_1$-NH$_2$; DMSO; vi) R$_2$-N=C=O; vii) HCl, toluene, 95°C;[343] viii) TFA;[344] ix) BrCN, TFA, CHCl$_3$/H$_2$O.[345]

97

1.7.4. Multidirectional cleavage strategies

Monofunctional resin cleavage procedures are well suited for the generation of targeted libraries where the key pharmacophoric group remaining constant forms the link between the resin and the molecules that are built up. Multidirectional cleavage offers the tremendous advantage that in the final cleavage step an additional element of diversity is incorporated. Thus, the amount of compounds is multiplied by the number of elements that can be incorporated (see *Figures 1.9.1 and 1.9.2, Chapter 1.9*). Most of these linkers which have been described for multidirectional cleavage procedures are termed also as "traceless" linkers as no element of the linker remains in the final molecules.

The main strategies include:

- Direct cleavage by nucleophilic substitution reactions
- Direct cleavage by electrophilic substitution reactions
- Activation of the linker group prior to cleavage ("safety catch" principle)

1.7.4.1. Direct cleavage by nucleophilic substitution

Figure 1.7.5 summarises the linkers and strategies that allow the final multidirectional cleavage from the resin using nucleophiles such as amines, alcoholates, thiolates and C-nucleophiles such as *Grignard*-reagents.

Figure 1.7.5

a:

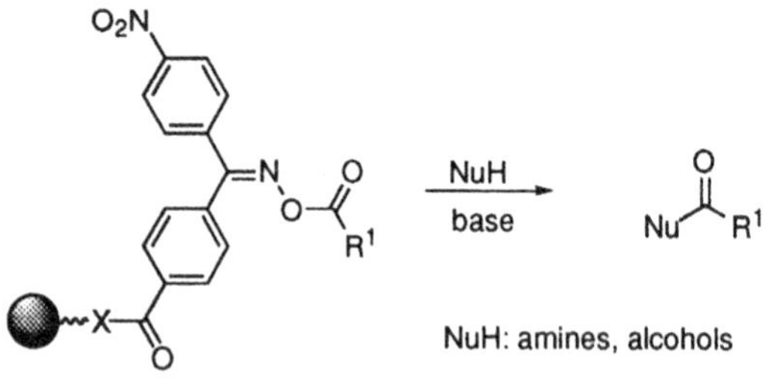

NuH: amines, alcohols

Refs. [106,346]

Figure 1.7.5 (continued)

b:

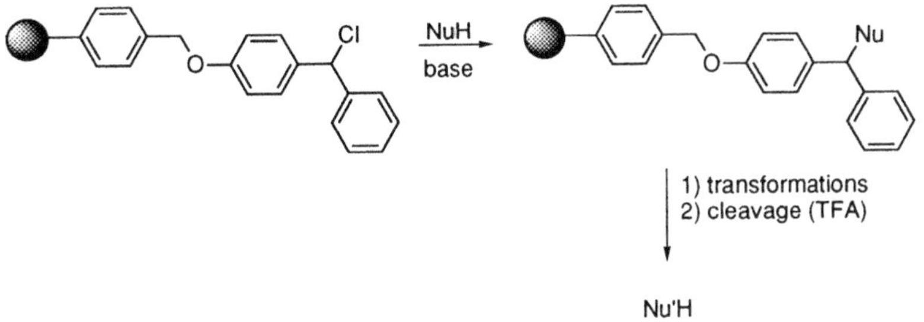
Cl
NuH
base
Nu
1) transformations
2) cleavage (TFA)
Nu'H

Ref. [322]

c: Ketones from *Grignard*-reagents and *Weinreb*-amides

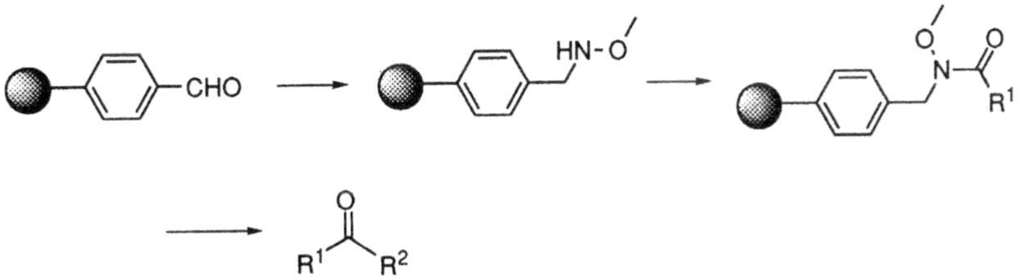
CHO
HN-O
O
N
O
R1
O
R1
R2

Ref. [346]

d:

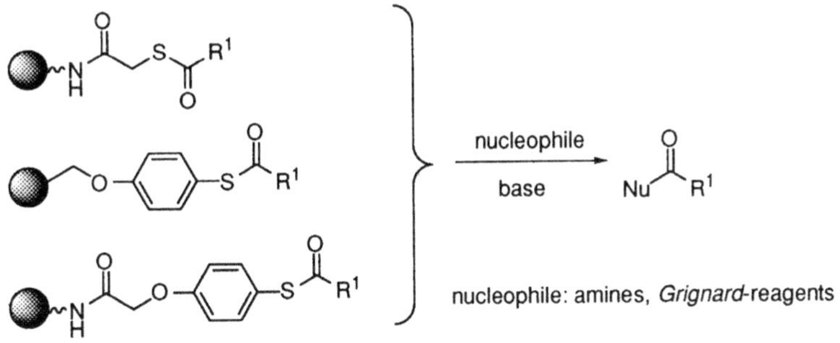
O
N
H
S
R1
O
O
S
R1
nucleophile
base
Nu
R1
nucleophile: amines, Grignard-reagents

Ref. [347]

e:

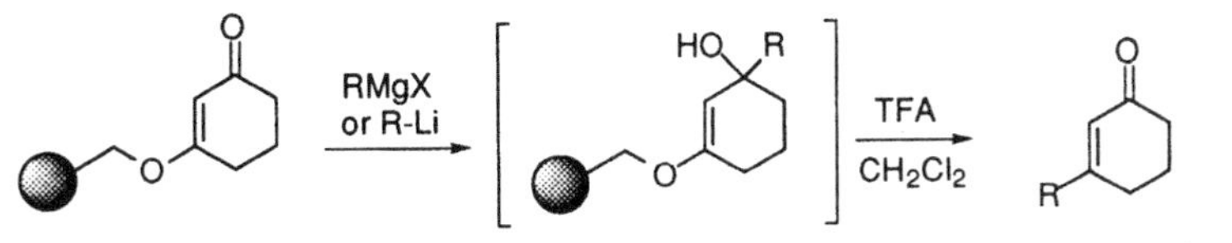
O
RMgX
or R-Li
HO
R
TFA
CH2Cl2
R

Ref. [348]

1.7.4.2. Multidirectional direct cleavage by electrophiles

Most of the linkers that allow multidirectional electrophilic cleavage from the resin are based on the chemistry of silicon. In *Figure 1.7.6* are summarised some of the most interesting examples.

Figure 1.7.6

a:

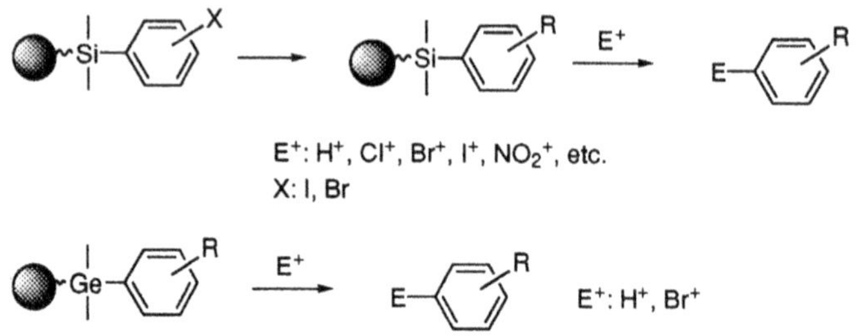

E$^+$: H$^+$, Cl$^+$, Br$^+$, I$^+$, NO$_2^+$, etc.
X: I, Br

E$^+$: H$^+$, Br$^+$

Refs. 349,350

b:

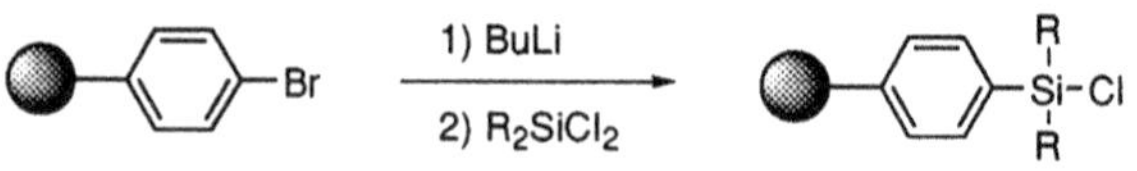

1) BuLi
2) R$_2$SiCl$_2$

Ref. 111

c:

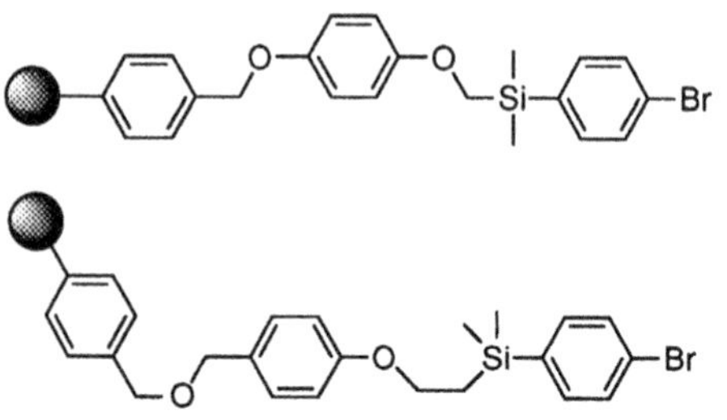

Ref. 351

d:

$-(CH_2)_3-Si-Cl$

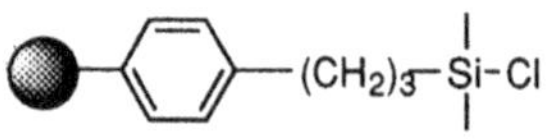

Ref. 352

1.7.4.3. "Safety-catch" linkers

Among the successful linker strategies that are being developed specifically for the solid-supported synthesis of small organic molecules, the "safety catch" principle has become increasingly popular. Thereby, a linker molecule is activated in the very last step before cleavage.

The major advantages are listed below:

- During synthesis of the library the linker moiety is completely stable to a wide range of reaction conditions.
- The linkage between the resin and the molecule can be specifically designed and planned in view of the structure and chemical stability of the final products.
- The linker group can often be reduced to a single atom such as sulfur, tin, silicon and others (*vide infra*). Thus, the rich chemistry and reactivities of these elements can be capitalised for the final activation and cleavage and can be integrated in the whole synthetic strategy.
- The "safety catch" principle generally leads to multidirectional resin cleavage which allows multiplication of the final library members both in terms of structural and functional diversity.

The safety catch principle has been first described by *Kenner et al.*[353] in the field of peptide chemistry and was originally based on the reactivity of the sulfonimide group. Meanwhile, this concept has been used successfully by several other groups. Examples are listed in *Figure 1.7.7*. Activation of the polymer-bound sulfonimido group in **1** (*Figure 1.7.7*, entry a) was achieved by alkylation with diazomethane or iodoacetonitrile. Subsequent substitution with nucleophiles such as amines and alkoholates resulted in multidirectional cleavage to form products of type **3** in high yield.[353-355] The same products can be obtained by using active esters of type **4** (*Figure 1.7.7*, entry b) which are activated via oxidation to the corresponding sulfones **5** followed by nucleophilic cleavage. This strategy was successfully used in an intramolecular version to form cyclic peptides.

Figure 1.7.7

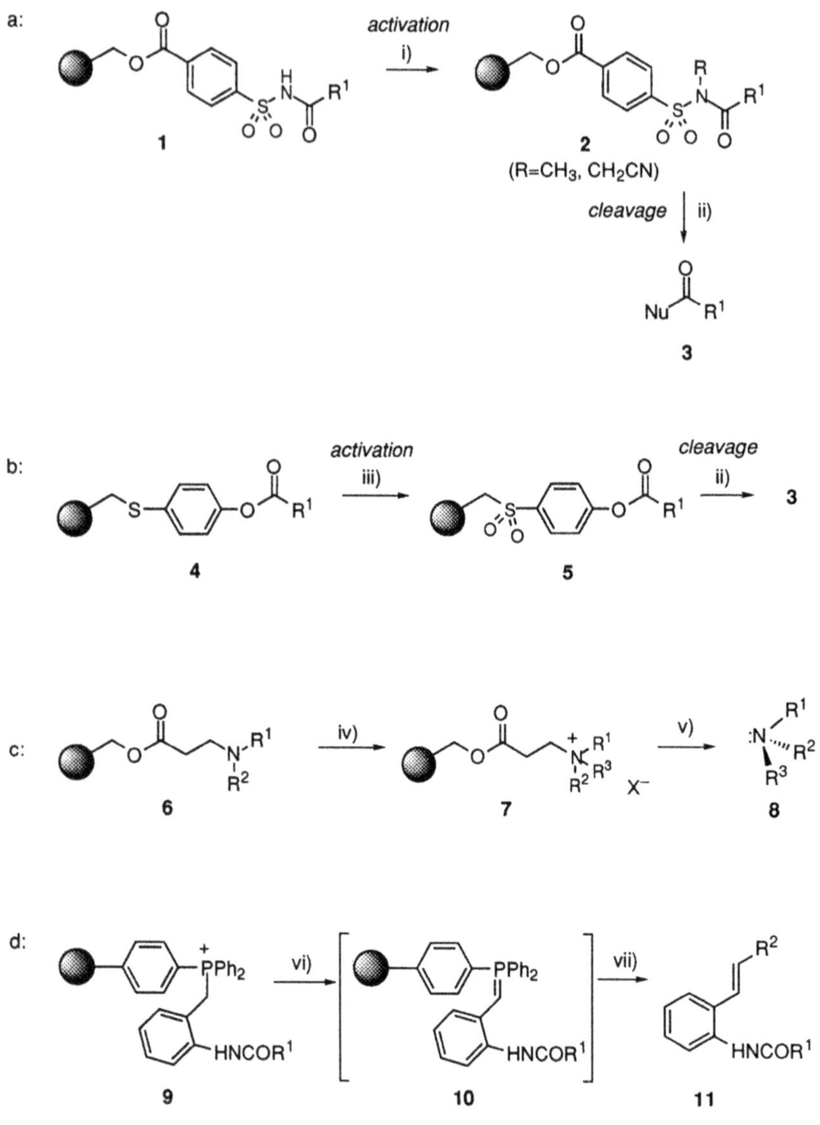
a:
activation
i)
1
2
(R=CH3, CH2CN)
cleavage
ii)
3
b:
activation
iii)
cleavage
ii)
3
4
5
c:
iv)
v)
X−
6
7
8
d:
vi)
vii)
PPh2
HNCOR1
9
10
11

e:

f:

g:

Reagents and conditions: i) CH_2N_2 or ICH_2CN, DBU; ii) Nu^- (amines or alcoholates); iii) *m*-CPBA, CH_2Cl_2; iv) R3-X, DMF; v) DIEA, DMF; vi) base; vii) R^2-CHO; viii) TFA; ix) Tf_2O, 2,6-di-*tert*-butyl-4-methylpyridine; CH_2Cl_2, -60 to -30°C; x) $Hg(OCOCF_3)_2$, CH_2Cl_2, H_2O, r.t.

The REM-linker can also be regarded as a safety-catch linker (*Figure 1.7.7*, entry c). Activation of polymer-bound **6** was achieved by alkylation to form **7**. Subsequent treatment with DIEA resulted in formation of the cleaved tertiary amines **8**.[327,328]

Polymer-bound phosphonium salts **9** form after activation with base the intermediate ylides **10** which can be multidirectionally cleaved by reaction with aldehydes R^2-CHO to form products of type **11**.[356]

An elegant biomimetic acyl-linker was designed and synthesised by *Frank et al.* (*Figure 1.7.7*, entry e) based on the chemistry of imidazoles. Thus, Boc-protected polymer-bound imidazole **12** was activated by cleavage of the Boc-protective group. Acyl transfer and subsequent nucleophilic cleavage led products of type **3**.[357,358]

A novel safety-catch principle for the multidirectional cleavage of pyrimidines and related heterocycles was first developed by *Obrecht et al.*[98,359] (*Figure 1.7.7*, entry f). Thus, polymer-bound thiopyrimidines of type **15** are activated by oxidation to sulfones **16** with *m*-CPBA. Multidirectional cleavage was performed with various nucleophiles such as amines, alcoholates, azide and C-nucleophiles like CN⁻ or malonates. Treatment of sulfones with ammonia or amines results in formation of pharmacologically interesting amino-pyrimidines. It is noteworthy that highly polar functional groups can be introduced in the very last cleavage step avoiding lengthy protective group manipulations. A similar strategy was recently described by *Gayo* and *Suto*.[360]

Another powerful sulfur-based safety-catch linker was developed by *Kahne et al.*[361-363] for the polymer-supported synthesis of oligosaccharides (*Figure 1.7.7*, entry g). Thio-linked saccharides can be efficiently cleaved from the resin in multidirectional fashion via sulfoxide **19** or by direct activation with $Hg(OCOCF_3)_2$.

These few recent examples impressively demonstrate the integration of the power of organic chemistry into the design and synthesis of linker and cleavage strategies for the creation of highly diverse, small "drug-like" molecule ensembles. Yet, there is much more to come in the future and it is easily imaginable that saftey-catch linkers will be the "winners" in this field.

1.8. Synthetic strategies in combinatorial and parallel synthesis

1.8.1. Solution- versus solid-phase chemistry

Since many advantages and disadvantages are associated with solution- and solid-phase chemistry we try to compare the pros and contras of the two strategies. We hope to be able to discern from this comparative study that the most efficient strategy to succesfully carry out parallel and combinatorial chemistry is to combine and integrate efficiently the strength and advantages of both methods in a given reaction sequence. As already pointed out in *Chapters 1.3, 1.5 and 1.6*, solid supported reagents and scavengers for excess of reagents are valuable tools for solution chemistry. In addition, reactions in solution can be complemented by additional solid-supported reaction steps.

solution chemistry

++ most reactions and reagents have been studied in solution

+ usually no excess of reagents have to be used

+ solvent effects can be studied and altered more readily

++ steric effects are usually less pronounced in solution and can be overcome more easily by using more drastic reaction conditions

++ reaction conditions are usually more easily adapted to a large variety of substituents

– – extensive and time consuming, chromatographic purifications are often necessary

–+ side products have to be separated and analysed (can be also an advantage in the first exploratory stage of a given project)

– – parallelisation and automation usually requires more initial effort

solution strategies will probably be more efficient for small libraries since less effort has to be put into method development

solid-phase chemistry

++ excess of reagents can be used to drive reactions to completion

++ purification procedures achieved by simple filtrations which can be easily automated

++ assuming complete spacial separation of the reactive sites on a given solid support, the principle of high dilution ("hyperentropic effect"[6]) can be used beneficially *e.g.* for intramolecular cyclisation reactions

++ the costs have to be compared with the labour intensive extraction and purification procedures usually necessary in solution

+/– linker molecules have to be designed which are compatible with the chemistry and the polymer matrix. Since the linker can be an integral part of the chemical strategy, this feature can also be advantageous.

– reaction conditions have to be established for each case

– the range of reactivity seems to be more restricted on solid supports than in solution

– chemistry on polymers is relatively expensive: costs for the polymers, robotic systems and for the solvents have to be considered

– reactions on solid support are more difficult to monitor

aspects of using highly loaded resins

++ especially critical for polymer-bound reagents and catalysts

++ irreversible binding of unpleasantly smelling and aggressive compounds

– high loading may lead to crosslinking reactions and thus interfere with the principle of high dilution

+ crosslinking reactions can be essentially completely suppressed by using correctly designed linkers.

+ high loading can be favourable for the reactivity of polymer-bound catalysts and reagents and also for intramolecular cyclisations and resin cleavage reactions

Solid-supported chemistry is very useful for the synthesis of larger series of compounds: investments in the development of methods will be paid off by faster production through simple robotic methods.

1.8.2. Compound mixtures versus single compound collections

As pointed out earlier, combinatorial and parallel synthesis are used to synthesise mixtures as well as single compounds. When evaluating the advantages and drawbacks of mixtures and single compound collections it is important to consider the following aspects:

- **Type of compounds:** - polypeptides, oligonucleotides, oligosaccharides
  - small carbocyclic and heterocyclic molecules

- **Type of library:** **random libraries**: large compound libraries consisting of highly diverse compounds

    (*mixtures and single compounds*)

    focused libraries: smaller compound libraries containing designed elements of diversity

    (*single compounds preferred*)

    lead optimisation libraries: small tailor-made compound collections centered around a lead structure

    (*single compounds essential*)

- **library format and assay throughput**

As discussed and schematically shown in *Figure 1.4.3* the lead discovery process is divided in several phases. In the very early phase of hit identification, where a large number of compounds are screened in the HTS, libraries usually consist of randomly selected compound collections where the structural and chemical diversity should be as large and random as possible.

Many tools are available to date in order to assess qualitatively the degree of diversity present in a given collection (some of those will be discussed in *Chapter 1.11*). In cases where a high throughput screen is available, single compounds or mixtures can be screened. Several pharmaceutical companies store and test their compound collections for general screening purposes as cocktails (typically 10-50 compounds per cocktail), others screen only single compounds. Screening cocktails has the advantage that less biological material is necessary to screen the whole library. There are also many problems associated with the screening of cocktails. Synergic and non-synergic effects between compounds in the cocktail often complicate an unambiguous interpretation of the biological results. A pioneering study performed at *Pfizer*[14] revealed that mixtures above twenty compounds are prone to a lot of false positive results. The storage of mixtures can also be problematic due to possible degradation and chemical modifications of the products. In addition, the high degree of miniaturisation of the screening format and the high throughput speaks rather for the single compound approach. Once one or several hits have been identified, focused libraries incorporating structural features around these hits can be planned and synthesised. Parallel and combinatorial chemistry approaches are especially valuable tools in this phase of lead discovery as they speed up the validation and identification process of potential hits and leads. If high throughput screens are available single compound collections will be the favourite choice.

1.8.3. Parallel and split-mixed synthesis

Reactions in solution or on solid supports can be carried out in a parallel format as schematically shown in *Figure 1.8.3.1*, where the two reactive building blocks **RBA¹** and **RBA²** are reacted with building blocks **RBB¹** and **RBB²** in four different vessels giving rise to the first generation of four intermediate products which are combined with **RBC¹** and **RBC²** to give the eight possible products in eight different reaction vessels. This protocol allows to unambiguously identify every product in each reaction vessel but requires all manipulations for each reaction. Efficient parallel processing and automation of the time consuming steps such as washing procedures is a prerequisite in this approach.

In the parallel approach **RBA¹** and **RBA²** can also be reacted with a mixture of building blocks **RBB¹** and **RBB²** giving rise to two mixtures of two compounds. Each mixture would further react with building blocks **RBC¹** and **RBC²** in two separate vessels to give two mixtures of four compounds as indicated in *Figure 1.8.3.2*.

A major problem associated with the synthesis of mixtures is that not all the reaction partners will couple with the same reaction rate and this will give rise to a non-equimolar distribution of the products. This problem could be solved more or less efficiently in the field of peptide[364,365] and oligonucleotide chemistry[366] by adjusting the concentrations of the amino acid- and nucleic acid components according to their respective reaction rates. In the field of combinatorial synthesis of small molecules this problem is much more severe.

Using the "split-mixed" or the "split and combine" technology developed by *Furka et al.*,[15,367] *Lam et al.*[368] and *Houghten et al.*[369] the problems of different reaction rates can be elegantly avoided as depicted schematically in *Figure 1.8.3.3*.

The polymer-bound reactive building blocks **RBA¹** and **RBA²** are in the first step mixed and then redivided in two equal portions of beads (**RBA¹** and **RBA²** are represented in roughly equimolar distribution) which are then reacted individually with building blocks **RBB¹** and **RBB²** to yield two pools (**A¹** and **B¹**) of beads. As only one building block is coupled to one pool of beads large excesses of reagents can be used. Since the two pools contain a mixture of two product beads the product identification has to be performed either by cleavage of the products and analysis by mass spectrometry or by deconvolution or tagging techniques (see *Chapter 1.10*). The two pools **A¹** and **B¹** are mixed, redivided and further individually coupled with building blocks **RBC¹** and **RBC²** to give two pools (**A²** and **B²**) of beads containing four products each. Using this technology, all eight products can be generated with only four reactions compared to twelve individual reactions necessary in parallel approach (*Figure 1.8.3.1*).

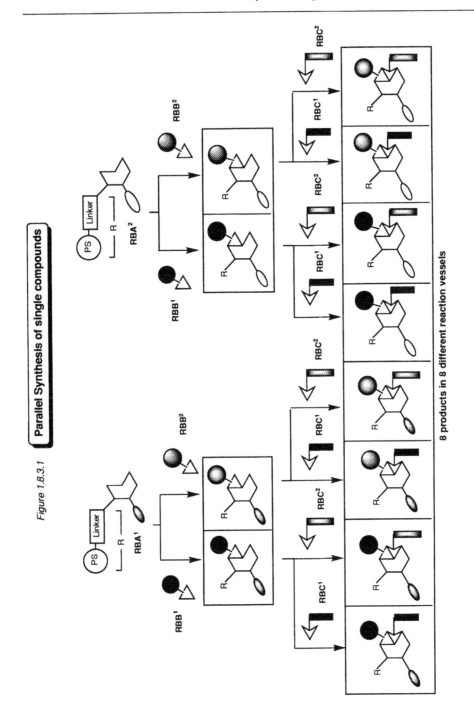

Figure 1.8.3.1

Parallel Synthesis of single compounds

8 products in 8 different reaction vessels

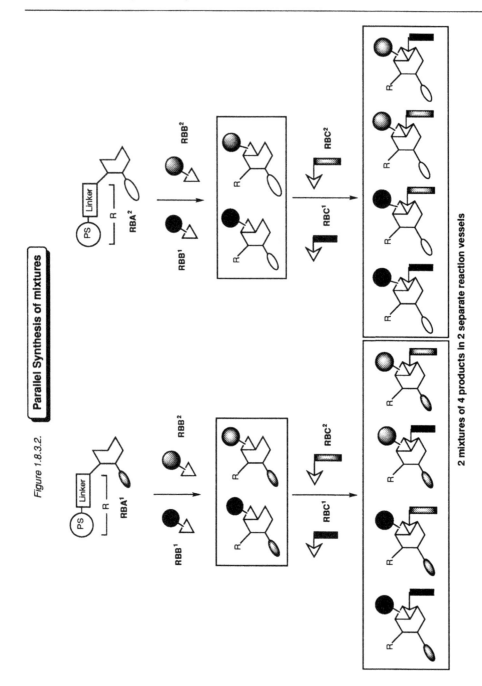

Figure 1.8.3.2. Parallel Synthesis of mixtures

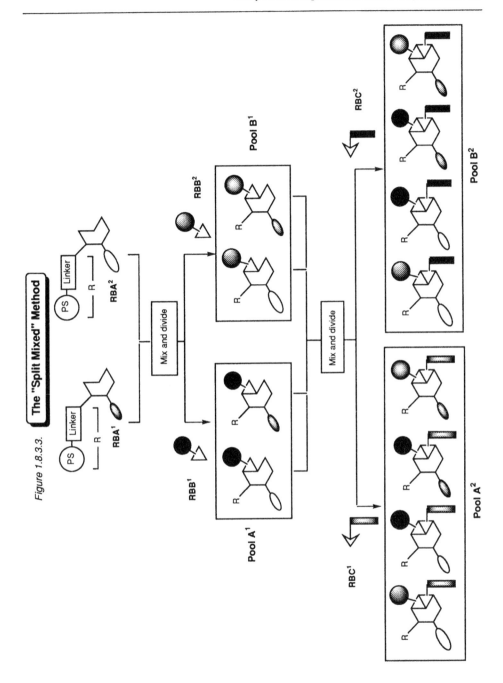

Figure 1.8.3.3.

Advantages of the split-mixed methodology

The split-mixed methodology is ideal for linear stategies like the synthesis of polypeptide libraries where the economisation of reaction steps gains importance. Large libraries can be synthesised, every bead containing only one single compound. Tagging or deconvolution are usually required in order to determine the structure of active compounds.

The substances can be either tested on bead or after cleavage depending on the availability of biological tests.

The methodology is limited, however, to relatively little material available on a single bead. Furthermore, resynthesis of the interesting members of the library in greater amounts will be necessary.

1.8.4. Linear versus convergent strategies

In *Chapters 1.8.1-1.8.3* we discussed the fundamental strategies of parallel and combinatorial synthesis. Whether reactions should be performed in solution or on solid supports or whether the compounds are obtained as singles or as mixtures from parallel or split-mixed synthesis, depends largely on the complexity of the underlying chemistry and the way compounds are identified and screened. The resulting structural diversity, however, is only determined by the way the molecules were assembled and the types of building blocks employed.

The basic idea of combinatorial chemistry is to synthesise (starting *e.g.* from building blocks A, B, C^1, C^2, and C^3) all possible combinations of products (P^1-P^3, etc. (*Figure 1.8.4.1*)) either as single compounds or as mixtures. Since products P^1-P^3 were created from similar building blocks using the same chemical strategy, they form an *ensemble* or a *library* of compounds having the same *core structure* with all possible combinations of appended substituents. Prerequisites for a successful combinatorial approach are sets of highly reactive building blocks of type A, B, and C which can be readily assembled in a **linear** or in a **convergent** fashion (*Figure 1.8.4.1*).

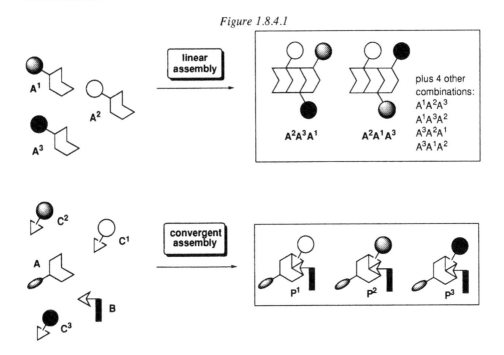

Figure 1.8.4.1

Linear assembly

In a linear assembly strategy (see also *Figure 1.3.2, Chapter 1.3*) building blocks of one single type, *e.g.* A, are combined using usually a well established bond forming reaction, *e.g.* amide bonds for peptides, phosphorester bonds for oligonucleotides and glycosidic linkage for oligosaccharides. Assembly of A^1-A^3 readily gives access to 6 possible combinations (assuming that for steric reasons not all three substituents can be adjacent to each other *(Figure 1.8.4.1*, upper part).

Not surprisingly, combinatorial chemistry originated from linear assembly strategies developed in peptide and oligonucleotide chemistry (cf. *Chapter 2*), since both building blocks and coupling chemistry have been extensively studied in solution as well as on solid support.

Convergent assembly

In a convergent assembly strategy the building blocks A, B, and C^1-C^3 ultimately can form products P^1-P^3. As can be seen from the simple sketch in *Figure 1.8.4.1* (lower part) the core structure differs significantly from linear strategies. A convergent strategy is usually more demanding both in terms of having access to the reactive building blocks A, B, C^1-C^3 ("reactophores") as well as elaborating the corresponding assembly chemistry. The whole plethora of organic reactions, *e.g.* condensation and pericyclic reactions, will be included. As will be discussed in *Chapter 1.8.5* and *1.8.6*, an exciting aspect of convergent strategies lies in the fact that building blocks A, B, and C^1-C^3 can potentially be assembled in different ways (see *Figures 1.8.5.1* and *1.8.5.2*).

There is currently a significant shift observable from linear to convergent strategies as mentioned in *Chapter 1.1*, probably due to the fact that convergent approaches are more likely than linear ones to generate small "drug-like" molecules.

1.8.5. Multicomponent and multigeneration reactions

When looking at multicomponent reactions we have to differentiate between *multicomponent one-pot* reactions, where the reactive components react to a product without giving rise to isolable intermediates, and *multigeneration* reactions, where the reactions proceed via intermediates which can be trapped and isolated. The two reaction schemes differ significantly in their scope and application for combinatorial organic synthesis as schematically depicted in *Figures 1.8.5.1* and *1.8.5.2*.

Lets suppose a case where polymer-supported reactive building block PS-A reacts with components B and C^1-C^3. A multicomponent one-pot reaction will always generate three polymer-bound products PS-P^1 to PS-P^3 and after cleavage three final products P^1 to P^3 *irrespective of the sequence of addition of the reagents (Figures 1.8.5.1* and *1.8.5.2* upper half). Conversely, in the multigeneration version the sequence of mixing the components is important and consequently we have to distinguish between two possible cases as shown in *Figure 1.8.5.1* (strategy A, lower half) and *Figure 1.8.5.2* (strategy B, lower half). In strategy A, the polymer-bound reagent PS-A reacts first with building block B to generate the stable intermediate PS-P^4, which can either be cleaved to give 1st generation products P^4 and/or be reacted with C^1-C^3 giving rise after cleavage to the products P^1-P^3 (2nd generation).

In strategy B, PS-A is first reacted with building blocks C^1-C^3 to generate the stable intermediates PS-P^4 to PS-P^6, which can be cleaved to give 1st generation products P^4-P^6 and/or be reacted with B to generate after cleavage from the resin again products P^1-P^3. It is important to note that the multicomponent one-pot reaction always generates P^1-P^3 whereas in strategy A (*Figure 1.8.5.1*) of the multicomponent cascade version four products P^1-P^4 with two different core structures are formed while in strategy B *(Figure 1.8.5.2)* six products P^1-P^6 with two different core structures are generated.

This simple sketch already illustrates the implications of multicomponent one-pot and multigeneration reactions on the number and on the diversity of the products that can be generated. While one-pot reactions allow the parallel generation of products with a minimum amount of manipulations and vessel transfers, the multigeneration versions offer the benefit of having access to a larger number of compounds and core structures, which ultimately results in greater diversity of the produced compound collections. In *Chapter 1.8.5.1* we will describe some of the most important multicomponent one-pot reactions, while in *Chapter 1.8.5.2* we will discuss some aspects of multigeneration reactions.

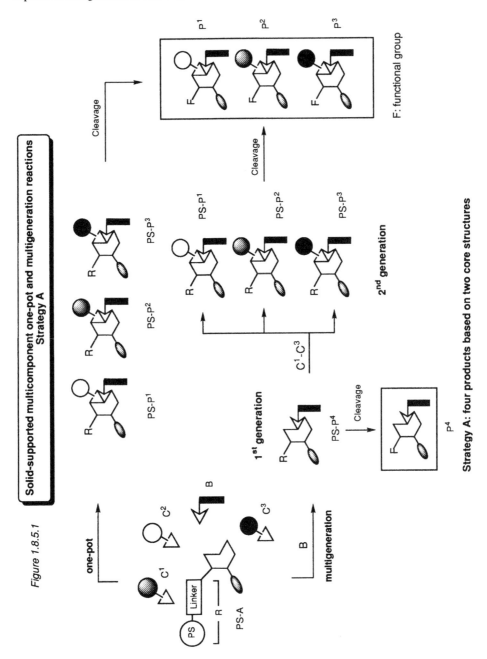

Figure 1.8.5.1

Solid-supported multicomponent one-pot and multigeneration reactions
Strategy A

Strategy A: four products based on two core structures

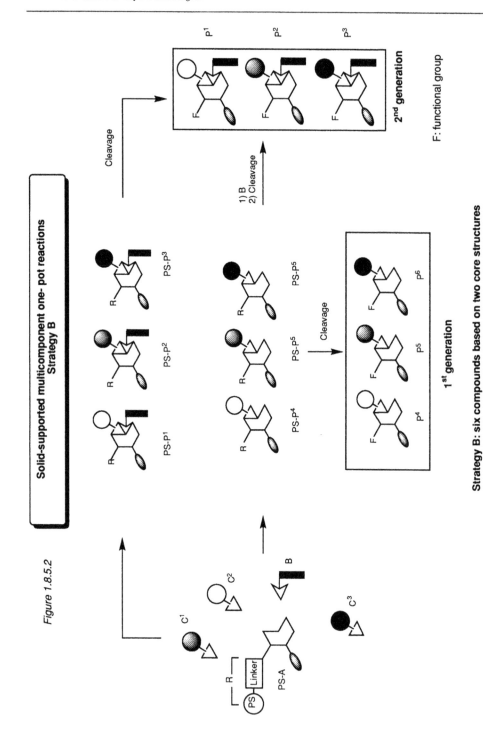

Figure 1.8.5.2

Solid-supported multicomponent one- pot reactions
Strategy B

Strategy B: six compounds based on two core structures

F: functional group

1.8.5.1. Multicomponent one-pot reactions

The pioneer of multicomponent reactions and one of the first scientists to extend multicomponent reactions to the generation of compound libraries was certainly *Ivar Ugi*. The two classical four component reactions combining an acid (R^1COOH, **1**), an aldehyde (R^2CHO, **2**), a primary amine (R^3NH_2, **3**) or a secondary amine (R^3R^4NH, **5**) and an isonitrile (R^4NC or R^5NC, **4**) are described in *Schemes 1.8.5.1* and *1.8.5.2*, respectively.

Scheme 1.8.5.1

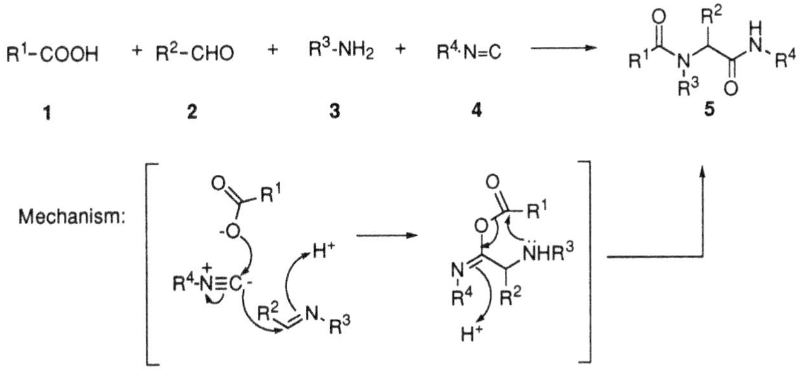

Scheme 1.8.5.2.

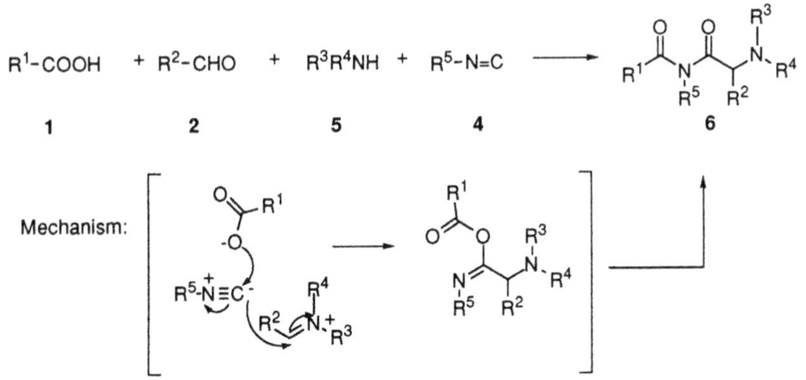

As typical for multicomponent one-pot reactions, the four components have potentially many possibilities to react in one way or another and many reversible equilibria will co-exist in the reaction mixture. After formation of the imine or immonium species which is still a reversible process, reaction of the isonitrile and the carboxylate with the imine generates a reactive acyloxy-imine, which in the case of a primary amine **3** can rearrange to form a *N*-alkylated dipeptide **5**

(*Scheme 1.8.5.1*) or, in the case of a secondary amine **5**, can rearrange to give imide **6**. Both products constitute the thermodynamically most stable species in the reaction mixture. Many variations of the *Ugi* four component condensation (*Ugi*-4CC) have been published and a short compilation is depicted in *Figure 1.8.5.3*.

Figure 1.8.5.3

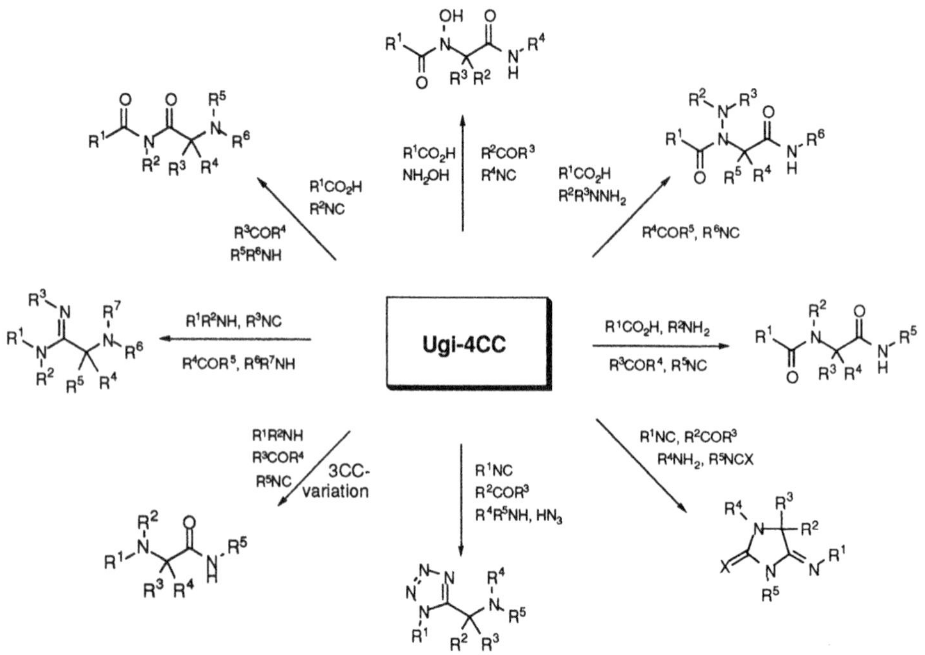

Related to the *Ugi* reaction is the *Passerini* three component variation[370] as shown in *Scheme 1.8.5.3*.

Scheme 1.8.5.3

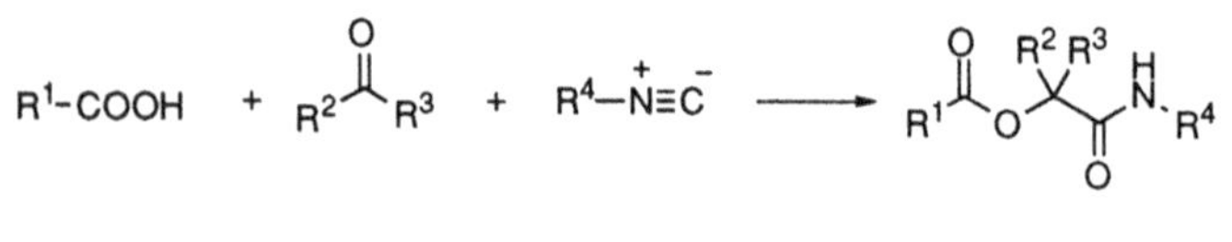

Several variations in which two of the functional groups are attached in the same molecule have been successfully used to synthesise anellated heterocyclic ring systems. Several interesting applications will be shown in *Chapter 3.3*.

118

In *Scheme 1.8.5.4*, three versions of the *Hantzsch* multicomponent one-pot reaction are shown leading to thiazoles, pyrroles and dihydropyridines.[371]

Scheme 1.8.5.4

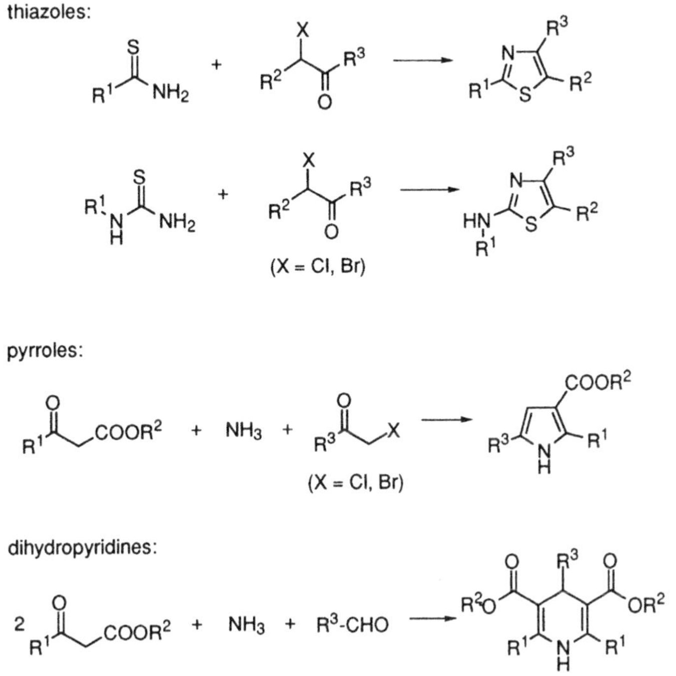

thiazoles:

(X = Cl, Br)

pyrroles:

(X = Cl, Br)

dihydropyridines:

One of the oldest multicomponent reactions is certainly the *Strecker*-synthesis,[372,373] in which an aldehyde is condensed with ammonia and HCN to form an aminonitrile as shown in *Scheme 1.8.5.5* (entry a).

Scheme 1.8.5.5

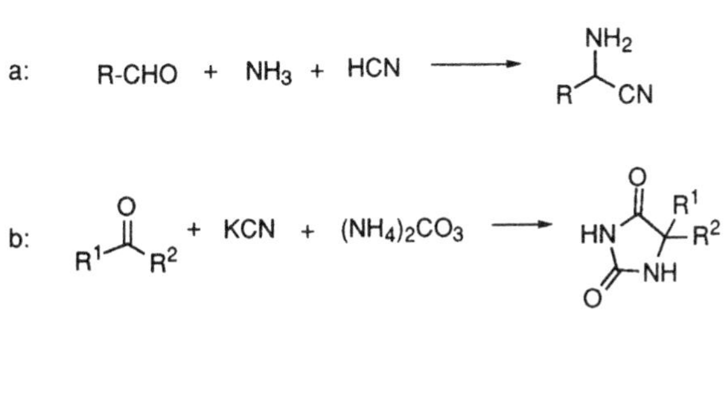

a: R-CHO + NH$_3$ + HCN $\longrightarrow$

b:

As many of the classical multicomponent reactions, the *Strecker* synthesis also takes advantage of the versatile chemistry of the initially formed imine. The formation of the amino nitrile, however, is reversible under the reaction conditions which usually results in lower yields. This problem was elegantly solved in the *Bucherer-Bergs* variation,[374-376] where the initially formed aminonitrile is irreversibly trapped by formation of a hydantoin as depicted in *Scheme 1.8.5.5* (entry b).

The *Mannich* reaction[377] (*Scheme 1.8.5.6*) involved in the biosynthesis of many alkaloids (*e.g.* tropane alkaloids, gramine) has been synthetically used in many elegant syntheses of alkaloids such as porantherine.[378]

Scheme 1.8.5.6

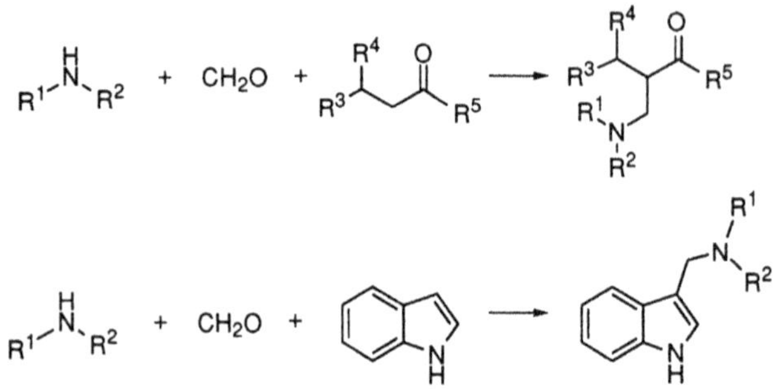

The *Erlenmeyer* azlactone synthesis (*Scheme 1.8.5.7*) combining a hippuric acid derivative, acetic anhydride and an aldehyde to form a didehydroazlactone. This azlactone approach has been extensively used for the synthesis of amino acids.[359]

Scheme 1.8.5.7

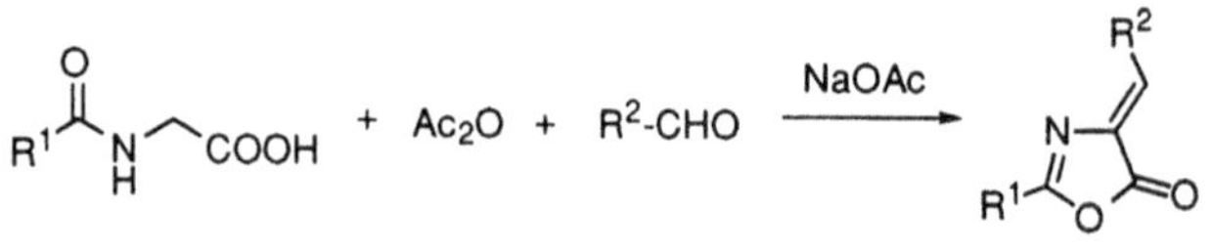

The *Biginelli* reaction[379,380] combines ureas with an aldehyde and a β-oxoacid derivative to form dehydropyrimidones as shown in *Scheme 1.8.5.8*.

Scheme 1.8.5.8

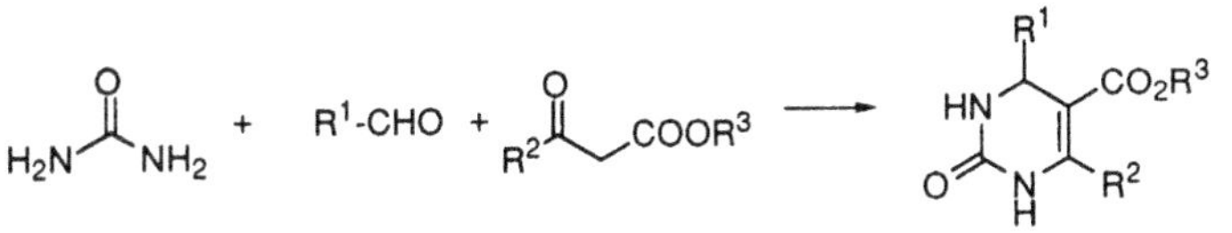

The *Atwal* modification of the *Hantzsch* synthesis consists of a condensation reaction between a thiouronium salt, an aldehyde and a β-oxoacid derivative to form dihydropyrimidine derivatives [13] as shown in *Scheme 1.8.5.9*.

Scheme 1.8.5.9

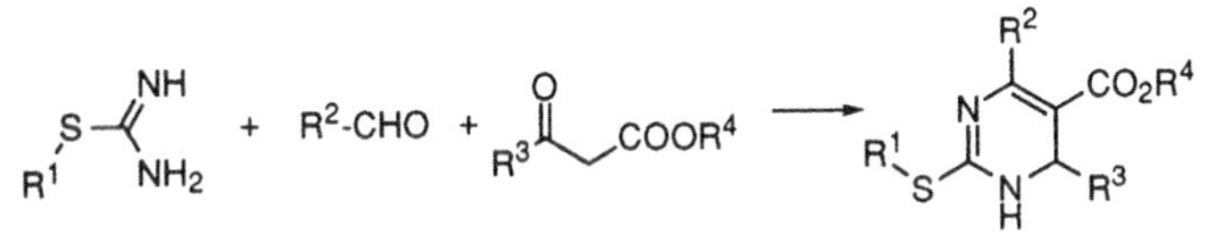

An interesting variation of the classical multicomponent reaction was recently described starting from a glyoxale and a tetrahydroisoquinoline generating in the first step an immonium species which underwent 1,3-dipolar cycloaddition with *N*-methyl maleimide as shown in *Scheme 1.8.5.10*.[381]

Scheme 1.8.5.10.

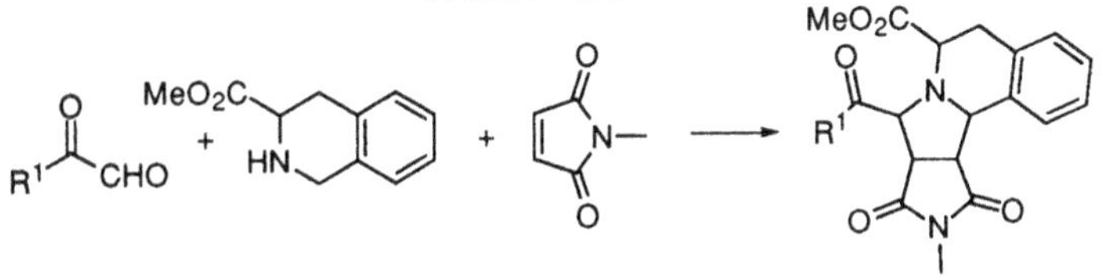

Combining a glyoxale, 2-(phenylthio)acetic acid and an isonitrile gave, in a modification of the *Passerini* reaction, substituted oxazoles[382] as depicted in *Scheme 1.8.5.11*.

Scheme 1.8.5.11.

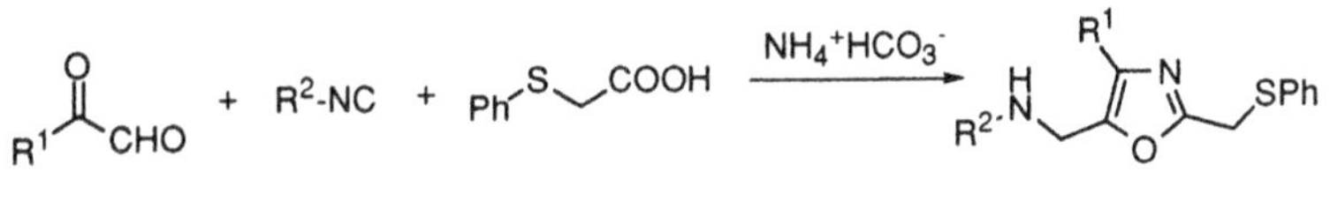

A novel three component reaction developed by *Hogan* and coworkers[383] involves condensation of esters, substituted hydrazines and epoxides to yield hydrazide derivatives as shown in *Scheme 1.8.5.12*.

Scheme 1.8.5.12

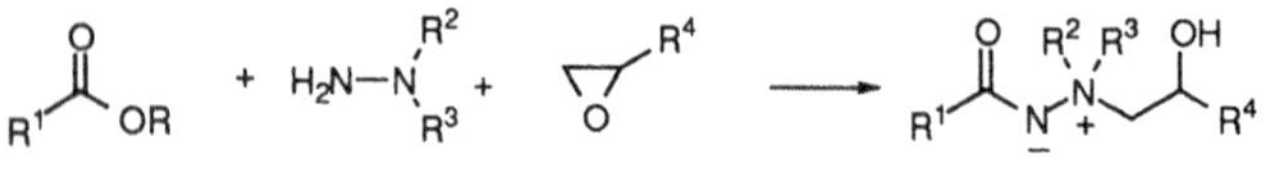

An interesting novel, *Lewis* acid catalysed three (or four) component reaction between a silyl enol ether, an enone and an imine leading to δ-lactams was recently reported by *Kobayashi et al.* [384]

Scheme 1.8.5.13

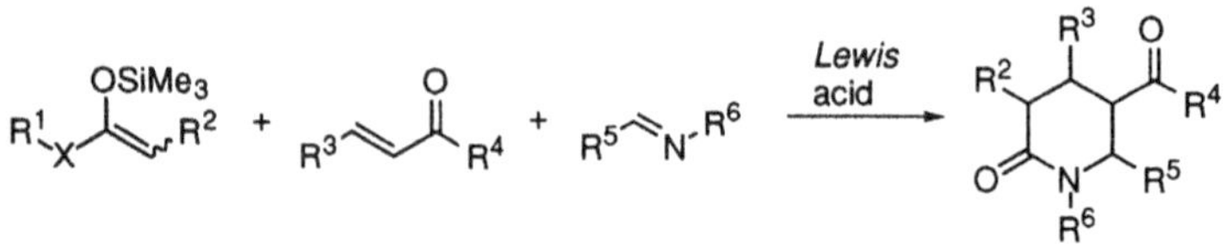

The *Pauson-Khand* reaction combines efficiently an alkyne, an alkene and carbon monoxide to form cyclopentenones (*Scheme 1.8.5.14*) and belongs to the very few purely C-C-coupling based multicomponent reactions.[385]

Scheme 1.8.5.14

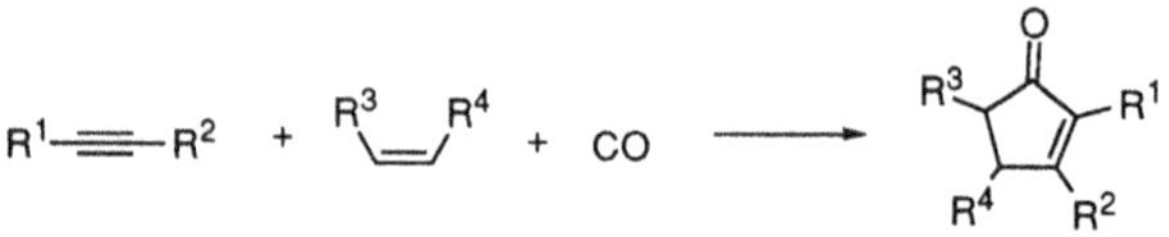

An interesting three component reaction combining anilines, olefins and aldehydes forming tetrahydroquinolines was described by *Grieco et al.*[386] and applied by *Armstrong et al.* for the construction of libraries based on this core structure[387]is presented in *Scheme 1.8.5.15*.

Scheme 1.8.5.15

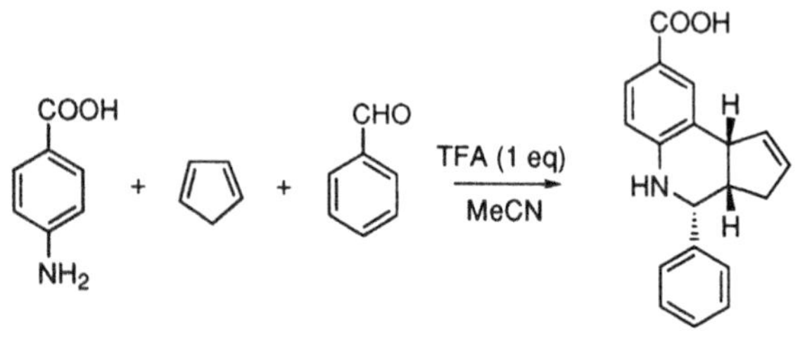

1.8.5.2. Multigeneration reactions

The following examples illustrate the important differences between multicomponent one-pot reactions and multigeneration reactions. Many of the previously mentioned multicomponent one-pot reactions proceeded via intermediately formed imines. Instead of forming these imines *in situ* they can be synthesised separately and engaged in a whole range of reactions as schematically shown in *Scheme 1.8.5.16*.

The utility of carrying out the reactions in a multigeneration mode is obvious. The number of accessible products and core structures increases substantially.

Scheme 1.8.5.16

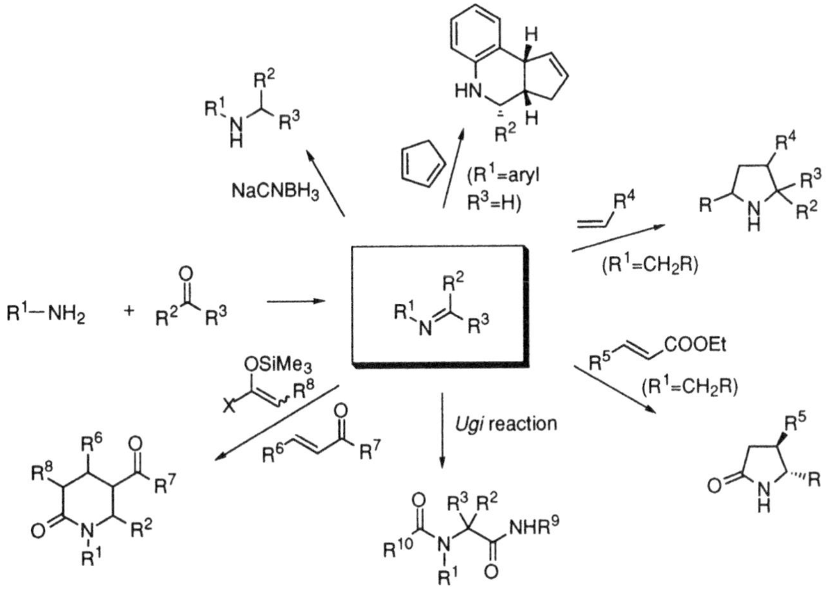

Indeed, most reaction sequences can be regarded as multigeneration reactions in which every reaction step generates a reactive species ready to undergo the next transformation. The perfection of this concept are the so-called "tandem"-, "domino"- or "cascade"-reactions in which the intermediates undergo the next transformation directly, i.e. without isolation or activation. Two examples will be discussed below to illustrate this concept.

Blechert et al.[388] developed a multicomponent cascade reaction for the synthesis of indole derivatives as depicted in *Scheme 1.8.5.17*. The first step of the sequence involves formation of a nitrone derivative starting from phenylhydroxylamines and aldehydes. The resulting nitrones were not isolated but captured by a cyanoallene in a 1,3-dipolar cycloaddition reaction followed by hetero-*Cope* rearrangement, ring-opening and condensation to yield an indole derivative.

Scheme 1.8.5.17

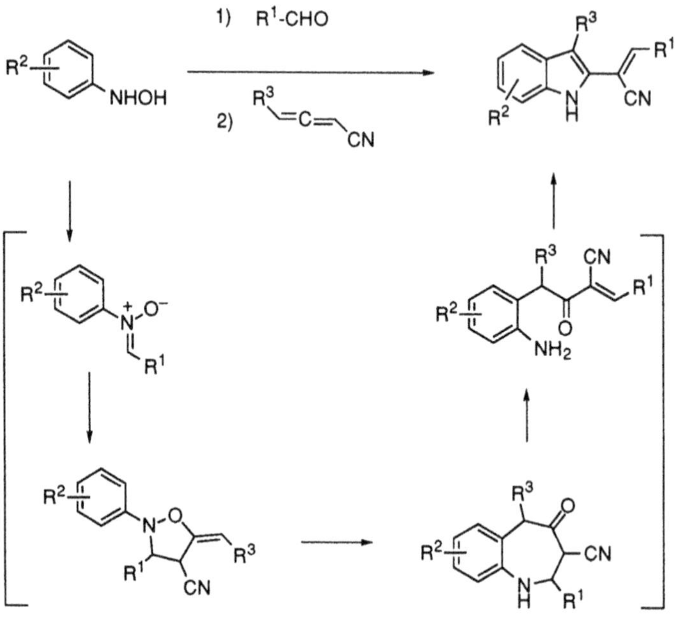

Variation of the different building blocks allowed subsequent anellation reactions. A broad range of indole alkaloid derivatives were generated using this strategy as shown in *Figure 1.8.5.4*.

Figure 1.8.5.4

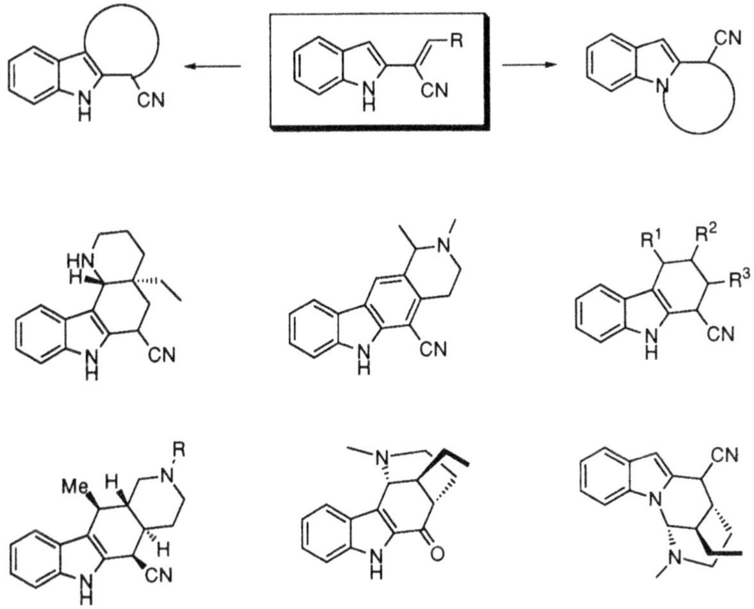

The tandem reaction studied by *Neier et al.*[389-391] is a combination of a *Diels-Alder* process and a *Ireland-Claisen* rearrangement. These pericyclic reactions usually proceed with good regio- and diastereoselectivities. In addition, the mechanisms of both reaction types are well known. Thus, *O*-butadienyl-*O*-TBDMS ketene acetals **1a-e** react with electron-deficient dienophiles such as *N*-phenylmaleimide at 60°C to yield initial *Diels-Alder* products *rac*-**2a-e**. These products are not isolated, but the newly created double bond in the cyclohexene moiety participates instantanously in an *Ireland-Claisen*-type rearrangement to form the substituted cyclohexenes *rac*-**3a-e** as single diastereoisomers. The products *rac*-**5a-e** were isolated after hydrolysis of the silyl ester and esterification with TMS-diazomethane in 43-77% yield (*Scheme 1.8.5.18*). The relative configuration of the products was controlled by the geometry of the starting ketene acetal **1**, the *endo*-selectivity of the *Diels-Alder* reaction, the facial selectivity of the sigmatropic rearrangement involving a boat-shaped transition state and was confirmed by an X-ray crystal structure determination of **4e**.

Scheme 1.8.5.18

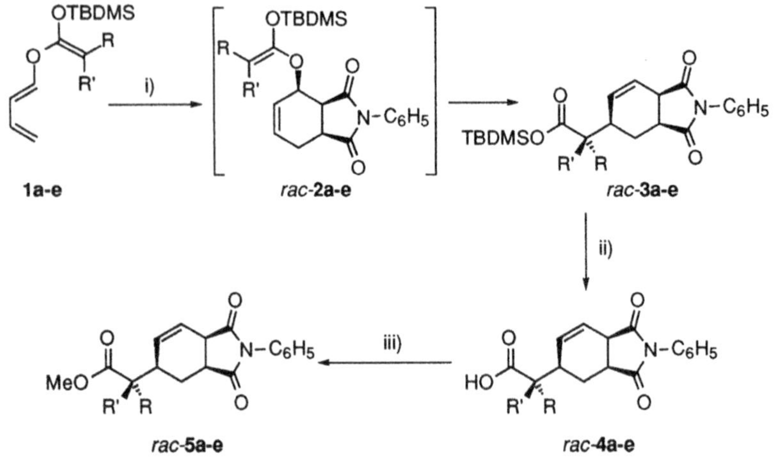

5a: R, R' = H, H (70%)
5b: R, R' = CH₃, H (77%)
5c: R, R' = CH₃, CH₃ (56%)
5d: R, R' = OCH₃, H (75%)
5e: R, R' = Phthalimido, H (65%)

Reagents and conditions: i) *N*-phenylmaleimide, THF, 60°C, r.t.; ii) 25% H_2SiF_6, DME, 30 min r.t.; iii) TMSCHN₂, THF, CH₃OH, 2h, r.t.

The cyclohexenes **5a-e** obtained from the tandem *Diels-Alder* reaction / *Ireland-Claisen* rearrangement sequence represent interesting building blocks with well defined geometry which are not easily accessible using other routes.

1.9. Classification of reactive building blocks

Despite the fact that more effort has to be put into the synthesis of suitable building blocks, we focus mainly on convergent strategies, which will lead more likely than linear strategies to small-molecular-weight "drug-like" molecules. *Figures 1.9.1* and *1.9.2* schematically show the construction of a library of molecules on solid support using a convergent multigeneration and a combined multigeneration/multicomponent approach coupled with monofunctional or multidirectional cleavage.

The first step usually involves coupling of a reactive building block **RB1** ("reactophor 1") via a linker molecule to a solid support. A reactophor is a reactive building block that consists essentially only of a reactive group and appended orthogonally protected functional groups which will serve as points for diversification. As previously discussed, the choice of the linker group is of prime importance as it determines the scope and limitations of the chemistry that can be employed to synthesise the compounds. Building block **RB1** consists of a reactive part that serves to construct the core structure of the target molecule, of one or several appended orthogonally protected functional groups G^1 and of an attachment site for a suitable linker group. The core structure, which will be most likely an interesting pharmacophor, is assembled by the reaction of **RB1** with **RB2**, which also contains orthogonally protected functional groups G^2 and G^3. Groups G^1-G^3 can now be converted into the final modified groups A, B, and C employing *l* building blocks of type A, *m* of type B and *n* of type C (*Figure 1.9.1*). In all these reactions the actual library is formed either as single compounds when the reactions are all carried out in separate vessels, or as mixtures when all reagents A(*l*), B(*m*) and C(*n*) are added at the same time to a suitably activated functional group. In *Figure 1.9.2* the modification of G^3 occurs by means of a multicomponent reaction (polymer- bound molecule + components C, D and E) leading to the generation of *l* x *m* x n^1 x n^2 x n^3 polymer-bound molecules.

This sketch emphasises again the importance of using multicomponent reactions in the diversification steps. Once the library is formed on the solid support, the molecules can be cleaved either by liberating *one functional group* or in a *multidirectional mode*. Note again, that the latter case generates *o* analogues of the built library in the final step.

Figure 1.9.1. **Convergent multigeneration assembly strategies**

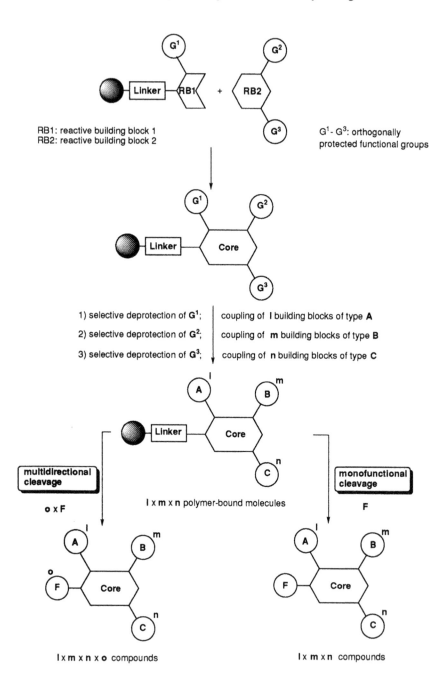

RB1: reactive building block 1
RB2: reactive building block 2

G^1- G^3: orthogonally
protected functional groups

1) selective deprotection of G^1;   coupling of l building blocks of type **A**

2) selective deprotection of G^2;   coupling of m building blocks of type **B**

3) selective deprotection of G^3;   coupling of n building blocks of type **C**

l x m x n polymer-bound molecules

multidirectional cleavage

o x F

monofunctional cleavage

F

l x m x n x o compounds

l x m x n compounds

Figure 1.9.2. **Convergent multigeneration/multicomponent assembly strategy**

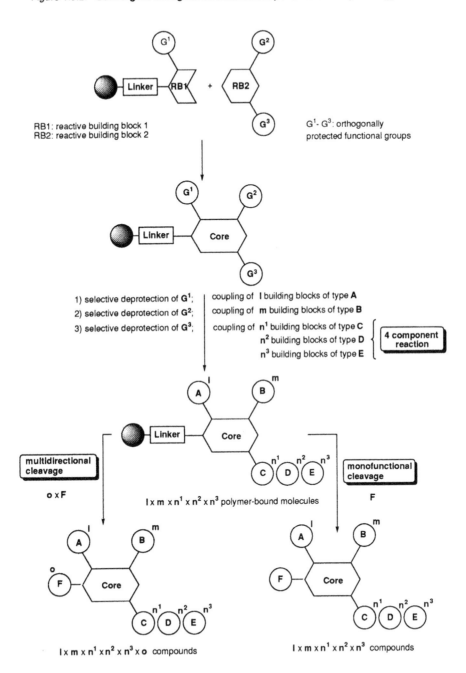

RB1: reactive building block 1
RB2: reactive building block 2

G^1- G^3: orthogonally protected functional groups

1) selective deprotection of G^1;   coupling of **l** building blocks of type **A**
2) selective deprotection of G^2;   coupling of **m** building blocks of type **B**
3) selective deprotection of G^3;   coupling of n^1 building blocks of type **C**
                                                            n^2 building blocks of type **D**   **4 component reaction**
                                                            n^3 building blocks of type **E**

multidirectional cleavage

monofunctional cleavage

$o \times F$

$l \times m \times n^1 \times n^2 \times n^3$ polymer-bound molecules

F

$l \times m \times n^1 \times n^2 \times n^3 \times o$ compounds

$l \times m \times n^1 \times n^2 \times n^3$ compounds

The important features of such an approach can be summarised as follows:

- The careful selection of the reactive building blocks ("reactophores") is a key issue as they predetermine the core structure and the appended functional groups. Orthogonally protected functional groups minimise the number of chemical transformations on the solid support and thus will increase the overall yield.

- In the actual diversification phase it is useful to incorporate multicomponent reactions for creating a large set of core structure analogues in a given step.

- Multidirectional cleavage reactions also amplify the number and the actual diversity of the synthesised compound library as shown in *Figures 1.9.1* and *1.9.2*.

- It was our aim to extract from the literature and from our own work the prototypical reactive building blocks that are especially well suited for such a multigeneration/multicomponent approach. We have classified the reactophores into three categories:

  - mono-, bis- and polydentate donor molecules (*Chapter 1.9.1*)

  - mono-, bis-, tris- and polyacceptor molecules (*Chapter 1.9.2*)

  - acceptor-donor molecules (*Chapter 1.9.3*)

The goals of classification:

In the initial phase the classification of reactophores was just an excercise to get an overview about the type of building blocks which were employed in solid-phase reactions. A careful analysis of those building blocks and reactions revealed that essentially all of them could be grouped into the above mentioned categories. In a second study we analysed the building blocks that were used in multicomponent one-pot reactions. Again, most of them fell into one of those categories.

We were especially interested in reactive building blocks that would allow to carry out multicomponent-cascade reactions giving rise to different core structures depending on the sequence of mixing ("combinatorics of building blocks"). Ultimately, this excercise can lead into the discovery of novel multicomponent reactions.

Requirements to a reactophore for COS

- Reactive and yet still storable at -20°C

- Should be synthesisable in gram quantities

- A convergent synthesis should be available that allows for an easy introduction of a wide range of orthogonally protected functional groups

- Each reactophor should lead into different core structures, depending on the reactive counterpart and the sequence of mixing.

- Orthogonally protected functional groups should allow a quick and easy diversification phase.

1.9.1. Mono- and polyvalent donor building blocks (nucleophiles)

In *Chart 1* are listed the most frequently employed mono-donor building blocks in SPOS such as tertiary amines, phosphines, phosphites, alcohols, boronic acids, stannanes, acetylenes, carboxylates and various organometallic nucleophiles. These building blocks are mostly used for the diversification steps to "decorate" the previously constructed core structures (see *Figures 1.9.1* and *1.9.2*) or for the final cleavage reactions.

In *Chart 2* are depicted various poly-valent donor building blocks such as thiols, secondary amines, phosphoryl imines and ylids which serve both the purposes of diversification and cleavage as well as of construction of the central core structures. Bidentate nucleophiles such as thioamides, thioureas, ureas, isothioureas, amidines, guanidines, 1,2-dihydroxybenzenes, 2-aminothiophenols and others (see *Chart 2*) were exclusively employed for the synthesis of the cores as exemplified in *Chapter 3 and 4* (the indicated references refer to these examples)

Chart 1: Monodentate donors

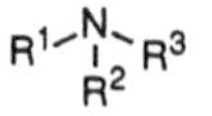 $R^1\text{-}N(R^2)\text{-}R^3$	1	392	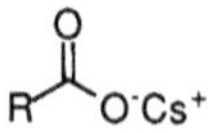 $R\text{-}C(=O)\text{-}O^-Cs^+$	8	
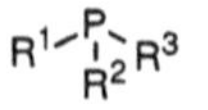 $R^1\text{-}P(R^2)\text{-}R^3$	2		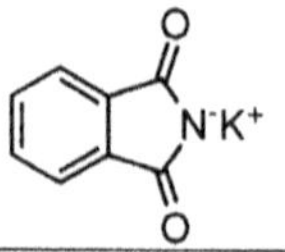 phthalimide N^-K^+	9	
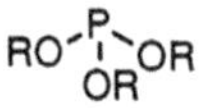 $RO\text{-}P(OR)\text{-}OR$	3		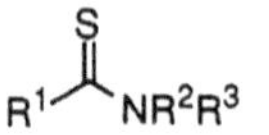 $R^1\text{-}C(=S)\text{-}NR^2R^3$	10	359, 393
$R\text{-}OH$	4	19, 394, 395	$R^1\text{-}Zn\text{-}R^2$	11	331
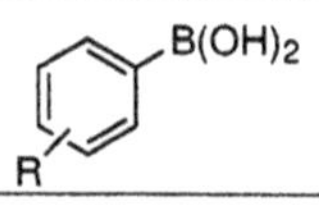 Ar-$B(OH)_2$	5	5,354	$R\text{-}Li$	12	
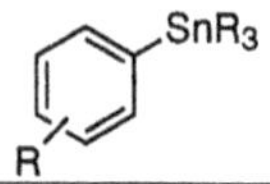 Ar-SnR_3	6	396-399	$R\text{-}MgX$	13	344, 400, 401
$R\text{≡}$	7	394, 402, 403	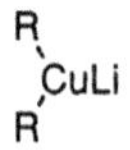 R_2CuLi	14	404

Chart 2: Polydentate donors

R–SH	1	61, 394, 405	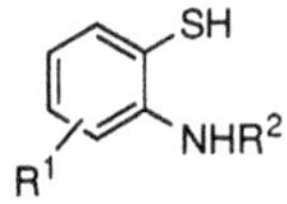 (aryl, SH, R^1, NHR^2)	11	
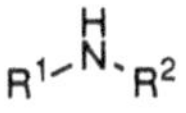 $R^1-\overset{H}{N}-R^2$	2	60, 61, 344, 364, 397, 406-412	$RO-NH_2$	12	
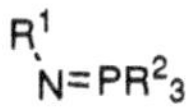 $\overset{R^1}{\underset{}{N}}=PR^2_3$	3	359, 406, 407, 409	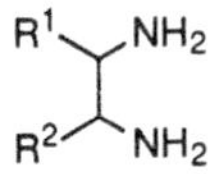 R^1—NH_2, R^2—NH_2	13	
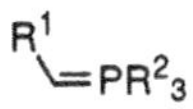 $R^1=PR^2_3$	4	4, 413, 414	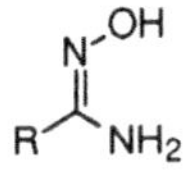 $\overset{N-OH}{\underset{R\quad NH_2}{}}$	14	393
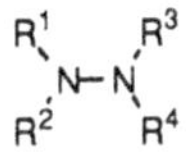 $\overset{R^1\quad R^3}{\underset{R^2\quad R^4}{N-N}}$	5		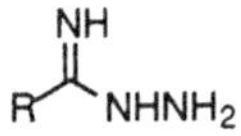 $\overset{NH}{\underset{R\quad NHNH_2}{}}$	15	
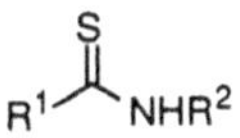 $\overset{S}{\underset{R^1\quad NHR^2}{}}$	6	359, 393	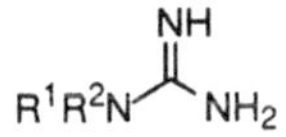 $\overset{NH}{\underset{R^1R^2N\quad NH_2}{}}$	16	
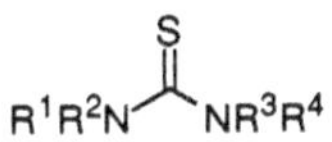 $\overset{S}{\underset{R^1R^2N\quad NR^3R^4}{}}$	7	359	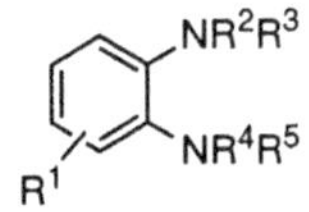 (aryl, NR^2R^3, R^1, NR^4R^5)	17	359, 415
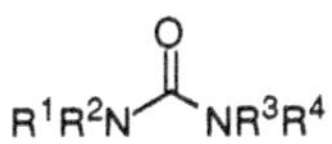 $\overset{O}{\underset{R^1R^2N\quad NR^3R^4}{}}$	8		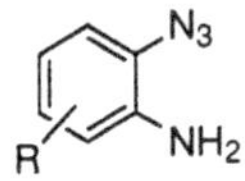 (aryl, N_3, R, NH_2)	18	359
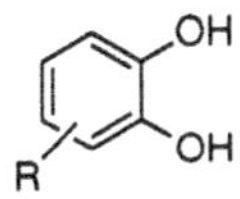 (aryl, OH, R, OH)	9	4, 344	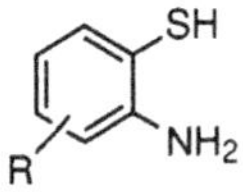 (aryl, SH, R, NH_2)	19	359
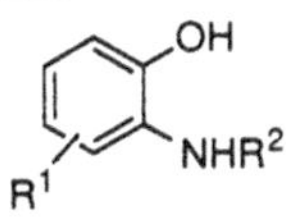 (aryl, OH, R^1, NHR^2)	10		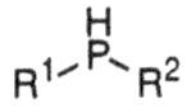 $R^1-\overset{H}{P}-R^2$	20	

1.9.2. Mono- and polyvalent acceptor molecules (electrophiles)

In *Chart 3* are shown first the mono-valent acceptor building blocks such as alkyl and aryl halides. These electrophiles are exclusively employed in the diversification steps or as final "scavenging" or "capping" reagents in cascade reactions. Acceptor molecules derived from aldehydes ketones, oximes, carboxylic acids and amides, nitriles, nitrones and others are used for both derivatisation reactions and the construction of central core structures. Bidentate acceptor reactophores such as α-bromoketones, α,β-unsaturated ketones and derivatives, glyoxals and α-alkynyl ketones (cf. *Chapter 1.9.4*) are the prototypes of key building blocks useful for SPOS as they allow the synthesis of a whole range of heterocyclic core structures. Some of the most interesting examples will be shown in *Chapters 3 and 4*.

Chart 3: Mono- and polyvalent acceptor molecules

Structure	No.	Reactions / Substituents
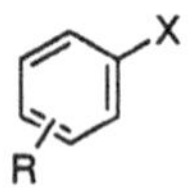 Aryl–X (R-substituted benzene with X)	1	X = Br, I[394, 396, 402, 403, 407, 416] X = OTf[331, 417] X = OMs[4,5]
R—X	2	N-Alkylation[60, 95, 344, 364, 397, 406-409, 411, 412] C-Alkylation[272, 354] S-Alkylation[61, 394, 405] O-Alkylation[19, 45, 394, 395, 418-420]
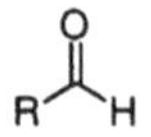 R–CHO (aldehyde)	3	Acetalisation[421, 422] Imine formation[344, 411, 423-427] *Wittig* reaction[413] *Horner-Emmons* reaction[414] Nitro-aldol reaction[428, 429]
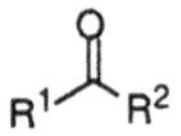 R^1–CO–R^2 (ketone)	4	401
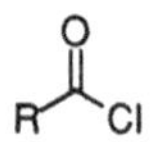 R–CO–Cl (acid chloride)	5	95, 311, 344, 398, 399
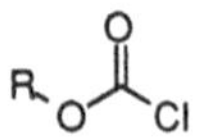 R–O–CO–Cl (chloroformate)	6	430, 431
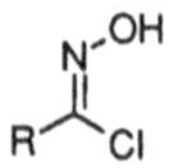 R–C(=N–OH)–Cl	7	432

Chart 3: Mono- and polyvalent acceptor molecules (continued)

Structure	№	Refs
R−C≡N	8	
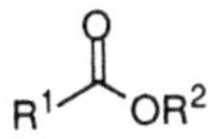 R¹–C(=O)–OR²	9	
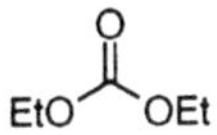 EtO–C(=O)–OEt	10	344
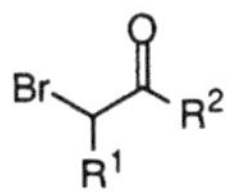 Br–CH(R¹)–C(=O)–R²	11	359, 393
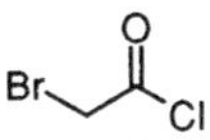 Br–CH₂–C(=O)–Cl	12	433
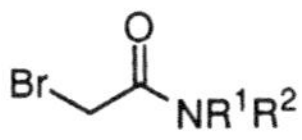 Br–CH₂–C(=O)–NR¹R²	13	61, 406, 407, 409
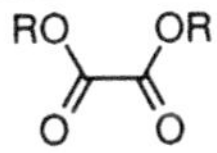 RO–C(=O)–C(=O)–OR	14	344
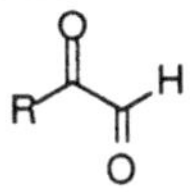 R–C(=O)–CH(=O)	15	434, 435
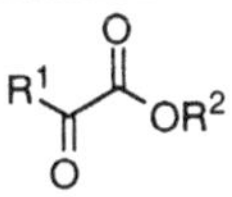 R¹–C(=O)–C(=O)–OR²	16	436
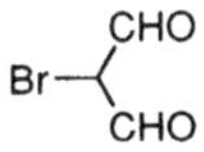 Br–CH(CHO)CHO	17	437
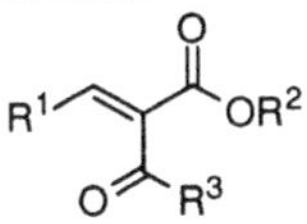 R¹–CH=C(C(=O)OR²)(C(=O)R³)	24	29, 344, 438-441
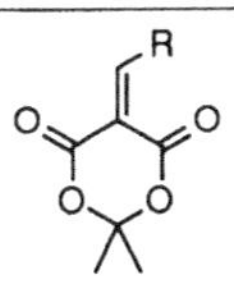 R-substituted dioxinone	25	442
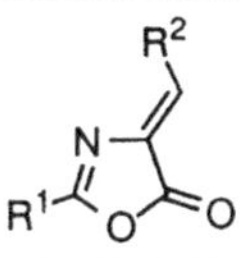 oxazolone R¹, R²	26	

Chart 3: Mono- and polyvalent acceptor molecules (continued)

Structure	No.	References
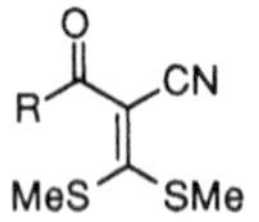 (R–C(=O)–C(CN)=C(SMe)(SMe))	27	
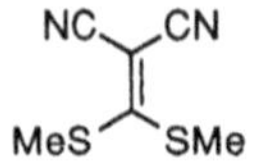 (NC–C(CN)=C(SMe)(SMe))	28	443
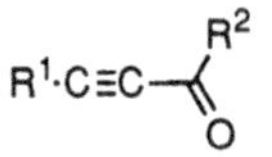 (R^1·C≡C–C(R²)=O)	29	98, 359, 444
 (Br–C≡C–C(R)=O)	30	

1.9.3. Acceptor-donor molecules

Combining nucleophilic and electrophilic centers, acceptor-donor molecules are highly versatile building blocks useful in many multicomponent and multigeneration strategies. After initial reaction with a donor molecule, these building blocks can potentially generate a novel donor center ready for reaction with a further acceptor-donor building block in a chain reaction or react with an acceptor species to terminate the cascade.

In *Chart 4* are listed some of the most popular and useful acceptor-donor reactophores that have been used in combinatorial and parallel synthesis such as epoxides, episulfides, imines, β-diketones. 2,2-Disubstituted 3-amino-2*H*-azirines constitute a highly reactive class of acceptor-donor reactophores that have been shown to be useful for the generation of a range of diverse heterocycles under mild reaction conditions. Furthermore, isocyanates, isothiocyanates and carbodiimides belong to the class of prototypical acceptor-donor reactophores as they exhibit high reactivity and are readily available.

Chart 4: Acceptor-donor molecules

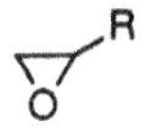	1	344, 383	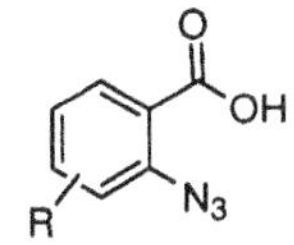	13	359, 412
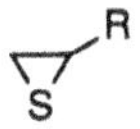	2		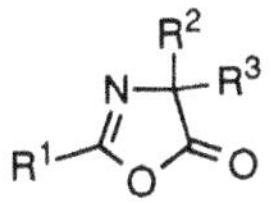	14	
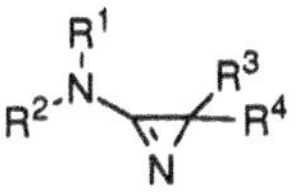	3	445	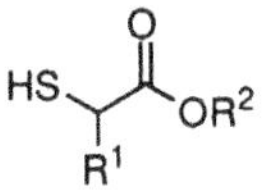	15	422, 424, 444, 446
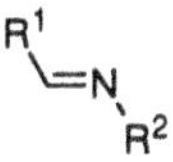	4	99, 344, 411, 423-427, 447	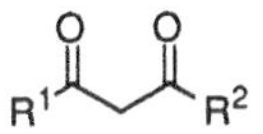	16	344, 438, 440, 448-451
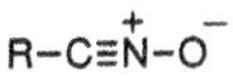 $R-C\equiv\overset{+}{N}-O^{-}$	5	428, 452	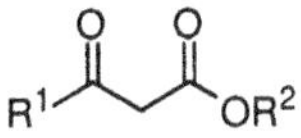	17	54, 393, 441, 448, 451, 453
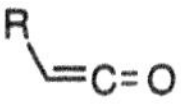	6	454	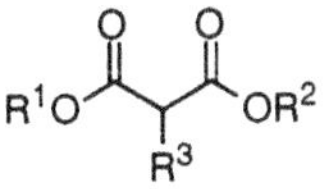	18	437
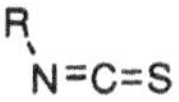	7	455	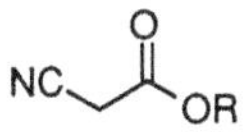	19	344, 455
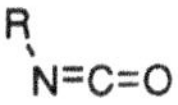	8	344, 427, 455-461	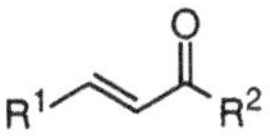	20	344, 413, 462
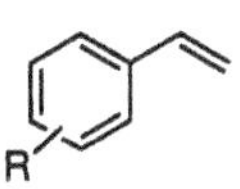	9	463	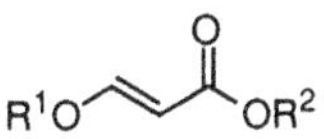	21	
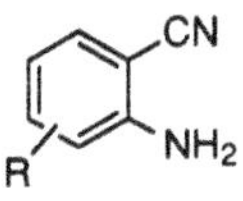	10	464, 465	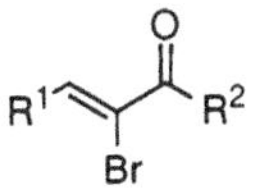	22	

Chart 4: Acceptor-donor molecules (continued)

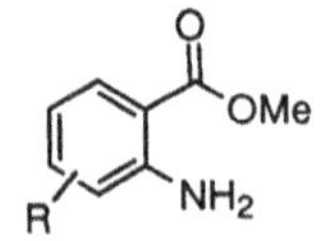 (structure: methyl 2-amino-R-benzoate with OMe, NH$_2$, R)	11	432	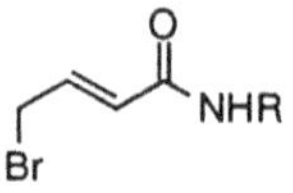 (structure: Br-CH$_2$-CH=CH-C(=O)-NHR)	23	407
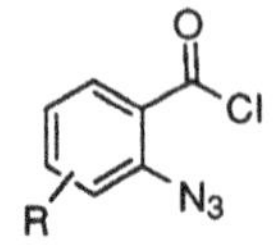 (structure: 2-azido-R-benzoyl chloride with Cl, N$_3$, R)	12	407	$R^1-N=C=N-R^2$	24	359

1.9.4. Acetylenic ketones as typical representatives of a bis-acceptor reactophore

As defined earlier, reactive building blocks that are useful for solid supported combinatorial and parallel approaches show a high reactivity, allowing the synthesis of several different core structures by muticomponent one-pot and multigeneration reactions. In addition they should be storable at -20°C and synthesisable in reasonable quantities (several grams) containing orthogonally protected functional groups for subsequent introduction of further elements of diversity. A classification of these reactive building blocks was presented in *Chapters 1.9.1-1.9.3*. An excellent example for such a class of reactive blocks constitute the α-alkynyl ketones (acetylenic ketones) as typical representatives of bis-acceptor molecules. They show the following features:

- α-Alkynyl ketones are formally analogues of β-dicarbonyl compounds
- In contrast to β-dicarbonyl compounds α-alkynyl ketones show good to excellent regioselectivities in the reaction with bidentate nucleophiles such as 2-amino thiophenols[466] and hydrazines (*Scheme 1.9.4.2*)
- They are synthetically easily available by simple acetylenic C,C-coupling reactions (*Scheme 1.9.4.1*)
- They tolerate a wide range of functional groups (*Scheme 1.9.4.1*)
- They react with mono- and bidentate nucleophiles to yield a wide range of mono- and bicyclic heterocycles (*Scheme 1.9.4.2*)
- They show a good balance between reactivity and stability (storage at 0°C possible)
- They can be easily synthesised on gram scale with a large variety of orthogonally protected functional groups.

α-Alkynyl ketones are readily available by the reaction of a lithium or magnesium acetylide with either an aldehyde followed by subsequent oxidation with MnO_2.[467-469] or by reaction with a *Weinreb* type amide[470] followed by hydrolysis as shown in *Scheme 1.9.4.1*.

Scheme 1.9.4.1

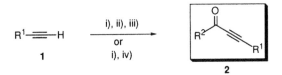

R^1: H, alkyl, aryl, SiMe$_3$, Br, CH$_2$OTHP, CH$_2$NHBoc, CH$_2$NHZ, CH$_2$NHTs
  CH$_2$NBoc$_2$, CH(OEt)$_2$, COOMe, COOtBu, SO$_2$Ph, CN, SnBu$_3$, CH$_2$C(Me)$_2$OH

R^2: H, alkyl, aryl, heteroaryl, CH$_2$OTHP, CH$_2$NHBoc, CH$_2$NHZ, COOEt

Reagents and conditions: i) iPrMgCl or nBuLi, THF, -78°C; ii) R^2-CHO; iii) MnO$_2$; iv) R^{2}CON(OCH$_3$)CH$_3$, H$_3$O$^+$

In *Scheme 1.9.4.2* are depicted some applications of α-alkynyl ketones in order to demonstrate their scope and utility for COS.

Thus, acid catalysed isomerisation reaction of acetylenic acetals **3** yield the pharmacologically interesting (*E*)-4-oxobut-2-enoic acids **4** in high yields[470] (entry a).

Acetylenic ketones **5**, **9** and **13** or acetals **8**, **12** and **16** gave by treatment with HBr in a regioselective manner 3-halofurans **6** and **7**[469] (entry b), 3-bromothiophenes **10** and **11**[467] (entry c), and N-protected bromopyrroles **14** and **15**[468] (entry d). These building blocks can be further functionalised, *e.g.* they have been used for the synthesis of optically pure amino acid analogues.[359]

Treatment of acetylenic ketones **2** with methyl thioglycolate gave in a tandem *Michael*-addition/ intramolecular *Knoevenagel*-condensation highly substituted thiophenes **18** in high yields[444] (entry e).

2,4-Disubstituted quinolines of type **21** were obtained from acetylenic ketones **2** as a result of a condensation reaction with 2-amino thiophenol **19** followed by sulfur extrusion[466] (entry f).
Acetylenic ketones have also been employed as dienophiles in Carbonyl-Alkyne-Exchange (CAE) reactions with dienes **22** (derived from 2,2-dialkyl-2,3-dihydro-4*H*-pyran-4-ones[471])(entry g) or **24**[472] (entry h) to yield aromatic compounds **23** and **25** in a highly regioselective way.

Reaction of **2** with alkyl- and arylhydrazines gave varying amounts of pyrazoles **27** and **28**[359,473] (entry i).

Acetylenic ketones **29** cyclised upon treatment with HBr to give flavones **30**[469] (entry k). Amino acid derivatives **32**[474,475] and **34**[476] were obtained by reaction with acetylenic ketones **31** and **33** (entries l and m).

Scheme 1.9.4.2

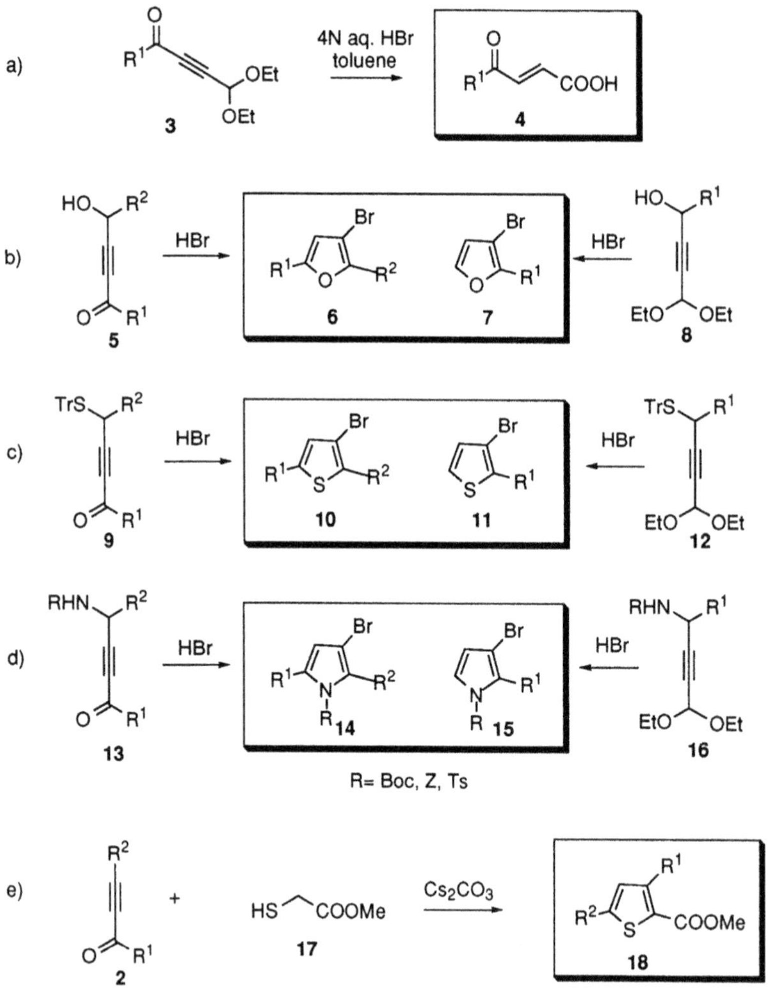

R= Boc, Z, Ts

Scheme 1.9.4.2 (continued)

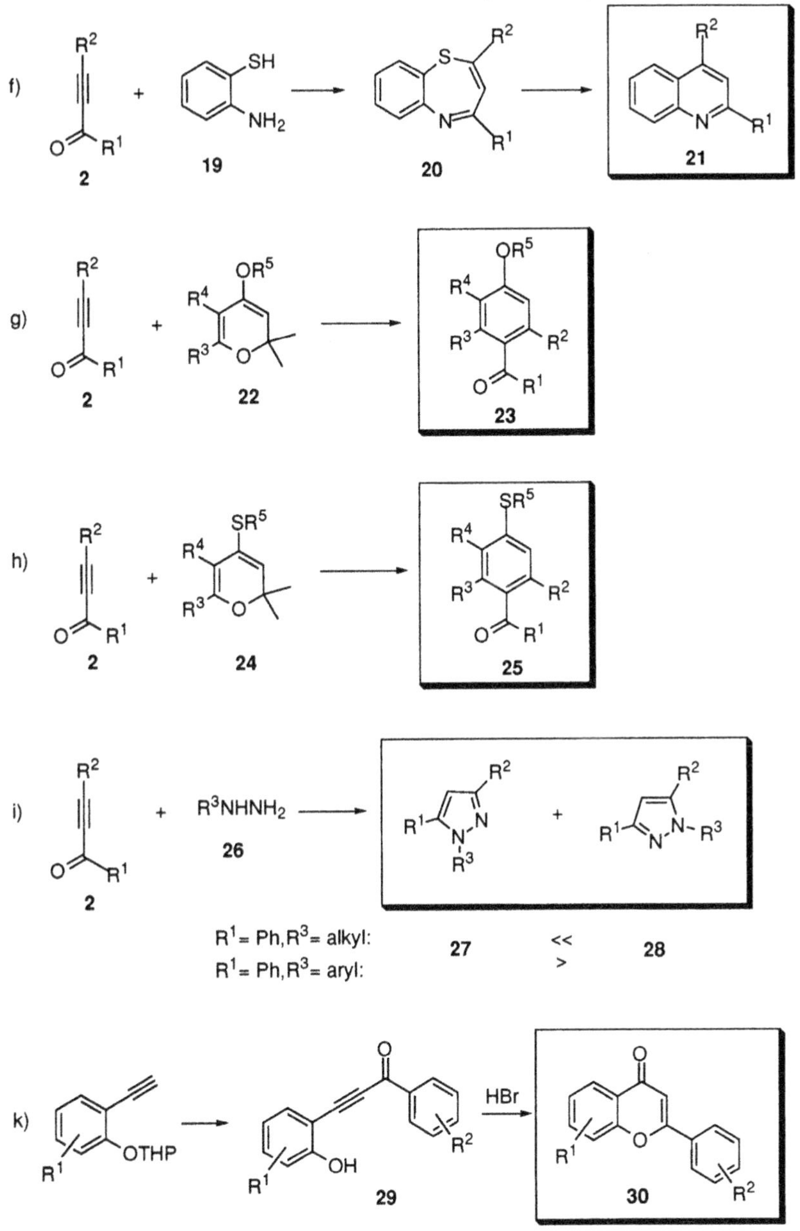

f)

g)

h)

i)

$R^1 = Ph, R^3 = alkyl$:   **27**   <<   **28**
$R^1 = Ph, R^3 = aryl$:   >

k)

Scheme 1.9.4.2 (continued)

l)

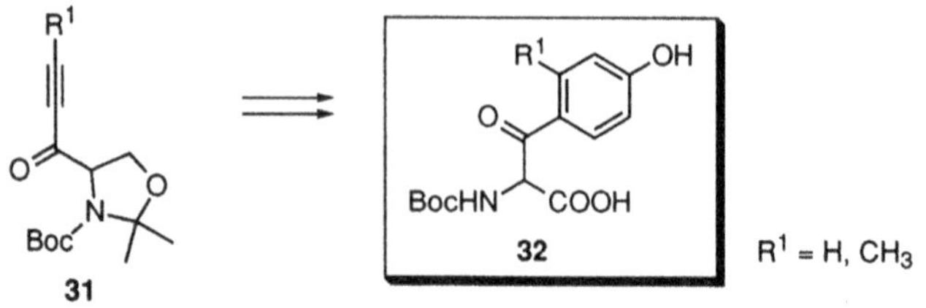

31

⟹

R^1 = H, CH$_3$

32

m)

33

⟶

34

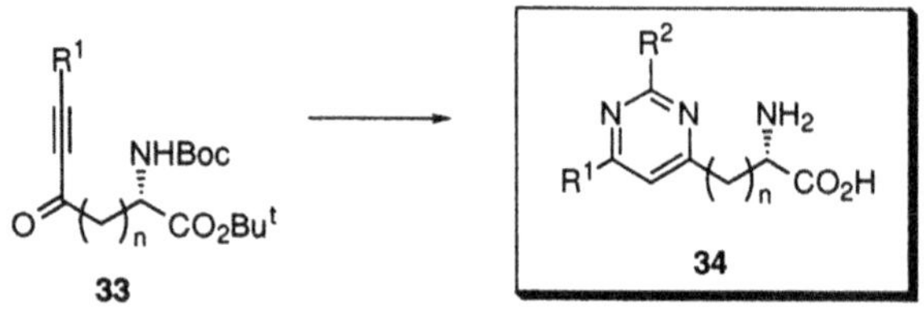

1.9.5. Combinatorics of reactive building blocks

As previously described, multicomponent one-pot and multigeneration reactions constitute the main strategies for the successful generation of small-molecular-weight compound libraries by parallel or combinatorial techniques. A pivotal role in all of these strategies play reactive building blocks which allow the rapid and efficient assembly of interesting pharmacologically meaningful core structures. In *Chapter 1.9.1-1.9.3* we have classified those building blocks into three families, whereas in *Chapter 1.9.4* we have demonstrated the scope of the highly versatile acetylenic ketones as bis-acceptor molecules for combinatorial synthesis.

In *Schemes 1.8.5.1 and 1.8.5.2* we have eluded to the fact that multicomponent cascade reactions usually lead to a higher number of compounds based on a higher number of core structures as compared to multicomponent one-pot reactions. Reactive building blocks which allow the construction of several different core structures depending on the sequence of mixing bear definitely a novel potential in combinatorial organic synthesis as shown in *Scheme 1.9.5.1*.

Suppose a case where the reactive building blocks **1-5** (three of each) would be available. Thioureas **1** are polyvalent donor molecules, isothiocyanates **2** and nitriles **3** can be viewed as acceptor-donor molecules, 2-bromoketones **4** are typical bis-acceptors whereas alkyl halides are mono-acceptors. In the typical *Hantzsch* synthesis of aminothiazoles,[359] **1** is condensed with **4** to give thiazoles **6** which subsequently can be alkylated with **5** to give the final products **7** (27 aminothiazoles).

Scheme 1.9.5.1

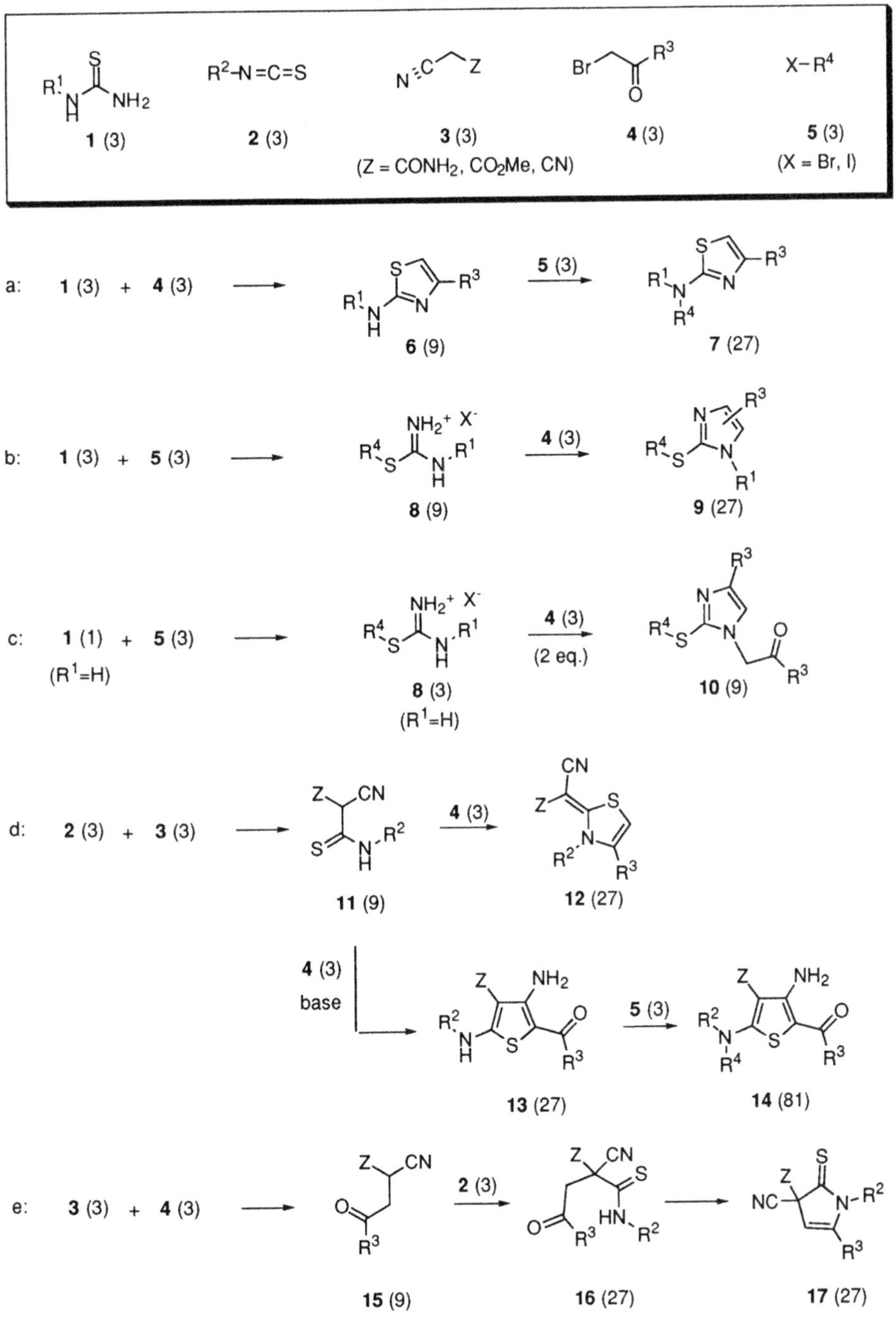

a: **1** (3) + **4** (3) ⟶ **6** (9) $\xrightarrow{\textbf{5 (3)}}$ **7** (27)

b: **1** (3) + **5** (3) ⟶ **8** (9) $\xrightarrow{\textbf{4 (3)}}$ **9** (27)

c: **1** (1) + **5** (3)
(R¹=H) ⟶ **8** (3)
(R¹=H) $\xrightarrow[\text{(2 eq.)}]{\textbf{4 (3)}}$ **10** (9)

d: **2** (3) + **3** (3) ⟶ **11** (9) $\xrightarrow{\textbf{4 (3)}}$ **12** (27)

4 (3) base ⟶ **13** (27) $\xrightarrow{\textbf{5 (3)}}$ **14** (81)

e: **3** (3) + **4** (3) ⟶ **15** (9) $\xrightarrow{\textbf{2 (3)}}$ **16** (27) ⟶ **17** (27)

143

Mixing **1** first with **5** (entry b) yields the isothiouronium salts[98] of type **8** which react with **4** in a *Hantzsch*-type condensation to yield imidazoles of type **9** (again 27 compounds). In a similar way, isothiourea **8** (R^1=H) reacts with two equivalents of **4** to give imidazoles **10** (9 compounds).[477] For the case where the reaction described in entry c is performed as a combinatorial one-pot synthesis in one reaction vessel, all possible 27 permutations of **10** can be obtained.

Conversely, mixing of isothiocyanates **2** with nitriles **3** in the presence of base yields intermediates **11** which undergo clean S-alkylation with **4**, followed by cyclisation yielding either methylene-thiazoles **12** (27 compounds) or, in the presence of base, aminothiophenes[455] of type **13** which subsequently can be alkylated with **5** to yield thiophenes of type **14** (27 compounds).

In a more speculative and intuitive way, **3** and **4** should react in the presence of a base to give intermediates of type **15** (9 compounds), which similarly to entry c should react with **2** to give intermediates **16** (27 compounds) followed by cyclodehydration to yield **17** (27 compounds).

Whereas a five component one-pot condensation would maximally yield 243 compounds with one single core structure, the decribed multicomponent cascade approach using reactive building blocks **1-5** as shown in *Scheme 1.9.5.1* gives rise to 288 compounds with 12 core structures. In addition, there are certainly several more combinations of reactive building blocks **1-5** possible, some of which might be novel multicomponent approaches. This simple experiment emphasises once more the importance of the chemical strategy in COS. Looking at the building blocks classified in *Chapters 1.9.1-1.9.3* it can be assumed that only a fraction of all possible combinations have been carried out so far. The *combinatorics of reactive building blocks* should certainly prove a successful approach to find novel multicomponent reactions useful for combinatorial and parallel synthesis.

1.10. Tagging and deconvolution

With the concept of combinatorial chemistry and the construction of chemical libraries the issue of the structural elucidaton of active leads emerged. If compounds are generated individually in a parallel synthesis, their structure is obviously predetermined. Similarly, compounds that are physically constrained to a location are accurately defined by their position. Resourceful strategies that have been employed include the spot-synthesis on paper,[478] the multipin,[8] Diversomer®,[427] photolitographic,[479] and the teabag technique.[9]

For compounds prepared in mixtures, the identification of the active compound(s) from these mixtures is approached in various ways. Methods comprise identification by analytical means, iterative or noniterative strategies, as well as the generation of indexed libraries, or elucidation of the structure by the aid of so-called "tags" that encode for the single building blocks and the synthetic history.

The nature of libraries and the strategy for the identification of active leads from these mixtures are closely linked. If tested on solid support, the identification of a hit can be fairly straightforward for one-bead-one-compound libraries. After the identification of the active compound (detection enabled *e.g.* by fluorescence or dye) the individual bead is isolated from the mixture and the structural elucidation is either accomplished by cleaving off and analysing the compound directly, or in encoded libraries indirectly by reading the tags. When dealing with mixtures in solution, the assignment is more challenging and different solutions for the identification have been created. These include deconvolution strategies or methods involving the partial release from the bead.

One method of identification is the **iterative deconvolution**. This method is based on the concept of a stepwise identification of the most favourable building blocks used during the course of constructing the library and involves the successive resynthesis of a less and less complex library reapplying this strategy until the active compound is identified. If for example a "tetrameric" product is to be constructed and four different building blocks are used in each step the resulting library will consist of $4^4 = 256$ members, represented as XXXX with X= A, B, C or D. In the first circle of this example a set of four sublibraries is constructed. Using for example the split synthesis method in the fourth synthetic step the aliquots are not *recombined* and are kept as separate pools, AXXX, BXXX, CXXX, and DXXX, thus defining one of the four variable components of the molecule. Screening reveals the most active compound*, in which sublibrary it is located and hence which building block in this first segment is preferred; in this example one of the 64 compounds of sublibrary **BXXX** is the most active one (*Figure 1.10.1*). Resynthesis of a smaller set of

---

* For a dicussion of the effectiveness of deconvolution approaches in finding the most active component and problems with cumulative activity see *Freier et al.* [480-482]

sublibraries is performed, with the optimal evaluated residue incorporated and the next segment to be identified defined by the new sublibraries with the general structure BAXX, BBXX, BCXX, and BDXX. Again these sublibraries are subjected to the screening procedure. This iterative process of screening and resynthesis of a less and less complex library is repeated until the structure of the compound is determined.

Houghten et al. used this approach in the deconvolution of hexapeptide libraries defining the last two positions jointly in the first cycle, creating 324 (18^2) subsets.[369,483,484] The strategy was likewise applied to deconvolute oligonucleotide,[366,485] peptidoid,[406] propanediol[429] and pyrrolidine[486] libraries.

Figure 1.10.1

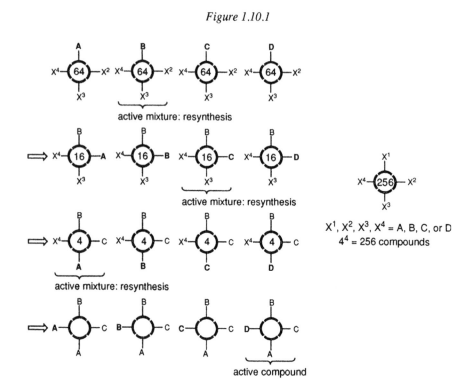

active mixture: resynthesis

active mixture: resynthesis

active mixture: resynthesis

active compound

$X^1, X^2, X^3, X^4 = $ A, B, C, or D
$4^4 = 256$ compounds

Illustration of iterative deconvolution. Sublibraries are screened for activity and less complex sublibraries with the identified building block incorporated are synthesised. The best building block for every position is identified stepwise through this process.

The **recursive deconvolution** is closely related to the iterative deconvolution. The difference lies within the construction of the library. After every split synthesis step a portion is stored. In the process of deconvolution this material can be used to establish the identity of the active compound having to perform only a number of synthetic steps rather than the repeated resynthesis of a whole new set of sublibraries. *Erb et al.* examined the recursive deconvolution in the construction and screening of a small pentapeptide library.[487] Evidently, the advantage is that a lengthy resynthesis is obsolete; though when dealing with a large library the handling of all these subsets can become rather complex.

Positional scanning uses a different approach to determine the most active molecule. Here the most favourable building block for each segment is evaluated. The rational is that by identifying the best building block in every segment, the structure of the most active compound can be easily deduced. In the case of the "tetrameric" product four separate sets of the library have to be constructed (*Figure 1.10.2*): OXXX (stands for the individual sets of AXXX, BXXX, CXXX, and DXXX), XOXX, XXOX and XXXO. A total of 16 sublibraries are thus produced each containing a mixture of 64 compounds with one building block at a given position defined. Screening discloses the preferential building block for every segment and hence the exact structure of the favoured compound.

This strategy was applied for the generation and screening of a hexapeptide library.[488] *Uebel et al.* used the approach for a detailed analysis of the peptide specificity and recognition of human TAP. Nonapeptide libraries were used to determine the effects of the individual residues at a given position upon binding. In a second set of studies the effect of backbone destabilisation by introducing D-amino acids at a given position was investigated.[489]

The major advantage of this method is the straightforwardness of the identification inasmuch as there is no need for any resynthesis and more assays. But if one subset shows no particular preference for any building blocks, resynthesis of all compounds with ambiguous segments is required. Additionally, the split synthesis cannot be readily applied, and this method relies on the stepwise reactions with defined mixtures rather than single building blocks.

Figure 1.10.2

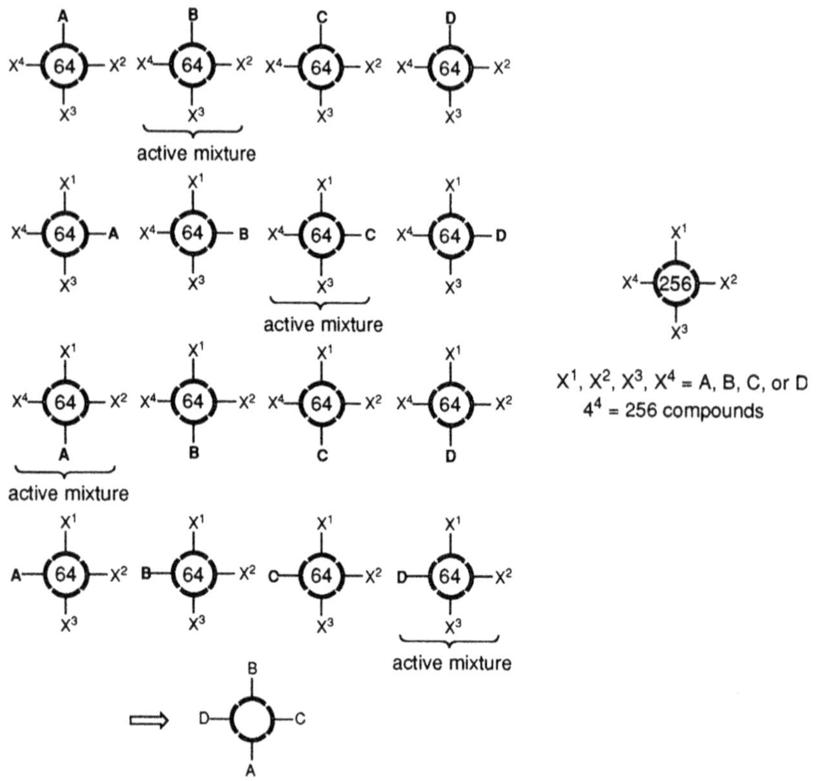

$X^1, X^2, X^3, X^4 = A, B, C, \text{ or } D$
$4^4 = 256$ compounds

Illustration of positional scanning. Four sets of sublibraries are generated with one building block for every given position of the molecule defined. Screening reveals the preferential building block for every single position. The active compound BCAD can be deduced and is synthesised for evaluation.

Another method uses **indexed** or **orthogonal** libraries for the structural elucidation. Two separate distinct sets of sublibraries are generated with the characteristic that for any compound in a set of the first sublibrary there is an identical one in every other subset of the second library. This orthogonal array permits the structural elucidation of an active lead by simple correlation as a consequence that any sublibrary from the first set and any sublibrary from the second set share only one identical compound. The idea can be illustrated as following: Reacting component X (X= A, B, C, D and E) and x (x= a, b, c, d and e) yield the product Xx, i.e. 25 different compounds. Two sets of sublibraries are created, one reacting every single component of X with a mixture of x (thus creating Ax, Bx, Cx, Dx, and Ex, each a mixture of 5 compounds) and the other one by reacting the mixture of components of X with every single component x (Xa, Xb, ...Xe). If screening reveals that the active compound is one of the Bx sublibrary and for the second set is one of the Xc, the structure of the active compound can be easily deduced to be Bc. Such orthogonal libraries offer the advantage, that false positives can be detected, since a hit has to appear in both libraries. The outlined strategy has been used for the synthesis and screening of carbamates,[464] tetrahydroacridines,[465] and arylpiperazines.[490] *Smith et al.* reacted 40 acid chlorides with a mixture of 40 amines and alcohols, and the 40 amines and alcohols with the mixture of 40 acid chlorides. The resulting 80 amide/ester libraries were tested in a variety of assays disclosing *e.g.* an inhibitor for MMP-1.[491] The concept allows the split synthesis approach and *Déprez et al.* demonstrated the power of this method in the structural elucidation of a tripeptide library consisting of 15625 trimers by generating two orthogonal sets of libraries each containing 125 sublibraries.[492] The orthogonal matrix in this example is set up by dividing the 25 amino acids into 5 groups, A1-A5, each containing 5 amino acids. The synthesis of the sublibraries is performed by incorporating the groups (i.e. the defined mixtures) into the single position of the tripeptide. A second set of 125 sublibraries is generated, now dividing the 25 amino acids into 5 groups, B1-B5, orthogonal to the first.

With the one bead one compound concept a further strategy to determine the structure of an active compound was employed; the notion of **encoding** the synthetic information. Methods to directly analyse the small amounts of compound present on the bead essentially exist for *sequenceable* structures. Additionally, analytical methods like mass spectrometry provide help in the structural elucidation of small organic molecules. Yet it is not always feasible and advantageous to directly analyse the compound from the solid support. To address the need for a sensitive, unambiguous, and fast structural elucidation and hence to circumvent possible laborious analyses the idea of encoding was adopted. Each synthetic step performed during the library formation is *paired* with the additional introduction of tags that encode for the synthetic transformation.

The first generation of encoding molecules were biopolymers. *Oligonucleotides* were used to encode for peptide libraries on solid support, the read-out method being PCR amplification and sequencing.[493-495] Another approach is the use of *peptides* to encode for libraries with nonsequenceable polymers such as peptides incorporating unnatural amino acids that cannot be distinguished from their natural counterparts in the *Edman* degradation.[496,497]

These methods rely on a orthogonal linker and protecting group strategy to introduce the tags and to construct the library independently, and that the two synthetic procedures are compatible. Though the usefulness of this encoding method was readily demonstrated its application is limited due to some severe constraints in the reaction conditions that can be performed.

A different approach was presented to address the problem of the stability of the biopolymeric tags by introducing *chemically inert small molecule tags* that produce a molecular barcode. The concept is intriguingly simple and elegant. Halophenol derivatives with varying hydrocarbon chain lengths are the nonsequential coding units. These molecules are easily separated and detected at a *subpicomolar* level by electron capture gas chromatography. As a consequence, the tags that make up the synthetic code can be used in very low amounts and can be introduced "cosynthetically" rather than by using orthogonal synthetic schemes. Attachment proceeds *via* a linker unit either by reacting the tag with the functional groups present on the solid support or independently by directly inserting it into the matrix. This linker also provides the handle for the selective release of the tags (either photolytically or oxidatively). Subsequent silylation allows the direct analysis of the encoding units by ECGC.[498,499]

The tags are used in such way that they form a **binary code** unique to every reagent and synthetic step. For the synthesis of a tripeptide using 7 building blocks (A, B, C, D, E, F, and G) during the three cycles of construction 9 tags would be required. Three tags for every step are used, T1, T2 and T3 for the first step, T4-T6 for the second and T7-T9 for the third to encode for the building blocks. T1 encodes for amino acid A, T2 for amino acid B, T3 for amino acid C, T1 *and* T2 for D, T1 and T3 for E, T2 and T3 for F and T1, T2 and T3 for amino acid G. Thus the *presence and absence* of the tags form the binary code. In this example the numerical description is 001, 010, 100, 011, 101, 110 and 111, read from the right to the left. This assignment is repeated with tags T4-T6 for the second and T7-T9 for the third step. A 9-bit binary code of 101 001 111 therefore encodes for the tripeptide with G incorporated in the first step, A incorporated in the second step and E incorporated in the final step.

Figure 1.10.3

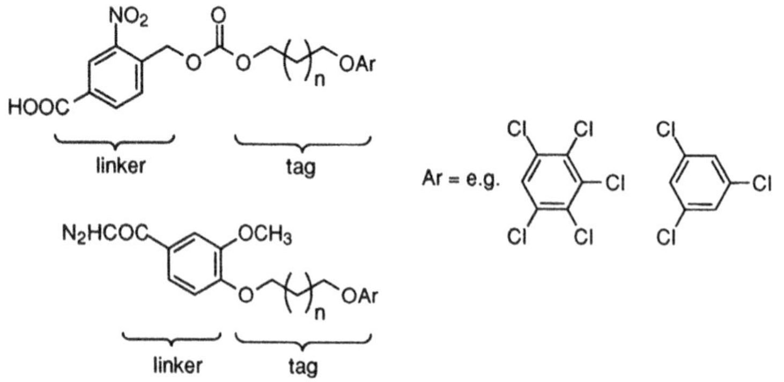

Schematic representation of the tags used for a binary encoded synthesis. The variation of the linker allows a different mode of attachment and release.

This binary encoding strategy was first used in the synthesis a peptide library that was screened against an anti-c-MYC monoclonal antibody.[498] Encoded combinatorial chemistry and screening has been successfully used by *Still* as a tool to evaluate the binding properties of synthetic receptors to large collections of peptides, disclosing remarkably selective receptors. The obtained insights were used in the design of new receptors and receptor libraries.[500-503] One of the receptors that has been designed and studied is the macrocyclic receptor A shown below.

A

Initial experiments with this receptor used α-amino acid derivatives to evaluate the properties of the receptor designed for enantioselective binding. The binding studies showed remarkable enantioselectivity (up to 99% e.e. for the L configuration) as well as side chain selectivity (up to 3.0 kcal/mol ΔΔG) for single amino acid derivatives. As the receptor displays additional amide groups that are not involved in binding the amino acid it was tested for its potential to sequence-selectively bind oligopeptides.[504,505] Here encoded combinatorial chemistry was employed to generate a 15^4 member tripeptide library on solid support and a dye-labelled analogue of the receptor was used for screening. In the assay only tight binding library members will display the colour of the dye-labelled receptor and the beads carrying the selective tripetides can be separated, their code analysed and the sequence hence determined. The receptor indeed disclosed high peptide-binding selectivity. The simplified macrocyclic receptor B displayed the most amazing selectivity in a series of modified receptors with less than 0.1% of the beads being coloured (the original bound about 1% of the library members). Decoding of the beads showed that 78% had the sequence (AA_3-AA_2-AA_1): D-Pro-L-Val-D-Gln and L-Lys-L-Val-D-Pro.[506]

Encoding and synthesis of a carbohydrate library,[507] and other small molecule libraries[508,509] are additionally reported (*vide infra*).

B

A different solution to the requirement of tag stability under a variety of reaction conditions was presented by *Ni et al.* Here the binary encoding strategy has been used alternatively with robust *secondary amine tags.* The tags encount for roughly 10% of the material on the bead and therefore the secondary amines are introduced sequentially, applying an orthogonal protecting group chemistry. For decoding the generated polyamide is subjected to acid hydrolysis and the amine tags are analysed as their dansyl derivatives by reversed phase HPLC equipped with a fluorescence monitor.[510]

This procedure was verified in the construction of a highly functionalised pyrrolidine library. The screening results were compared to those formerly obtained for the same compounds using an iterative deconvolution approach. Both strategies revealed the same new, highly potent ACE inhibitor among close to 500 library members. The encoded screening strategy proved to be somewhat superior as it disclosed a second and third most inhibitory compound that had been discarded during the deconvolution process.[486,511]

Figure 1.10.4

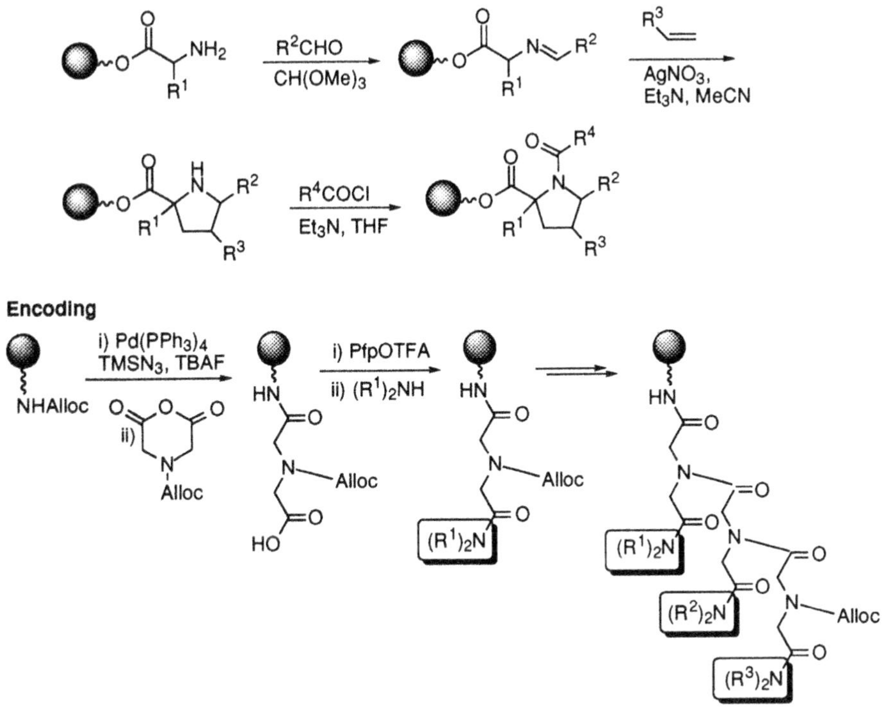

Encoding

Above, the outline of the synthesis of a pyrrolidine library. Below, the orthogonal protecting group chemistry used to introduce the tags.

The coding of *encapsulated* beads by the use of radiofrequency encodable microchips recording the synthetic history of the library members (peptides) was reported. The container that includes the beads and the chip can be regarded as a macroscopic version of a single bead (one-"capsule"-one-compound) and the concept is related to *Houghten's* teabag method.[512,513] *Xiao et al.* used the method to encode a taxoid library.[514] The synthetic strategy was based on the introduction of a glutamic acid handle to attach the template to solid phase. The C-7, C-2' position and the amino group of the glutamic acid were used to introduce diversity (*Figure 1.10.5*).

Figure 1.10.5

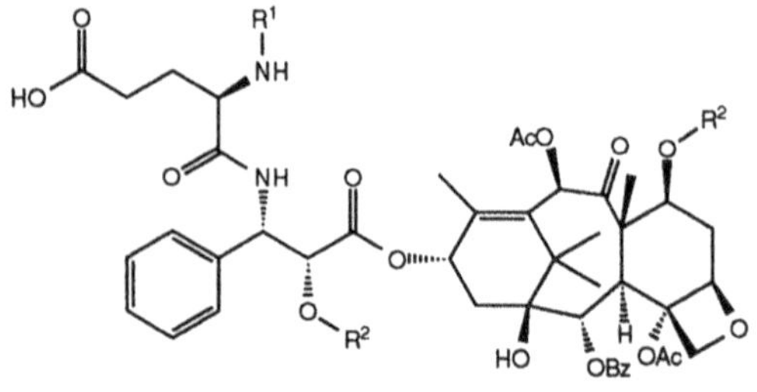

Based on the outlined template Xiao et al. *synthesised a 400-member radiofrequency encoded taxoid library.*

Other methods take advantage of the readiness of assays on solid support – the confinement of the substance to the bead (one-bead-one-compound) that permits a simultaneous screening of individual discrete compounds.

Figure 1.10.6

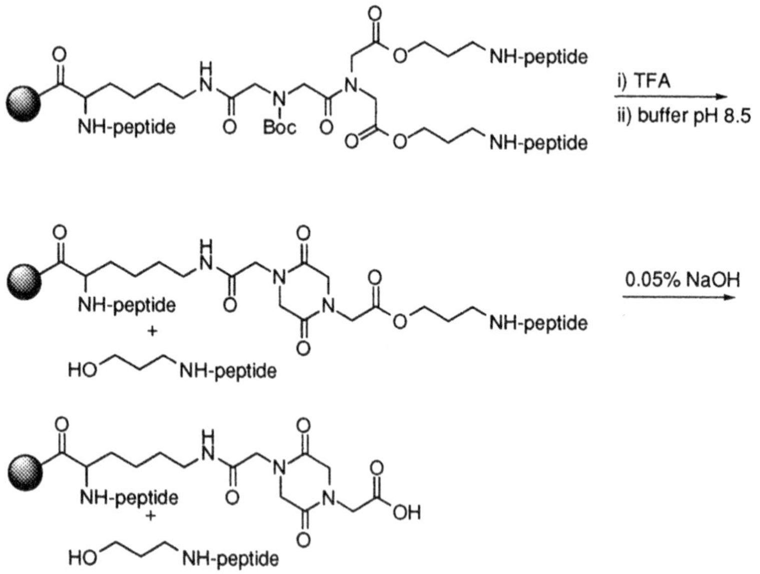

Double releasable linkers by Kocis et al.[515]

This strategy is based on a **multiple release** of the compound from the solid support for solution based assays. The first cycle starts from *pools of beads*. One part of the compounds bound to the beads is released into solution and is removed for subsequent screening. If the screening in solution reveals activity for one of these pools the original beads are recovered and used in the next cycle. Now *single beads* are tested, releasing a second fraction that is subsequently transferred and screened. At this point the structure of the active compound at issue can be determined either directly by liberating the last part from the bead or indirectly by cleaving off the tags that encode for the structure. This approach has been used in the construction and screening of peptide libraries, introducing linkers that allow a sequential *orthogonal* release in conjunction with an uncleavable linker for sequencing positives[515-517] (*Figure 1.10.6*). A photocleavable *o*-nitrobenzyl linker allowing the controlled *partial* release of compounds has been used alternatively in the synthesis of an encoded small molecule library.[508] In addition the synthesis of a 1143-member dihydrobenzopyran library and a 6727-member acylpiperidine library are reported.[509]

1.11. Diversity (planning/assessment)

Combinatorial chemistry has been developed as a method of rapidly producing large numbers of compounds for screening in biological assays. However, the number of reaction products which could theoretically be synthesised by all the possible combinations of appropriate reagents using a given reaction will exceed by far the synthesis and screening resources of any organisation. The term "virtual library" has been used to denote all compounds accessible by parallel synthesis. The goal is therefore to design a much smaller but "smarter" library for actual synthesis to achieve maximum diversity with a minimal number of compounds.

Even if there is no mechanistic understanding of the target at all, compound classes for library synthesis can be selected by choosing "privileged" structures, where the display of a number of functionalities on a certain core structure has already provided drugs or drug candidates.[518] Molecules can be grouped together using structure-based parameters such as molecular weight, shape, flexibility, lipophilicity, charge distribution and dipole moment. The definition and measurement of diversity[519-521] of a given library are important for the choice of the starting materials.[522,523] However, it is difficult to decide which parameters are relevant for biological activity.[524,525]

If more structural information about the receptor-ligand binding becomes available, the number of methods for "rational" selection of appropriate compounds increases. Even limited amount of information about the target can have a profound effect on the design strategy. Site-directed mutagenesis can identify key residues in the receptor giving clues which functionalities a potential ligand should contain. Furthermore, the design can be based on molecular recognition motifs in biological systems such as secondary structural motifs of proteins or transition state mimetics.

If the molecular structure of the intended target is known, ligands can be designed *de novo* by molecular modelling to fit the receptor binding site.[526] However, cavity-filling algorithms often suggest molecules hard to synthesise and therefore should be restricted to virtual libraries accessible by parallel synthesis.

Instead of designing ligands *de novo*, a pharmacophore model can be used in a flexible search of the virtual library to determine molecules for synthesis and screening. A number of diversity assessing programs such as Selector,[527] Jarpat,[528] C²-Diversity,[529] Chem-X[530] and HARPick[531] allow an intelligent compound selection to ensure maximum value from the synthesis and screening effort made.

Important issues for the design of the library synthesis sequence are:[404] The molecule should incorporate many different sets of building blocks providing maximal element variability. The

156

ability to utilise families of different building blocks means opportunity for introduction of diversity and is governed by the methods employed. The chemistry may involve linear, convergent or multi-component reactions.[532]

For a rapid library synthesis, the building blocks used in the synthesis should be commercially available or readily accessible. The selection of reagents can be facilitated by search of chemical databases such as ACD.[533]

Furthermore, the reaction conditions required for the synthesis should be compatible with the display of as many functionalities as possible, including reactive functional groups commonly found in drugs. Support in the selection or optimisation of reactions can be given by reaction searching databases such as REACCS[533] or CROSSFIRE.[534]

The best strategy for the synthesis of a library is dictated by the available building blocks, the target structure and the library format (whether mixtures or individual compounds).

1.12. Analytical methods

1.12.1. Analysis of single compounds in solution

The structure elucidation of the active compounds depends critically on the library format. A library of discrete compounds, prepared by multiple parallel synthesis, can be characterised routinely by standard means such as TLC, IR, NMR and MS. As for conventionally synthesised compounds, the identity of these products is determined by its site of synthesis.

The situation is somewhat different if a compound library is prepared by solid-phase combinatorial synthesis by the split method (one bead-one compound). If the active library members are non-sequenceable but coded, the structure is determined indirectly by reading the code after cleavage from the solid support. If the substances are non-coded, product analysis can be achieved by traditional characterisation after cleavage. Since the amount of substance bound to a single bead is very small (in the order of 100 pmol), indications on the structure can be best obtained by mass spectrometric determination of the molecular weight. Apart from its sensitivity, MS has also a short sampling time, allowing rapid characterisation of a large number of samples. Cleavage from the solid support is performed either in a separate step prior to measurement with TFA or photochemically[535] or directly in the spectrometer under matrix-assisted, laser-desorption-ionisation time-of-flight (MALDI-TOF) conditions.[536] This was demonstrated with peptides,[535-540] peptoids,[541] and with low molecular weight, non-oligomeric "small" molecules.[542,543]

1.12.2. Analysis of single compounds on the bead

Compared to solution-phase organic synthesis, solid-phase reactions can not be followed by TLC or HPLC and are therefore more difficult to monitor. Product analysis can be achieved by cleavage from the solid support followed by traditional characterisation. Nevertheless, a large variety of analytical methods are available for product analysis on the bead.

A number of classical reactions are used to monitor solid-phase reactions, including the titration of reactive functionalities such as amines, carboxylic acids, phenols or thiophenols.[61,544] Resin-bound peptides can be analysed directly on the support by *Edman* degradation. Coupling yields of peptides can be determined rapidly by photometric Fmoc determination.[545-548] UV-active by-products resulting from the cleavage of other protecting groups can be quantified as well.[549-551] Resin loading are typically calculated using *Volhard* chloride titration[552] or elemental analysis. These methods tend to be inaccurate due to low loadings of the substrates on the support.

Support-bound products can be analysed on the bead by IR or NMR spectroscopy in a non-destructive manner. But due to the dominating strong resonance signals of the polymer backbone, interpretations of the analytical results are difficult. Using PEG-PS resins, gel-phase ^{13}C-NMR spectra of resin-bound compounds can give a sufficient resolution to monitor reactions.[553-558] The combination of *magic-angle spinning* (MAS) and *gel-phase NMR* minimises the line broadening observed in conventional solid-phase ^{1}H-spectra.[559-562]

FT-IR is a very convenient method for the evaluation of resin-bound intermediates and can give information on the completeness of reactions.[5,169,400,563,564] Resin-bound substrates are also suitable for KBr pellet IR analysis and for photoacoustic IR.[565]

1.12.3. Analysis of compound mixtures

A reliable analysis of multicomponent mixtures is particularly difficult and the situation becomes more complicated and less meaningful with increasing mixture size. Nevertheless, the analysis of peptide libraries is possible by mass spectroscopic means.[566-572] As mixture size increases, the main aim of any analysis is to ensure that the vast majority of expected library members are represented in the mixture. The use of MS in conjunction with other analytical techniques (HPLC-MS, GC-MS, MS-MS) have been used successfully in library characterisation.[413,425,573] Electronspray-MS is a particularly mild method for ionisation.[566] *Matrix-assisted Laser Desorption Ionisation – time-of-flight* (MALDI-TOF) mass spectrometry is a tool of increasing importance in multicomponent analysis. Its reduced tendency to preferential ionisation results in an increased likelihood of observing ions from all components.[574,575]

159

1.13. Robotic systems and workstations

Combinatorial and parallel methods have made it possible to speed up the synthesis of molecule ensembles using linear or convergent strategies leading to a large array of structures such as peptides, oligosaccharides, peptoids and also to an increasing number of highly diverse heterocyclic "drug-like" small organic molecules.

The shift towards small molecules, synthesised in a parallel single compound array was initiated on one hand by an increasing number of useful solid-supported synthetic protocols that are emerging almost daily in the literature but also by a rapidly growing number of useful workstations and robotic systems suited for solid-supported and liquid phase parallel synthesis. As schematically depicted in *Chapter 1.8* (cf. *Figures 1.8.3.1-1.8.3.3*), libraries of compounds can be synthesised in parallel as real mixtures (cf. *Figure 1.8.3.1*), as single compounds (cf. *Figure 1.8.3.2*) and as pools using the "split-mixed" methodology (cf. *Figure 1.8.3.3*). While parallel synthesis necessitates more individual chemical operations to be performed simultaneously, it offers the benefit of avoiding tagging strategies (required in the "split-mixed" methodology) and of the possibility to synthesise flexible quantities required in a given project. The appearance of suitable workstations and robotic systems which carry out the laborious and "boring" simultaneous operations have certainly motivated many research groups and companies to shift towards the parallel synthesis of single compounds.

These robotic systems and workstations can be grouped into mainly three categories as schematically shown in *Table 1.13*:

- Manual devices: typically used for process development

- Semi-automatic workstations: libraries of 100-500 compounds

- Fully automated robots: large libraries

Table 1.13: Categories of robotic systems

Type of system	Manual handling	Type and complexity of library	Library size	costs	building blocks
1 manual devices	intense	focused libraries process development	small (10-100)	small	1-10g
2 semi-automated workstations	manual addressing of reagents possible	focused libraries flexible adaptation of novel processes	medium (100-500)	medium 10'000- 100'000 US$	10-50g
3 fully automated workstations	require operators for maintenance	• random to focused libraries • Processes with high reproducibility required	large (> 1000)	high > 100'000 US$	>50g

Whereas the format, size and degree of automation vary considerably in these different workstations and robotic systems, the basic operations remain the same. Therefore, with the aid of a schematic representation of a medium sized workstation an illustration of the basic modes of action should be achieved (*Figure 1.13*).

Figure 1.13

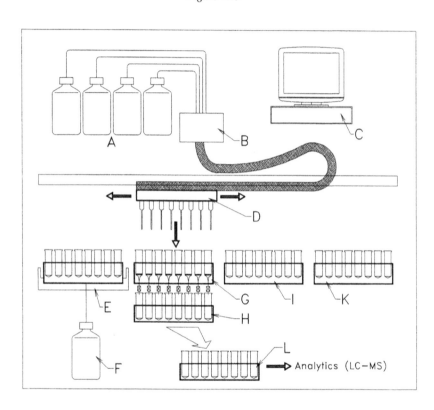

Legend:

A   Reservoir to address solvents for washing cycles or for dilutions

B   Valve system controlling the addition of solvents and reagents

C   Computer system handling the operation of the robotic arm **D**, valve system **B**, adressing of reagents and solvents. In addition, the data (structural and spatial) of the products synthesised in a parallel array is handled electronically (typically with ISIS Base or similar data base)

D   Robotic arm containing several automatic syringes (typically 1-8) which address reagents or solvents (for dilution or washing cycles) to the reaction vessels. The robotic arm is programmed to move to reaction blocks **E, G, I** and **K**

E   Reaction block containing reaction vessels (typically 8-96) suitable for solid-phase and solution chemistry. The reaction block (made from Teflon® or aluminium) can be vortexed as well as cooled and heated (typical temperature range -40 to 150°C)

F   Reservoir to collect solvents from washing cycles (waste)

G   Cleavage block

H   Compounds are collected in parallel array from cleavage block.

I   Block for liquid/liquid extractions in solution chemistry

K   Block containing reagents as normalised solutions or automatically weighed.

L   Solutions collected from cleavage block are evaporated in parallel and analysed by LC/MS.

References

(1) A. Eschenmoser and M. V. Kisakürek, *Helv. Chim. Acta* **1996**, *79*, 1249.

(2) N. K. Mathur, C. K. Narang and R. E. Williams, *Polymers as Aids in Organic Chemistry*, Academic Press: New York (1980).

(3) R. B. Merrifield, *J. Am. Chem. Soc.* **1963**, *85*, 2149.

(4) C. C. Leznoff, *Acc. Chem. Res.* **1978**, *11*, 327.

(5) J. M. J. Fréchet, *Tetrahedron* **1981**, *37*, 663.

(6) J. I. Crowley and H. Rapoport, *Acc. Chem. Res.* **1976**, *9*, 135.

(7) K. C. Nicolaou and R. K. Guy, *Angew. Chem. Int. Ed. Engl.* **1995**, *34*, 2079.

(8) H. M. Geysen, R. H. Meloen and S. J. Barteling, *Proc. Natl. Acad. Sci. USA* **1984**, *81*, 3998.

(9) R. A. Houghten, *Proc. Natl. Acad. Sci. USA* **1985**, *82*, 5131.

(10) E. M. Gordon, R. W. Barrett, W. J. Dower, S. P. Fodor and M. A. Gallop, *J. Med. Chem.* **1994**, *37*, 1385.

(11) M. A. Gallop, R. W. Barrett, W. J. Dower, S. P. Fodor and E. M. Gordon, *J. Med. Chem.* **1994**, *37*, 1233.

(12) P. H. H. Hermkens, H. C. J. Ottenheijm and D. Rees, *Tetrahedron* **1996**, *52*, 4527.

(13) F. Balkenhohl, C. von dem Bussche-Hünnefeld, A. Lansky and C. Zechel, *Angew. Chem. Int. Ed. Engl.* **1996**, *35*, 2288.

(14) N. K. Terrett, M. Gardner, D. W. Gordon, R. J. Koblecki and J. Steele, *Tetrahedron* **1995**, *51*, 8135.

(15) A. Furka, F. Sebestyen, M. Asgedom and G. Dibo, *Int. J. Pept. Prot. Res.* **1991**, *37*, 487.

(16) R. L. Letsinger and V. Mahadevan, *J. Am. Chem. Soc.* **1965**, *87*, 3526.

(17) R. Arshady, G. W. Kenner and A. Ledwith, *Makromol. Chem.* **1976**, *177*, 2911.

(18) R. Arshady, E. Atherton, M. J. Gait, K. Lee and R. C. Sheppard, *J. Chem. Soc., Chem. Commun.* **1979**, 423.

(19) S. J. Danishefsky, K. F. McClure, J. T. Randolph and R. B. Ruggeri, *Science* **1993**, *260*, 1307.

(20) A. Cheminat, C. Benezra, M. J. Farrall and J. M. J. Fréchet, *Can. J. Chem.* **1981**, *59*, 1405.

(21) L. D. Arnold, H. I. Assil and J. C. Vederas, *J. Am. Chem. Soc.* **1989**, *111*, 3973.

(22) K. Akaji and Y. Kiso, *Tetrahedron Lett.* **1997**, *38*, 5185.

(23) D. Cordier and M. W. Hosseini, *New J. Chem.* **1990**, *14*, 611.

(24) C. Oefner, A. D'Arci, F. K. Winkler, B. Eggimann and M. Hosang, *EMBO J.* **1992**, *11*, 3921.

(25) R. Ross, E. W. Raines and D. F. Bowden-Pope, *Cell* **1986**, *46*, 155.

(26) C. F. Ibañez, T. Ebendal, G. Barbany, J. Murray-Rust, T. Blundell and H. Perrson, *Cell* **1992**, *69*, 329.

(27) R. A. Kramer, M. D. Schaber, A. M. Skalka, K. Ganguly, F. Wong-Staal and E. P. Reddy, *Science* **1986**, *231*, 1580.

(28) A. H. Davidson, A. H. Drummond, W. Galloway and M. Whittaker, *Chemistry & Industry* **1997**, *7*, 258.

(29) A. MacDonald, R. Ramage, S. H. DeWitt and E. Hogan, *Diversomer Technology: Recent Advances in the Generation of Chemical Diversity*, conference: *"Combinatorial Synthesis Symposium"*, Exeter, 1995.

(30) S. R. Hubbard, L. Wei, L. Ellis and W. A. Hendrickson, *Nature* **1994**, *372*, 746.

(31) J. P. Griffith, J. L. Kim, E. E. Kim, M. D. Sintchak, J. A. Thompson, M. J. Fitzgibbon, M. A. Fleming, P. R. Caron, K. Hsiao and M. A. Navia, *Cell* **1995**, *82*, 507.

(32) H.-J. Böhm, G. Klebe and H. Kubinyi, *Wirkstoffdesign*, Spektrum: Heidelberg (1996).

(33) I. D. Kuntz, *Science* **1992**, *257*, 1078.

(34) K. Müller, D. Obrecht, A. Knierzinger, C. Stankovic, C. Spiegler, W. Bannwarth, A. Trzeciak, G. Englert, A. M. Labhardt and P. Schönholzer, in: *Perspectives in Medicinal Chemistry*, B. Testa, E. Kyburz, W. Fuhrer and R. Giger (Ed.), p. 513, Verlag Helv. Chim. Acta: Basel (1993).

(35) D. S. Kemp, *TIBTECH* **1990**, *8*, 249.

(36) J. P. Schneider and J. W. Kelly, *Chem. Rev.* **1995**, *95*, 2169.

(37) D. Obrecht, M. Altorfer and J. A. Robinson, *Novel Peptide Mimetic Building Blocks and Strategies for Efficient Lead Finding*, in: *Advances in Medicinal Chemistry*, B. E. Maryanoff and A. B. Reitz (Ed.), in press, p. JAI Press Inc.: Greenwich, Connecticut (1998).

(38) K. Hostettmann, *Chimia* **1998**, *52*, 1.

(39) J. W. Westley, *Polyether Antibiotics: Naturally occurring acid ionophors*, Vol. I&II, Marcel Dekker, Inc.: New York (1982).

(40) G. Albers-Schoenberg, B. H. Arison, Y. Chabala, A. W. Douglas, P. Eskola, M. H. Fisher, A. Lusi, H. Mrozik, J. L. Smith and R. L. Tolman, *J. Am. Chem. Soc.* **1981**, *103*, 4216.

(41) T. Hata, Y. Sano, R. Sugwara, A. Matsume, K. Kanamori, T. Shima and T. Hoshi, *J. Antibiot., Ser. A* **1956**, *9*, 141.

(42) T. Fukuyama, F. Nakasubo, J. Coccuzza and J. Kishi, *Tetrahedron Lett.* **1977**, *18*, 4295.

(43) J. Kishi, *J. Nat. Prod.* **1979**, *42*, 549.

(44) B. E. Bierer, B. S. Matila, R. Standaert, L. A. Herzenberg, S. J. Burakoff, G. Grabtree and S. L. Schreiber, *Proc. Natl. Acad. Sci. USA* **1990**, *87*, 9231.

(45) S. P. Douglas, D. M. Whitfield and J. J. Krepinsky, *J. Am. Chem. Soc.* **1995**, *117*, 2116.

(46) H. Han, M. Wolfe, S. Brenner and K. D. Janda, *Proc. Natl. Acad. Sci. USA* **1995**, *92*, 6419.

(47) R. H. Andreatta and H. Rink, *Helv. Chim. Acta* **1973**, *56*, 1205.

(48) H. Hayatsu and H. G. Khorana, *J. Am. Chem. Soc.* **1967**, *89*, 3880.

(49) H. Hayatsu and H. G. Khorana, *J. Am. Chem. Soc.* **1966**, *88*, 3182.

(50) M. M. Shemyakin, Y. A. Ovchinnikov, A. A. Kiryushkin and I. V. Kozhevnikova, *Tetrahedron Lett.* **1965**, *6*, 2323.

(51) M. Mutter, H. Hagenmaier and E. Bayer, *Angew. Chem. Int. Ed. Engl.* **1971**, *10*, 811.

(52) W. Rapp, *PEG grafted polystyrene tentacle polymers*, in: *Combinatorial Peptide and Nonpeptide Libraries*, G. Jung (Ed.), p. 425, VCH: Weinheim (1996).

(53) D. R. Cody, S. H. DeWitt, J. C. Hodges, J. S. Kiely, W. H. Moos, M. R. Pavia, B. D. Roth, M. C. Schroeder and C. J. Stankovic, US 5324483, 1994.

(54) S. H. DeWitt, *Diversomer Technology: Recent Advances in the Generation of Chemical Diversity*, conference: *"Chemical Lead Generation in Drug Discovery"*, Cambridge, 1995.

(55) S. H. DeWitt and A. W. Czarnik, *Acc. Chem. Res.* **1996**, *29*, 114.

(56) S. Hobbs-DeWitt, J. S. Kiely, C. J. Stankovic, M. C. Schroeder, D. M. Reynolds Cody and M. R. Pavia, *Proc. Natl. Acad. Sci. USA* **1993**, *90*, 6909.

(57) N. J. Maeji, R. M. Valerio, A. M. Bray, R. A. Campbell and H. M. Geysen, *Reactive Polymers* **1994**, *22*, 203.

(58) H. M. Geysen, S. J. Rodda, T. J. Mason, G. Tribbick and P. G. Schoofs, *J. Immun. Meth.* **1987**, *102*, 259.

(59) A. M. Bray, D. S. Chiefari, R. M. Valerio and N. J. Maeji, *Tetrahedron Lett.* **1995**, *36*, 5081.

(60) B. A. Bunin, M. J. Plunkett and J. A. Ellman, *Proc. Natl. Acad. Sci. USA* **1994**, *91*, 4708.

(61) A. A. Virgilio and J. A. Ellman, *J. Am. Chem. Soc.* **1994**, *116*, 11580.

(62) R. M. Valerio, A. M. Bray and H. Patsiouras, *Tetrahedron Lett.* **1996**, *37*, 3019.

(63) R. Frank, *Bioorg. Med. Chem. Lett.* **1993**, *3*, 425.

(64) B. Blankemeyer-Menge and R. Frank, *Tetrahedron Lett.* **1988**, *29*, 5871.

(65) P. Diddams, *Inorganic supports and catalysts-an overview*, in: *Solid Supports and Catalysts in Organic Synthesis*, K. Smith (Ed.), p. 3, Ellis Horwood Ltd.: Chichester (1992).

(66) H. U. Blaser, *Mod. Synth. Meth.* **1995**, *7*, 179.

(67) J. M. Maud, *Organic Supports – an Overview*, in: *Solid Supports and Catalysts in Organic Synthesis*, K. Smith (Ed.), p. 171, Ellis Horwood Ltd.: Chichester (1992).

(68) W. P. Hohenstein and H. Mark, *J. Polymer Science* **1946**, *1*, 127.

(69) F. H. Winslow and W. Matreyek, *Ind. Eng. Chem.* **1951**, *43*, 1108.

(70) H. Logemann, in: *Houben-Weyl, Methoden der Organischen Chemie*, E. Müller (Ed.), Vol. 14/1, p. 839, Georg Thieme Verlag: Stuttgart (1961).

(71) W. Heitz and R. Michels, *Macromol. Chem.* **1971**, *148*, 9.

(72) A. Guyot, *Pure & Appl. Chem.* **1988**, *60*, 365.

(73) D. H. Rich and S. K. Gurwara, *J. Am. Chem. Soc.* **1975**, *97*, 1575.

(74) E. Bayer, *Angew. Chem. Int. Ed. Engl.* **1991**, *30*, 113.

(75) T. Kusama and H. Hayatsu, *Chem. Pharm. Bull.* **1970**, *18*, 319.

(76) A. Guyot and M. Bartholin, *Design and properties of polymers as materials for fine chemistry*, in: *Progress in Polymer Science*, A. D. Jenkins and V. T. Stannett (Ed.), Vol. 8, p. 277, Pergamon Press: Oxford (1982).

(77) R. L. Albright, *Reactive Polymers* **1986**, *4*, 155.

(78) K. W. Pepper, H. M. Paisley and M. A. Young, *J. Chem. Soc.* **1953**, 4097.

(79) T. J. Nieuwstad, A. P. G. Kieboom, A. J. Breijer, J. van der Linden and H. van Bekkum, *Rec. Trav. Chim. Pays-Bas* **1976**, *95*, 225.

(80) R. P. Pinnell, G. D. Khune, N. A. Khatri and S. L. Manatt, *Tetrahedron Lett.* **1984**, *25*, 3511.

(81) R. S. Feinberg and R. B. Merrifield, *Tetrahedron* **1974**, *30*, 3209.

(82) J. T. Sparrow, *Tetrahedron Lett.* **1975**, *16*, 4637.

(83) W. T. Ford and S. A. Yacoub, *J. Org. Chem.* **1981**, *46*, 819.

(84) T. Balakrishnan and W. T. Ford, *J. Appl. Polymer Science* **1982**, *27*, 133.

(85) S. Mohanraj and W. T. Ford, *Macromolecules* **1986**, *19*, 2470.

(86) A. R. Mitchell, S. B. H. Kent, M. Engelhard and R. B. Merrifield, *J. Org. Chem.* **1978**, *43*, 2845.

(87) A. R. Mitchell, S. B. H. Kent, B. W. Erickson and R. B. Merrifield, *Tetrahedron Lett.* **1976**, *17*, 3795.

(88) A. R. Mitchell, B. W. Erickson, M. N. Ryabtsev, R. S. Hodges and R. B. Merrifield, *J. Am. Chem. Soc.* **1976**, *98*, 7357.

(89) J. T. Sparrow, *J. Org. Chem.* **1976**, *41*, 1350.

(90) J. P. Tam, F. S. Tjoeng and R. B. Merrifield, *J. Am. Chem. Soc.* **1980**, *102*, 6117.

(91) N. M. Weinshenker and C.-M. Shen, *Tetrahedron Lett.* **1972**, *13*, 3281.

(92) N. M. Weinshenker, C. M. Shen and J. Y. Wong, *Org. Synth.* **1977**, *56*, 95.

(93) H. Ito, N. Takamatsu and I. Ichikizaki, *Chem. Lett.* **1975**, 577.

(94) A. Einhorn, *Ann.* **1908**, *361*, 113.

(95) C. C. Zikos and N. G. Ferderigos, *Tetrahedron Lett.* **1995**, *36*, 3741.

(96) S.-S. Wang, *J. Org. Chem.* **1975**, *40*, 1235.

(97) R. Deans and V. M. Rotello, *J. Org. Chem.* **1997**, *62*, 4528.

(98) D. Obrecht, C. Abrecht, A. Grieder and J.-M. Villalgordo, *Helv. Chim. Acta* **1997**, *80*, 65.

(99) S. Kobayashi, I. Hachiya, S. Suzuki and M. Moriwaki, *Tetrahedron Lett.* **1996**, *37*, 2809.

(100) X. Beebe, N. E. Schore and M. J. Kurth, *J. Org. Chem.* **1995**, *60*, 4196.

(101) J. M. J. Fréchet and C. Schuerch, *J. Am. Chem. Soc.* **1971**, *93*, 492.

(102) H. M. Relles and R. W. Schluenz, *J. Am. Chem. Soc.* **1974**, *96*, 6469.

(103) B.-D. Park, H.-I. Lee, S.-J. Ryoo and Y.-S. Lee, *Tetrahedron Lett.* **1997**, *38*, 591.

(104) D. C. Neckers, D. A. Kooistra and G. W. Green, *J. Am. Chem. Soc.* **1972**, *94*, 9284.

(105) M. Tomoi, N. Kori and H. Kakiuchi, *Reactive Polymers* **1985**, *3*, 341.

(106) W. F. DeGrado and E. T. Kaiser, *J. Org. Chem.* **1980**, *45*, 1295.

(107) A. Ajayaghosh and V. N. Rajasekharan Pillai, *Tetrahedron Lett.* **1995**, *36*, 777.

(108) F. Cramer and H. Köster, *Angew. Chem. Int. Ed. Engl.* **1968**, *7*, 473.

(109) F. Helfferich, *Ion Exchange*, Mc Graw-Hill: New York (1962).

(110) L. M. Dowling and G. R. Stark, *Biochemistry* **1969**, *8*, 4728.

(111) M. J. Farrall and J. M. J. Fréchet, *J. Org. Chem.* **1976**, *41*, 3877.

(112) F. Camps, J. Castells, M. J. Ferrando and J. Font, *Tetrahedron Lett.* **1971**, *12*, 1713.

(113) N. M. Weinshenker, G. A. Crosby and J. Y. Wong, *J. Org. Chem.* **1975**, *40*, 1966.

(114) G. A. Crosby, N. M. Weinshenker and H.-S. Uh, *J. Am. Chem. Soc.* **1975**, *97*, 2232.

(115) M. Bernard and W. T. Ford, *J. Org. Chem.* **1983**, *48*, 326.

(116) T. M. Fyles and C. C. Leznoff, *Can. J. Chem.* **1976**, *54*, 935.

(117) E. Atherton, M. J. Gait, R. C. Sheppard and B. J. Williams, *Bioorg. Chem.* **1979**, *8*, 351.

(118) E. Atherton, D. L. J. Clive and R. C. Sheppard, *J. Am. Chem. Soc.* **1975**, *97*, 6584.

(119) E. Atherton, E. Brown and R. C. Sheppard, *J. Chem. Soc., Chem. Commun.* **1981**, 1151.

(120) P. W. Small and D. C. Sherrington, *J. Chem. Soc., Chem. Commun.* **1989**, 1589.

(121) M. Meldal, *Tetrahedron Lett.* **1992**, *33*, 3077.

(122) M. Meldal, F.-I. Auzanneau, O. Hindsgaul and M. M. Palcic, *J. Chem. Soc., Chem. Commun.* **1994**, 1849.

(123) M. Meldal, I. Svendsen, K. Breddam and F.-I. Auzanneau, *Proc. Natl. Acad. Sci. USA* **1994**, *91*, 3314.

(124) M. Meldal and I. Svendsen, *J. Chem. Soc., Perkin Trans. 1* **1995**, 1591.

(125) M. Renil and M. Meldal, *Tetrahedron Lett.* **1996**, *37*, 6185.

(126) M. Kempe and G. Barany, *J. Am. Chem. Soc.* **1996**, *118*, 7083.

(127) D. Su and F. M. Menger, *Tetrahedron Lett.* **1997**, *38*, 1485.

(128) A. Akelah and D. C. Sherrington, *Chem. Rev.* **1981**, *81*, 557.

(129) D. C. Bailey and S. H. Langer, *Chem. Rev.* **1981**, *81*, 109.

(130) J. M. J. Fréchet, G. D. Darling, S. Itsuno, P.-Z. Lu, M. Vivas de Meftahi and W. A. Rolls Jr., *Pure & Appl. Chem.* **1988**, *60*, 353.

(131) E. C. Blossey and W. T. Ford, in: *Comprehensive Polymer Science*, G. C. Eastmond, A. Ledwith, S. Russo and P. Sigwalt (Ed.), Vol. 6, p. 81, Pergamon Press: Oxford (1989).

(132) S. J. Shuttleworth, S. M. Allin and P. K. Sharma, *Synthesis* **1997**, 1217.

(133) W. T. Ford and E. C. Blossey, in: *Preparative Chemistry Using Supported Reagents*, P. Laszlo (Ed.), p. 193, Academic Press: New York (1987).

(134) G. Gelbard, in: *Preparative Chemistry Using Supported Reagents*, N. K. Mathur, C. K. Narang and R. E. Williams (Ed.), p. 213, Academic Press: New York (1987).

(135) J. Licto and H. Marrakchi, in: *Preparative Chemistry Using Supported Reagents*, N. K. Mathur, C. K. Narang and R. E. Williams (Ed.), p. 235, Academic Press: New York (1987).

(136) K. Smith, *Solid Supports and Catalysts in Organic Synthesis*, Ellis Horwood: Chichester (1992).

(137) J. Rebek Jr., *Tetrahedron* **1979**, *35*, 723.

(138) G. M. Blackburn, M. J. Brown and M. R. Harris, *J. Chem. Soc., Chem. Commun.* **1966**, 611.

(139) J. P. Collman, K. M. Kosydar, M. Bressan, W. Lamanna and T. Garrett, *J. Am. Chem. Soc.* **1984**, *106*, 2569.

(140) S. L. Regen, *J. Org. Chem.* **1977**, *42*, 875.

(141) F. Minutolo, D. Pini and P. Salvadori, *Tetrahedron Lett.* **1996**, *37*, 3375.

(142) C. E. Song, J. W. Yang, H. J. Ha and S. Lee, *Tetrahedron: Asymmetry* **1996**, *7*, 645.

(143) P. B. Rheiner, H. Sellner and D. Seebach, *Helv. Chim. Acta* **1997**, *80*, 2027.

(144) B. J. Cohen, M. A. Kraus and A. Patchornik, *J. Am. Chem. Soc.* **1981**, *103*, 7620.

(145) S. Mazur and P. Jayalekshmy, *J. Am. Chem. Soc.* **1979**, *101*, 677.

(146) C. C. Leznoff and J. Y. Wong, *Can. J. Chem.* **1972**, *50*, 2892.

(147) C. C. Leznoff and D. M. Dixit, *Can. J. Chem.* **1977**, *55*, 3351.

(148) T. M. Fyles, C. C. Leznoff and J. Weatherston, *Can. J. Chem.* **1978**, *56*, 1031.

(149) J. M. Goldwasser and C. C. Leznoff, *Can. J. Chem.* **1978**, *56*, 1562.

(150) J. Y. Wong, C. Manning and C. C. Leznoff, *Angew. Chem. Int. Ed. Engl.* **1974**, *13*, 666.

(151) J. M. J. Fréchet and L. J. Nuyens, *Can. J. Chem.* **1976**, *54*, 926.

(152) P. Hodge and J. Waterhouse, *J. Chem. Soc., Perkin Trans. 1* **1983**, 2319.

(153) S. C. Story and J. V. Aldrich, *Int. J. Pept. Prot. Res.* **1994**, *43*, 292.

(154) B. J. Cohen, M. A. Kraus and A. Patchornik, *J. Am. Chem. Soc.* **1977**, *99*, 4165.

(155) G. Cainelli, M. Contento, F. Manescalchi and R. Regnoli, *J. Chem. Soc., Perkin Trans. 1* **1980**, 2516.

(156) M. Bessodes and K. Antonakis, *Tetrahedron Lett.* **1985**, *26*, 1305.

(157) J. Rebek Jr and F. Gaviña, *J. Am. Chem. Soc.* **1975**, *97*, 3453.

(158) J. Rebek Jr, D. Brown and S. Zimmerman, *J. Am. Chem. Soc.* **1975**, *97*, 454.

(159) J. Rebek Jr, D. Brown and S. Zimmerman, *J. Am. Chem. Soc.* **1975**, *97*, 4407.

(160) A. Warshawsky, R. Kalir and A. Patchornik, *J. Am. Chem. Soc.* **1978**, *100*, 4544.

(161) J. J. Parlow, *Tetrahedron Lett.* **1995**, *36*, 1395.

(162) H. Frank and H. Hagenmeier, *Experimentia* **1975**, *31*, 131.

(163) G. Heusel, G. Bovermann, W. Göhring and G. Jung, *Angew. Chem. Int. Ed. Engl.* **1977**, *16*, 642.

(164) H. Han and K. D. Janda, *Angew. Chem. Int. Ed. Engl.* **1997**, *36*, 1731.

(165) D. E. Bergbreiter and R. Chandran, *J. Am. Chem. Soc.* **1985**, *107*, 4792.

(166) E. Seymour and J. M. J. Fréchet, *Tetrahedron Lett.* **1976**, *17*, 3669.

(167) L. A. Carpino, J. R. Williams and A. Lopusinski, *J. Chem. Soc., Chem. Commun.* **1978**, 450.

(168) L. A. Carpino, E. M. E. Mansour and J. Knapczyk, *J. Org. Chem.* **1983**, *48*, 666.

(169) J. I. Crowley and H. Rapoport, *J. Am. Chem. Soc.* **1970**, *92*, 6363.

(170) C. C. Leznoff and P. I. Svirskaya, *Angew. Chem. Int. Ed. Engl.* **1978**, *17*, 947.

(171) J. J. Parlow, *Tetrahedron Lett.* **1996**, *37*, 5257.

(172) M. G. Siegel, P. J. Hahn, B. A. Dressman, J. E. Fritz, J. R. Grunwell and S. W. Kaldor, *Tetrahedron Lett.* **1997**, *38*, 3357.

(173) L. M. Gayo and M. J. Suto, *Tetrahedron Lett.* **1997**, *38*, 513.

(174) R. M. Lawrence, S. A. Biller, O. M. Fryszman and M. A. Poss, *Synthesis* **1997**, 553.

(175) T. A. Keating and R. W. Armstrong, *J. Am. Chem. Soc.* **1996**, *118*, 2574.

(176) S. W. Kaldor, M. G. Siegel, J. E. Fritz, B. A. Dressman and P. J. Hahn, *Tetrahedron Lett.* **1996**, *37*, 7193.

(177) R. J. Booth and J. C. Hodges, *J. Am. Chem. Soc.* **1997**, *119*, 4882.

(178) T. L. Deegan, O. W. Gooding, S. Baudart and J. A. Porco Jr., *Tetrahedron Lett.* **1997**, *38*, 4973.

(179) S. W. Kaldor, J. E. Fritz, J. Tang and E. R. McKinney, *Bioorg. Med. Chem. Lett.* **1996**, *6*, 3041.

(180) A. A. Virgilio, S. C. Schürer and J. A. Ellman, *Tetrahedron Lett.* **1996**, *37*, 6961.

(181) T. Brunelet, C. Jouitteau and G. Gelbard, *J. Org. Chem.* **1986**, *51*, 4016.

(182) C. R. Harrison and P. Hodge, *J. Chem. Soc., Perkin Trans. 1* **1976**, 605.

(183) G. Gelbard and S. Colonna, *Synthesis* **1977**, 113.

(184) G. Cardillo, M. Orena, S. Sandri and C. Tomasini, *Chemistry & Industry* **1983**, 643.

(185) G. Cainelli, F. Manescalchi and M. Panunzio, *Synthesis* **1976**, 472.

(186) S. Colonna, A. Re, G. Gelbard and E. Cesarotti, *J. Chem. Soc., Perkin Trans. 1* **1979**, 2248.

(187) C. R. Harrison and P. Hodge, *Synthesis* **1980**, 299.

(188) G. Cainelli, F. Manescalchi and M. Panunzio, *Synthesis* **1979**, 141.

(189) R. Yamada, *Mem. Def. Acad., Math., Phys., Chem. Eng., Yokosuka, Jap.* **1973**, *13*, 411.

(190) R. Yamada, *Mem. Def. Acad., Math., Phys., Chem. Eng., Yokosuka, Jap.* **1969**, *9*, 667.

(191) J. V. Weber, P. Faller, G. Kirsch and M. Schneider, *Synthesis* **1984**, 1044.

(192) M. M. Salunkhe, D. G. Salunkhe, A. S. Kanade, R. B. Mane and P. P. Wadgaonkar, *Synth. Commun.* **1990**, *20*, 1143.

(193) G. Cardillo, M. Orena and S. Sandri, *J. Org. Chem.* **1986**, *51*, 713.

(194) G. Cardillo, M. Orena, G. Porzi and S. Sandri, *J. Chem. Soc., Chem. Commun.* **1982**, 1309.

(195) A. Bongini, G. Cardillo, M. Orena, S. Sandri and C. Tomasini, *J. Chem. Soc., Perkin Trans. 1* **1986**, 1339.

(196) M. M. Salunkhe, A. R. Sande, A. S. Kanade and P. P. Wadgaonkar, *Synth. Commun.* **1997**, *27*, 2885.

(197) G. Cardillo, M. Orena, G. Porzi and S. Sandri, *Synthesis* **1981**, 793.

(198) W. Xu, R. Mohan and M. M. Morrissey, *Tetrahedron Lett.* **1997**, *38*, 7337.

(199) M. J. Kurth, L. A. Ahlberg Randall and K. Takenouchi, *J. Org. Chem.* **1996**, *61*, 8755.

(200) H. Stephensen and F. Zaragoza, *J. Org. Chem.* **1997**, *62*, 6096.

(201) G. Cainelli, G. Cardillo, M. Orena and M. Sandri, *J. Am. Chem. Soc.* **1976**, *98*, 6737.

(202) L. G. Wade Jr. and L. M. Stell, *J. Chem. Educ.* **1980**, *57*, 438.

(203) G. Cardillo, M. Orena and S. Sandri, *Tetrahedron Lett.* **1976**, *17*, 3985.

(204) J. M. J. Fréchet, J. Warnock and M. J. Farrall, *J. Org. Chem.* **1978**, *43*, 2618.

(205) J. M. J. Fréchet, P. Darling and M. J. Farrall, *J. Org. Chem.* **1981**, *46*, 1728.

(206) G. Cainelli, M. Contento, F. Manescalchi and L. Plessi, *Synthesis* **1989**, 45.

(207) G. Cainelli, M. Contento, F. Manescalchi and L. Plessi, *Synthesis* **1989**, 47.

(208) C. R. Harrison and P. Hodge, *J. Chem. Soc., Perkin Trans. 1* **1982**, 509.

(209) B. Hinzen and S. V. Ley, *J. Chem. Soc., Perkin Trans. 1* **1998**, 1.

(210) B. Hinzen and S. V. Ley, *J. Chem. Soc., Perkin Trans. 1* **1997**, 1907.

(211) C. R. Harrison and P. Hodge, *J. Chem. Soc., Chem. Commun.* **1974**, 1009.

(212) C. R. Harrison and P. Hodge, *J. Chem. Soc., Perkin Trans. 1* **1976**, 2252.

(213) A. R. Sande, M. H. Jagadale, R. B. Mane and M. M. Salunkhe, *Tetrahedron Lett.* **1984**, *25*, 3501.

(214) N. M. Goudagaon, P. P. Wadgaonkar and G. W. Kabalka, *Synth. Commun.* **1989**, *19*, 805.

(215) G. W. Kabalka, P. P. Wadgaonkar and N. Chatla, *Synth. Commun.* **1990**, *20*, 293.

(216) H. W. Gibson and F. C. Bailey, *J. Chem. Soc., Chem. Commun.* **1977**, 815.

(217) G. Brunow, L. Koskinen and P. Urpilainen, *Acta Chemica Scandinavica* **1981**, 53.

(218) A. Nag, A. Sarkar, S. K. Sarkar and S. K. Palit, *Synth. Commun.* **1987**, *17*, 1007.

(219) B. Tamami and M. M. Lakouraj, *Synth. Commun.* **1995**, *25*, 3089.

(220) G. A. Crosby, US 3,928,239, 1975.

(221) J. M. J. Fréchet, E. Bald and P. Lecavalier, *J. Org. Chem.* **1986**, *51*, 3462.

(222) R. ter Halle, E. Schulz and M. Lemaire, *Synlett* **1997**, 1257.

(223) N. M. Weinshenker and C.-M. Shen, *Tetrahedron Lett.* **1972**, *13*, 3285.

(224) M. C. Desai and L. M. Stephens Stramiello, *Tetrahedron Lett.* **1993**, *34*, 7685.

(225) M. Adamczyk and J. R. Fishpaugh, *Tetrahedron Lett.* **1996**, *37*, 4305.

(226) W. Graf, *Chemistry & Industry* **1987**, 232.

(227) A. Bongini, G. Cainelli, M. Contento and F. Manescalchi, *J. Chem. Soc., Chem. Commun.* **1980**, 1278.

(228) A. Bongini, G. Cainelli, M. Contento and F. Manescalchi, *Synthesis* **1980**, 143.

(229) S. Cacchi and L. Caglioti, *Synthesis* **1979**, 64.

(230) F. Rahaingoson, B. Kimpiobi-Ninafiding, Z. Mouloungui and A. Gaset, *Synth. Commun.* **1992**, *22*, 1923.

(231) K. Smith, D. M. James, I. Matthews and M. R. Bye, *J. Chem. Soc., Perkin Trans. 1* **1992**, 1877.

(232) C. Yaroslavsky, A. Patchornik and E. Katchalski, *Tetrahedron Lett.* **1970**, *11*, 3629.

(233) M. Metelko and M. Zupan, *Synth. Commun.* **1988**, *18*, 1821.

(234) A. J. Biloski and B. Ganem, *Synthesis* **1983**, 537.

(235) G. Rosini, R. Ballini and M. Petrini, *Synthesis* **1986**, 46.

(236) R. Ballini, M. Petrini and G. Rosini, *Synthesis* **1987**, 711.

(237) R. Schwesinger, *Chimia* **1985**, *39*, 269.

(238) G. Cainelli, G. Cardillo and M. Orena, *J. Chem. Soc., Perkin Trans. 1* **1979**, 1597.

(239) J. M. Miller, K.-H. So and J. H. Clark, *J. Chem. Soc., Chem. Commun.* **1978**, 466.

(240) G. Cardillo, M. Orena, G. Porzi, S. Sandri and C. Tomasini, *J. Org. Chem.* **1984**, *49*, 701.

(241) B. A. Kulkarni and A. Ganesan, *Angew. Chem. Int. Ed. Engl.* **1997**, *36*, 2454.

(242) F. Camps, J. Castells, J. Font and F. Vela, *Tetrahedron Lett.* **1971**, *12*, 1715.

(243) S. V. McKinley and J. W. Rakshys, *J. Chem. Soc., Chem. Commun.* **1972**, 134.

(244) J. Castells, J. Font and A. Virgili, *J. Chem. Soc., Perkin Trans. 1* **1979**, 1.

(245) M. Bernard, W. T. Ford and E. C. Nelson, *J. Org. Chem.* **1983**, *48*, 3164.

(246) J. J. Landi and H. R. Brinkman, *Synthesis* **1992**, 1093.

(247) S. L. Regen and D. P. Lee, *J. Org. Chem.* **1975**, *40*, 1669.

(248) R. Caputo, C. Ferreri, S. Noviello and G. Palumbo, *Synthesis* **1986**, 499.

(249) R. Caputo, C. Ferreri and G. Palumbo, *Synthesis* **1987**, 386.

(250) A. A. El-Shehawy, M. Y. Abdelaal, K. Watanabe, K. Ito and S. Itsuno, *Tetrahedron: Asymmetry* **1997**, *8*, 1731.

(251) K. Soai, S. Niwa and M. Watanabe, *J. Org. Chem.* **1988**, *53*, 927.

(252) K. Soai, T. Suzuki and T. Shono, *J. Chem. Soc., Chem. Commun.* **1994**, 317.

(253) M. Watanabe and K. Soai, *J. Chem. Soc., Perkin Trans. 1* **1994**, 837.

(254) D. Seebach, R. E. Marti and T. Hintermann, *Helv. Chim. Acta* **1996**, *79*, 1710.

(255) S. Kobayashi and S. Nagayama, *J. Am. Chem. Soc.* **1996**, *118*, 8977.

(256) S. Kobayashi and S. Nagayama, *Synlett* **1997**, 653.

(257) S. Kiyooka, Y. Kido and Y. Kaneko, *Tetrahedron Lett.* **1994**, *35*, 5243.

(258) S. Itsuno, K. Kamahori, K. Watanabe, T. Koizumi and K. Ito, *Tetrahedron: Asymmetry* **1994**, *5*, 523.

(259) B. M. Kim and K. B. Sharpless, *Tetrahedron Lett.* **1990**, *31*, 3003.

(260) B. B. Lohray, A. Thomas, P. Chittari, J. R. Ahuja and P. K. Dhal, *Tetrahedron Lett.* **1992**, *33*, 5453.

(261) B. B. Lohray, E. Nadanan and V. Bhushan, *Tetrahedron Lett.* **1994**, *35*, 6559.

(262) E. Nandanan, A. Sudalai and T. Ravindranathan, *Tetrahedron Lett.* **1997**, *38*, 2577.

(263) D. Pini, A. Petri, A. Nardi, C. Rosini and P. Salvadori, *Tetrahedron Lett.* **1991**, *32*, 5175.

(264) D. Pini, A. Petri and P. Salvadori, *Tetrahedron: Asymmetry* **1993**, *4*, 2351.

(265) D. Pini, A. Petri and P. Salvadori, *Tetrahedron* **1994**, *50*, 11321.

(266) C. E. Song, E. J. Roh, S. Lee and I. Kim, *Tetrahedron: Asymmetry* **1995**, *6*, 2687.

(267) M. J. Farall, T. Durst and J. M. J. Fréchet, *Tetrahedron Lett.* **1979**, *20*, 203.

(268) R. Kalir, A. Warshawsky, M. Fridkin and A. Patchornik, *Eur. J. Biochem.* **1975**, *59*, 55.

(269) W. Huang and A. G. Kalivretenos, *Tetrahedron Lett.* **1995**, *36*, 9113.

(270) G. A. Olah, X.-Y. Li, Q. Wang and G. K. S. Prakash, *Synthesis* **1993**, 693.

(271) H. S. Moon, N. E. Schore and M. J. Kurth, *J. Org. Chem.* **1992**, *57*, 6088.

(272) H. S. Moon, N. E. Schore and M. J. Kurth, *Tetrahedron Lett.* **1994**, *35*, 8915.

(273) M. Reggelin and V. Brenig, *Tetrahedron Lett.* **1996**, *37*, 6851.

(274) A. V. Purandare and S. Natarajan, *Tetrahedron Lett.* **1997**, *38*, 8777.

(275) K. Burgess and D. Lim, *J. Chem. Soc., Chem. Commun.* **1997**, 785.

(276) S. M. Allin and S. J. Shuttleworth, *Tetrahedron Lett.* **1996**, *37*, 8023.

(277) M. Kawana and S. Emoto, *Tetrahedron Lett.* **1972**, *13*, 4855.

(278) M. Kawana and S. Emoto, *Bull. Chem. Soc. Jpn.* **1974**, *47*, 160.

(279) P. M. Worster, C. R. McArthur and C. C. Leznoff, *Angew. Chem. Int. Ed. Engl.* **1979**, *18*, 221.

(280) C. R. McArthur, P. M. Worster, J.-L. Jiang and C. C. Leznoff, *Can. J. Chem.* **1982**, *60*, 1836.

(281) A. R. Colwell, L. R. Duckwall, R. Brooks and S. P. McManus, *J. Org. Chem.* **1981**, *46*, 3097.

(282) J. S. Früchtel and G. Jung, *Angew. Chem. Int. Ed. Engl.* **1996**, *35*, 17.

(283) C. Blackburn, F. Albericio and S. A. Kates, *Drugs Fut.* **1997**, *22*, 1007.

(284) Novabiochem, *The combinatorial catalogue*, (1997).

(285) G. Barany and R. B. Merrifield, *Solid-Phase Peptide Synthesis*, in: *The Peptides*, E. Gross and J. Meienhofer (Ed.), Vol. 2, p. 1, Academic Press: New York (1979).

(286) E. Atherton, C. J. Logan, R. C. Sheppard, *J. Chem. Soc., Perkin Trans. 1* **1981**, 538.

(287) S. S. Wang, *J. Am. Chem. Soc.* **1973**, *95*, 1328.

(288) M. Mergler, R. Tanner, J. Gosteli and P. Grogg, *Tetrahedron Lett.* **1988**, *29*, 4005.

(289) H. Rink, *Tetrahedron Lett.* **1987**, *28*, 3787.

(290) K. Barlos, D. Gratos, J. Kallitis, G. Papaphotion, Y. Wenghing and W. Schäfer, *Tetrahedron Lett.* **1989**, *30*, 3943.

(291) K. Barlos, O. Chatzi, D. Gratos and G. Stavropulos, *Int. J. Pept. Prot. Res.* **1991**, *37*, 513.

(292) D. E. Graitanopoulos and J. Weinstock, WO 9534813, 1995.

(293) S. B. Katli, *J. Chem. Soc., Chem. Commun.* **1992**, 843.

(294) C. Garcia-Echeverria, *Tetrahedron Lett.* **1997**, *38*, 8933.

(295) F. Albericio, E. Giralt and R. Eritja, *Tetrahedron Lett.* **1991**, *32*, 1515.

(296) R. Ramage, C. A. Barron, S. Bidecki and D. W. Thomas, *Tetrahedron Lett.* **1987**, *28*, 4105.

(297) A. Routledge, T. Stock, S. L. Flitsch and N. J. Turner, *Tetrahedron Lett.* **1997**, *38*, 8287.

(298) D. G. Mullen and G. Barany, *J. Org. Chem.* **1988**, *53*, 5240.

(299) H. Kuntz, B. Dombo and W. Kosch, *Peptides*, conference: *Proc. 20th Eur. Pept. Symp.*, 1988.

(300) F. S. Tjoeng and G. A. Heavner, *J. Org. Chem.* **1983**, *48*, 355.

(301) S. W. Wang, *J. Org. Chem.* **1976**, *41*, 3258.

(302) R. P. Hammer, F. Albericio, E. Giralt and G. Barany, *Int. J. Pept. Prot. Res.* **1990**, *28*, 31.

(303) J. P. Tam, *J. Org. Chem.* **1985**, *50*, 5291.

(304) J. P. Tam, *Tetrahedron Lett.* **1981**, *22*, 2854.

(305) G. R. Matsueda and J. M. Stewart, *Peptides* **1981**, *2*, 45.

(306) F. Albericio and G. Barany, *Int. J. Pept. Prot. Res.* **1987**, *30*, 206.

(307) D. Sarantakis and J. J. Bicksler, *Tetrahedron Lett.* **1997**, *38*, 7325.

(308) M. Patek and M. Lebl, *Tetrahedron Lett.* **1991**, *32*, 3891.

(309) P. Sieber, *Tetrahedron Lett.* **1987**, *28*, 2107.

(310) Y. Han, S. L. Bontems, P. Hegyes, M. C. Munson, C. Minor, M. C. Kates, F. Albericio and G. Barany, *J. Org. Chem.* **1996**, *61*, 6326.

(311) A. Ayayagosh and V. N. R. Pillai, *Tetrahedron Lett.* **1995**, *36*, 777.

(312) S. J. Teague, *Tetrahedron Lett.* **1996**, *37*, 5751.

(313) P. G. Sammes, *Sulfonamides and Sulfones*, in: *Comprehensive Medicinal Chemistry*, C. Hansch, P. G. Sammes and J. B. Taylor (Ed.), Vol. 2, p. 255, Pergamon Press: Oxford (1990).

(314) A. M. Fivush and T. M. Willson, *Tetrahedron Lett.* **1997**, *38*, 7151.

(315) B. Raju and T. P. Kogan, *Tetrahedron Lett.* **1997**, *38*, 7325.

(316) S. M. Dankwardt, P. B. Smith, J. A. Porco and C. H. Nguyen, *Synlett* **1997**, 854.

(317) K. A. Beaver, A. C. Siegmund and K. L. Spear, *Tetrahedron Lett.* **1996**, *37*, 1145.

(318) L. Richter and M. C. Desai, *Tetrahedron Lett.* **1997**, *38*, 321.

(319) S. L. Mellor, C. McGuire and W. C. Chan, *Tetrahedron Lett.* **1997**, *38*, 3311.

(320) T. Vorherr, **1996**, personal communication.

(321) C. D. Floyd, C. N. Lewis, S. R. Patel and M. Whittaker, *Tetrahedron Lett.* **1996**, *37*, 8045.

(322) R. S. Garigipath, *Tetrahedron Lett.* **1997**, *38*, 6807.

(323) E. E. Swayze, *Tetrahedron Lett.* **1997**, *38*, 8465.

(324) P. Conti, D. Domont, J. Cals, H. C. J. Ottenheijm and D. Leysen, *Tetrahedron Lett.* **1997**, *38*, 2915.

(325) C. Ho and M. J. Kukla, *Tetrahedron Lett.* **1997**, *38*, 2799.

(326) C. Kay, P. J. Murray, L. Sandow and A. B. Holmes, *Tetrahedron Lett.* **1997**, *38*, 6941.

174

(327) A. R. Brown, D. C. Rees, Z. Rankovic and J. R. Morphy, *J. Am. Chem. Soc.* **1997**, *119*, 3288.

(328) J. R. Morphy, Z. Rankovic and D. C. Rees, *Tetrahedron Lett.* **1996**, *37*, 3209.

(329) K. Kaljuste and A. Undén, *Tetrahedron Lett.* **1996**, *37*, 3031.

(330) W. Bannwarth, J. Huebscher and R. Barner, *Bioorg. Med. Chem. Lett.* **1996**, *6*, 1525.

(331) L. A. Thompson and J. A. Ellman, *Tetrahedron Lett.* **1994**, *35*, 9333.

(332) J. Swistok, J. W. Tilley, W. Danho, R. Wagner and K. Mulkerins, *Tetrahedron Lett.* **1989**, *30*, 5045.

(333) W. Neugebauer and E. Escher, *Helv. Chim. Acta* **1989**, *72*, 1319.

(334) J. W. Apsimon and D. M. Dixit, *Synth. Commun.* **1982**, *12*, 113.

(335) Kobayashi and e. al., *Tetrahedron Lett.* **1996**, *37*, 5569.

(336) J. Alsina, C. Chiva, M. Ortiz, F. Rabanal, E. Giralt and F. Albericio, *Tetrahedron Lett.* **1997**, *38*, 883.

(337) M. Bonnet, M. Bradley and J. D. Kilburn, *Tetrahedron Lett.* **1996**, *37*, 5409.

(338) H. M. Zhong, M. N. Greco and B. E. Maryanoff, *J. Org. Chem.* **1997**, *62*, 9326.

(339) J. Nielsen and P. H. Rasmussen, *Tetrahedron Lett.* **1996**, *37*, 3351.

(340) F. Camps, J. Castells and J. Pi, *Ann. Quim.* **1974**, *70*, 848.

(341) G. Massimo, F. Zani, E. Coghi, A. Belloti and P. Mazza, *Farmaco* **1990**, *45*, 439.

(342) A. W. Czarnik, *Diversomer Technology, a practical approach to parallel simultaneous organic synthesis*, conference: *Synthetic chemical libraries in drug discovery*, London, 1995.

(343) S. A. Kolodziej and B. C. Hamper, *Tetrahedron Lett.* **1996**, *37*, 5277.

(344) D. M. Renolds Cody, S. H. Hobbs DeWitt, J. C. Hodges, J. S. Kiely, W. M. Moos, M. R. Pavia, B. D. Roth, M. C. Schroeder and C. J. Stankovic, US 5,324,483, 1994.

(345) D.-H. Ko, D. J. Kim, C. S. Lyu, I. K. Min and H.-S. Moon, *Tetrahedron Lett.* **1998**, *39*, 297.

(346) R. Mohan, Y.-L. Chou and M. M. Morrissey, *Tetrahedron Lett.* **1996**, *37*, 3963.

(347) I. Vlattas, J. Dellureficio, R. Dunn, I. I. Sytwu and J. Stanton, *Tetrahedron Lett.* **1997**, *38*, 7321.

(348) M. E. Frayley and R. S. Rubino, *Tetrahedron Lett.* **1997**, *38*, 3365.

(349) Y. Han, S. D. Walker and R. S. Young, *Tetrahedron Lett.* **1996**, *37*, 2703.

(350) M. J. Plunkett and J. A. Ellman, *J. Org. Chem.* **1997**, *62*, 2885.

(351) B. Chenera, J. A. Finkelstein and D. F. Veber, *J. Am. Chem. Soc.* **1995**, *117*, 11999.

(352) F. X. Woolard, J. Partsch and J. A. Ellman, *J. Org. Chem.* **1997**, *62*, 6102.

(353) G. W. Kenner, J. R. Mc Dermott and R. C. Sheppard, *J. Chem. Soc., Chem. Commun.* **1971**, 636.

(354) J. A. Backes and J. A. Ellman, *J. Am. Chem. Soc.* **1994**, *116*, 11171.

(355) B. J. Backes, A. A. Virgilio and J. A. Ellman, *J. Am. Chem. Soc.* **1996**, *118*, 3055.

(356) I. Hughes, *Tetrahedron Lett.* **1996**, *37*, 7595.

(357) S. Hoffmann and R. Frank, *Tetrahedron Lett.* **1994**, *35*, 7763.

(358) G. Panke and R. Frank, *Tetrahedron Lett.* **1998**, *39*, 17.

(359) A. Chucholowski, T. Masquelin, D. Obrecht, J. Stadlwieser and J. M. Villalgordo, *Chimia* **1996**, *50*, 530.

(360) L. M. Gayo and M. J. Suto, *Tetrahedron Lett.* **1997**, *38*, 211.

(361) L. Yan, C. M. Taylor, R. Goodnow Jr. and D. Kahne, *J. Am. Chem. Soc.* **1994**, *116*, 6953.

(362) S.-H. Kim, D. Augeri, D. Yang and D. Kahne, *J. Am. Chem. Soc.* **1994**, *116*, 1766.

(363) D. Kahne, S. Walker, Y. Cheng and D. Van Engen, *J. Am. Chem. Soc.* **1989**, *111*, 6881.

(364) J. M. Ostresh, G. M. Husar, S. E. Blondelle, B. Dorner, P. A. Weber and R. A. Houghten, *Proc. Natl. Acad. Sci. USA* **1994**, *91*, 11138.

(365) H. M. Geysen, S. J. Rodda and T. J. Mason, *Mol. Immunol.* **1986**, *23*, 709.

(366) D. J. Ecker, T. A. Vickers, R. Hanecak, V. Driver and K. Anderson, *Nucleic Acids Res.* **1993**, *21*, 1853.

(367) F. Sebestyen, G. Dibo, A. Kovacs and A. Furka, *Bioorg. Med. Chem. Lett.* **1993**, *3*, 413.

(368) K. S. Lam, S. E. Salmon, E. M. Hersh, V. J. Hruby, W. M. Kamiersky and R. J. Knapp, *Nature* **1991**, *354*, 82.

(369) R. A. Houghten, C. Pinilla, S. E. Blondelle, J. R. Appel, C. T. Dooley and J. H. Cuervo, *Nature* **1991**, *354*, 84.

(370) H. Passerini, *Gazz. Chim. Ital.* **1921**, *51*, 126.

(371) A. Hantzsch, *Ber. Dtsch. Chem. Ges.* **1890**, *23*, 1474.

(372) A. Strecker, *Justus Liebigs Ann. Chem.* **1854**, *91*, 349.

(373) A. Strecker, *Justus Liebigs Ann. Chem.* **1850**, *75*, 27.

(374) H. Bergs, Ger. Pat. 566094, 1929.

(375) H. T. Bucherer and H. T. Fischbeck, *J. Pract. Chem.* **1934**, *140*, 69.

(376) H. T. Bucherer and H. T. Fischbeck, *J. Pract. Chem.* **1934**, *140*, 28.

(377) C. Mannich and W. Krosche, *Arch. Pharm.* **1921**, *250*, 647.

(378) E. J. Corey and R. D. Ballanson, *J. Am. Chem. Soc.* **1974**, *96*, 6516.

(379) P. Biginelli, *Ber. Dtsch. Chem. Ges.* **1893**, *26*, 447.

(380) P. Biginelli, *Ber. Dtsch. Chem. Ges.* **1891**, *24*, 2962.

(381) R. Grigg, Z. Rankovic, M. Thornton-Pett and A. Somasunderam, *Tetrahedron* **1993**, *49*, 8679.

(382) R. Bossio, S. Marcaccini and R. Pepino, *J. Chem. Res. (S)* **1991**, *15*, 320.

(383) E. Peisach, D. Casebier, S. D. Gallion, P. Furth, G. A. Petsko, J. C. Hogan Jr. and D. Ringe, *Science* **1995**, *269*, 66.

(384) S. Kobayashi, R. Akiyama and M. Moriwaki, *Tetrahedron Lett.* **1997**, *38*, 4819.

(385) I. U. Khand, G. R. Knox, P. L. Pauson, W. E. Watts and M. I. Foremen, *J. Chem. Soc., Perkin Trans. 1* **1973**, *9*, 977.

(386) P. Grieco and A. Bahsas, *Tetrahedron Lett.* **1988**, *29*, 5855.

(387) A. Kiselyov and R. W. Armstrong, *Tetrahedron Lett.* **1997**, *38*, 6163.

(388) S. Blechert, R. Knier, H. Schroers and T. Wirth, *Synthesis* **1995**, 592.

(389) M. Baak, Y. Rubin, A. Franz, H. Stoeckli-Evans, L. Bigler, J. Nachbauer and R. Neier, *Chimia* **1993**, *47*, 233.

(390) A. Franz, P.-Y. Eschler, M. Tharin and R. Neier, *Tetrahedron* **1996**, *52*, 11643.

(391) J. Velker, V. Linder, P.-Y. Eschler, A. Franz and R. Neier, *A new stereocontrolled Tandem Reaction of butadienyl Ester Enolates*, conference: *Herbstversammlung der NSCG*, Lausanne, CH, 1997.

(392) J. R. Morphy, Z. Rankovic and D. C. Rees, *Tetrahedron Lett.* **1996**, *37*, 3209.

(393) A. Merritt, *Solution phase multiple synthesis in drug discovery*, conference: *"Molecular diversity & combinatorial chemistry"*, San Diego, 1996.

(394) M. R. Pavia, G. M. Whitesides, D. G. Hangauer and M. E. Hediger, WO 95/04277, 1995.

(395) T. A. Rano and K. T. Chapman, *Tetrahedron Lett.* **1995**, *36*, 3789.

(396) M. S. Despande, *Tetrahedron Lett.* **1994**, *35*, 5613.

(397) B. A. Bunin and J. A. Ellman, *J. Am. Chem. Soc.* **1992**, *114*, 10997.

(398) M. J. Plunkett and J. A. Ellman, *J. Am. Chem. Soc.* **1995**, *117*, 3306.

(399) M. J. Plunkett and J. A. Ellman, *J. Org. Chem.* **1995**, *60*, 6006.

(400) J. R. Hauske and P. Dorff, *Tetrahedron Lett.* **1995**, *36*, 1589.

(401) J. R. Petersen, *A Simple Approach to Small Synthetic Libraries*, conference: *"Small Molecular Libraries for Drug Discovery"*, La Jolla, 1995.

(402) K.-L. Yu, M. S. Desphande and D. M. Vyas, *Tetrahedron Lett.* **1994**, *35*, 8919.

(403) J. K. Young, J. C. Nelson and J. S. Moore, *J. Am. Chem. Soc.* **1994**, *116*, 10841.

(404) J. A. Ellman, *Acc. Chem. Res.* **1996**, *29*, 132.

(405) I. Sucholeiki, *Tetrahedron Lett.* **1994**, *35*, 7307.

(406) R. N. Zuckermann, E. J. Martin, D. C. Spellmeyer, G. B. Stauber, K. R. Shoemaker, J. M. Kerr, G. M. Figliozzi, D. A. Goff, M. A. Siani, R. J. Simon, S. C. Banville, E. G. Brown, L. Wang, L. S. Richter and W. H. Moos, *J. Med. Chem.* **1994**, *37*, 2678.

(407) D. A. Goff and R. N. Zuckermann, *J. Org. Chem.* **1995**, *60*, 5744.

(408) D. A. Goff and R. N. Zuckermann, *J. Org. Chem.* **1995**, *60*, 5748.

(409) R. N. Zuckermann, J. M. Kerr, S. B. H. Kent and W. H. Moos, *J. Am. Chem. Soc.* **1992**, *114*, 10646.

(410) S. M. Dankwardt, S. R. Newman and J. L. Krstenansky, *Tetrahedron Lett.* **1995**, *36*, 4923.

(411) J. Green, *J. Org. Chem.* **1995**, *60*, 4287.

(412) C. G. Boojamra, K. M. Burow and J. A. Ellman, *J. Org. Chem.* **1995**, *60*, 5742.

(413) C. Chen, L. A. Ahlberg, R. B. Miller, A. D. Jones and M. J. Kurth, *J. Am. Chem. Soc.* **1994**, *116*, 2661.

(414) D. A. Willard and V. Jammalamadaka, *Chemistry and Biology* **1995**, *2*, 45.

(415) D. S. Dhanoa, *"Lead Generation and Optimisation"*, conference: New Orleans, 1996.

(416) M. Hiroshige, J. R. Hauske and P. Zhou, *Tetrahedron Lett.* **1995**, *36*, 4567.

(417) F. W. Forman and I. Sucholeiki, *J. Org. Chem.* **1995**, *60*, 523.

(418) D. A. Campbell, J. C. Bermak, T. S. Burkoth and D. V. Patel, *J. Am. Chem. Soc.* **1995**, *117*, 5381.

(419) V. Krchnak, Z. Flegelova, A. S. Weichsel and M. Lebl, *Tetrahedron Lett.* **1995**, *36*, 6193.

(420) L. S. Richter and T. R. Gadek, *Tetrahedron Lett.* **1994**, *35*, 4705.

(421) H. Ritter and R. Sperber, *Macromolecules* **1994**, *27*, 5919.

(422) M. Patek, B. Drake and M. Lebl, *Tetrahedron Lett.* **1995**, *36*, 2227.

(423) P. T. Ho, D. Chang, J. W. X. Zhong and G. F. Musso, *Peptide Research* **1993**, *6*, 10.

(424) C. P. Holmes, J. P. Chinn, C. P. Look, E. M. Gordon and M. A. Gallop, *J. Org. Chem.* **1995**, *60*, 7328.

(425) D. W. Gordon and J. Steele, *Bioorg. Med. Chem. Lett.* **1995**, *5*, 47.

(426) M. A. Gallop and M. A. Murphy, WO 95/35278, 1995.

(427) S. H. DeWitt, J. S. Kiely, C. J. Stankoviv, M. C. Schroeder, D. M. Renolds Cody and M. R. Pavia, *Proc. Natl. Acad. Sci. USA* **1993**, *90*, 6909.

(428) X. Beebe, C. L. Chiappani, M. M. Olmstaed, M. J. Kurth and N. E. Schore, *J. Org. Chem.* **1995**, *60*, 4204.

(429) M. J. Kurth, L. A. Ahlberg, R. C. Chen, C. Melander, R. B. Miller, K. McAllister, G. Reitz, R. Kang, T. Nakatsu and C. Green, *J. Org. Chem.* **1994**, *59*, 5862.

(430) S. M. Hutchins and K. T. Chapman, *Tetrahedron Lett.* **1994**, *35*, 4055.

(431) S. M. Hutchins and K. T. Chapman, *Tetrahedron Lett.* **1995**, *36*, 2583.

(432) A. Dornow and K. Fischer, *Chem. Ber.* **1966**, *99*, 72.

(433) J. C. Phelan, B. K. Blackburn and T. R. Gadek, *Solid- phase Synthesis of Fibrinogen-GP IIb/IIIa Binding Inhibitors*, conference: *209th ACS National Meeting*, Anaheim, 1995.

(434) C. Zhang, E. J. Moran, T. F. Woiwode, K. M. Short and A. M. M. Mjalli, *Tetrahedron Lett.* **1996**, *37*, 751.

(435) A. M. M. Mjalli, S. Sarshar and T. J. Baiga, *Tetrahedron Lett.* **1996**, *37*, 2943.

(436) A. A. McDonald, S. H. DeWitt, E. M. Hogan and R. Ramage, *Tetrahedron Lett.* **1996**, *37*, 4815.

(437) B. Chenera, J. Elliott and M. Moore, WO 95/16712, 1995.

(438) M. C. Desai, *Strategy for the Synthesis of Heterocyclic Libraries for the Discovery of Lead Structures*, conference: *"Small Molecule Libraries for Drug Discovery"*, San Diego, 1996.

(439) M. F. Gordeev, D. V. Patel, J. Wu and E. M. Gordon, *Tetrahedron Lett.* **1996**, *37*, 4643.

(440) D. V. Patel, *Solid-Phase Combinatorial Synthesis of Small Molecule Libraries*, conference: *"Solid Phase Synthesis: Developing Small Molecule Libraries"*, San Diego, 1996.

(441) R. Ramage, *Diversomer Technology: Recent Advances in the Generation of Chemical Diversity*, conference: *"Synthesis of Small Molecules on Solid Phase"*, Basel, 1996.

(442) L. F. Tietze, A. Steinmetz and F. Balkenhohl, *Bioorg. Med. Chem. Lett.* **1997**, *7*, 1303.

(443) T. Masquelin, R. Bär, F. Gerber, D. Sprenger and Y. Mercadal, *Helv. Chim. Acta* **1998**, in press.

(444) D. Obrecht, F. Gerber, D. Sprenger and T. Masquelin, *Helv. Chim. Acta* **1997**, *80*, 531.

(445) H. Heimgartner, *Angew. Chem. Int. Ed. Engl.* **1991**, *30*, 238.

(446) C. P. Holmes, WO 96/00148, 1996.

(447) M. Cushman and E. J. Madaj, *J. Org. Chem.* **1987**, *52*, 907.

(448) L. D. Robinett, *Solid Phase Synthesis of Dehydropyrimidines by the Atwal Modification of the Biginelli Reaction*, conference: *211th ACS National Meeting*, New Orleans, 1996.

(449) L. F. Tietze and A. Steinmetz, *Synlett* **1996**, 667.

(450) A. L. Marzinzik and E. R. Felder, *Tetrahedron Lett.* **1996**, *37*, 1003.

(451) P. Wipf and A. Cunningham, *Tetrahedron Lett.* **1995**, *36*, 7819.

(452) Y. Pei and W. H. Moos, *Tetrahedron Lett.* **1994**, *35*, 5825.

(453) N. Q. McDonald, J. Murray-Rust and T. Blundell, *Structure* **1995**, *3*, 1.

(454) B. Ruhland, A. M. Bhandari, E. M. Gordon and M. A. Gallop, *J. Am. Chem. Soc.* **1996**, *118*, 253.

(455) F. Zaragoza, *Tetrahedron Lett.* **1996**, *37*, 6213.

(456) H. V. Meyers, G. J. Dilley, T. L. Durgin, T. S. Powers, N. A. Winsinger, W. Zhu and M. R. Pavia, *Molecular Diversity* **1995**, *1*, 13.

(457) E. K. Kick and J. A. Ellman, *J. Med. Chem.* **1995**, *38*, 1427.

(458) S. Hanessian and R.-Y. Yang, *Tetrahedron Lett.* **1996**, *37*, 5835.

(459) H. Han, M. M. Wolfe, S. Brenner and K. D. Janda, *Proc. Natl. Acad. Sci. USA* **1995**, *92*, 6419.

(460) S. Brenner, WO 95/16918, 1995.

(461) A. J. Speziale and L. R. Smith, *Org. Synth.* **1973**, *Coll. Vol V*, 204.

(462) S. V. Ley, D. M. Mynett and W.-J. Koot, *Synlett* **1995**, 1017.

(463) D. R. Tortolani and S. A. Biller, *Tetrahedron Lett.* **1996**, *37*, 5687.

(464) M. C. Pirrung and J. Chen, *J. Am. Chem. Soc.* **1995**, *117*, 1240.

(465) M. C. Pirrung and J. H.-L. Chan, *Chem. Biol.* **1995**, *2*, 621.

(466) T. Masquelin and D. Obrecht, *Tetrahedron* **1997**, *53*, 641.

(467) T. Masquelin and D. Obrecht, *Tetrahedron Lett.* **1994**, *35*, 9387.

(468) T. Masquelin and D. Obrecht, *Synthesis* **1995**, 276.

(469) D. Obrecht, *Helv. Chim. Acta* **1989**, *72*, 447.

(470) D. Obrecht and B. Weiss, *Helv. Chim. Acta* **1989**, *72*, 117.

(471) D. Obrecht, *Helv. Chim. Acta* **1991**, *74*, 27.

(472) C. Zumbrunn, Ph.D. thesis, University of Zürich (1998).

(473) G. Coispeau, J. Elguero and F. Jaquier, *Bull. Soc. Chim. Fr.* **1970**, 689.

(474) R. K. Olsen and X. Feng, *Tetrahedron Lett.* **1991**, *32*, 5721.

(475) X. Feng and R. K. Olsen, *J. Org. Chem.* **1992**, *57*, 5811.

(476) R. M. Adlington, J. E. Baldwin, D. Catterick and G. J. Pritchard, *J. Chem. Soc., Chem. Commun.* **1997**, 1757.

(477) R. M. Dodson, *J. Am. Chem. Soc.* **1948**, *70*, 2753.

(478) R. Frank, *Tetrahedron* **1992**, *48*, 9217.

(479) S. P. A. Fodor, J. L. Read, M. C. Pirrung, L. Stryer, A. T. Lu and S. Solas, *Science* **1991**, *251*, 767.

(480) S. M. Freier, D. A. M. Konings, J. R. Wyatt and D. J. Ecker, *J. Med. Chem.* **1995**, *38*, 344.

(481) D. A. M. Konings, J. R. Wyatt, D. J. Ecker and S. M. Freier, *J. Med. Chem.* **1996**, *39*, 2710.

(482) L. Wilson-Lingardo, P. W. Davis, D. J. Ecker, N. Hébert, O. Acevedo, K. Sprankle, T. Brennan, L. Schwarcz, S. M. Freier and J. R. Wyatt, *J. Med. Chem.* **1996**, *39*, 2720.

(483) C. T. Dooley, N. N. Chung, B. C. Wilkes, P. W. Schiller, K. M. Bidlack, G. W. Pasternak and R. A. Houghten, *Science* **1994**, *266*, 2019.

(484) C. T. Dooley, N. N. Chung, P. W. Schiller and R. A. Houghten, *Proc. Natl. Acad. Sci. USA* **1993**, *90*, 10811.

(485) J. R. Wyatt, T. A. Vickers, J. L. Roberson, R. W. Buckheit, T. Klimkait, E. DeBaets, P. W. Davis, B. Rayner, J. L. Imbach and D. J. Ecker, *Proc. Natl. Acad. Sci. USA* **1994**, *91*, 1356.

(486) M. M. Murphy, J. R. Schulleck, E. M. Gordon and M. A. Gallop, *J. Am. Chem. Soc.* **1995**, *117*, 7029.

(487) E. Erb, K. D. Janda and S. Brenner, *Proc. Natl. Acad. Sci. USA* **1994**, *91*, 11422.

(488) C. T. Dooley and R. A. Houghten, *Life Sci.* **1993**, *52*, 1509.

(489) S. Uebel, W. Kraas, S. Kienle, K.-H. Wiesmüller, G. Jung and R. Tampé, *Proc. Natl. Acad. Sci. USA* **1997**, *94*, 8976.

(490) L. Neuville and J. Zhu, *Tetrahedron Lett.* **1997**, *38*, 4091.

(491) P. W. Smith, J. Y. Q. Lai, A. R. Whittington, B. Cox, J. G. Houston, C. H. Stylli, M. N. Banks and P. R. Tiller, *Bioorg. Med. Chem. Lett.* **1994**, *4*, 2821.

(492) B. Déprez, X. Williard, L. Bourel, H. Coste, F. Hyafil and A. Tartar, *J. Am. Chem. Soc.* **1995**, *117*, 5405.

(493) S. Brenner and R. A. Lerner, *Proc. Natl. Acad. Sci. USA* **1992**, *89*, 5381.

(494) J. Nielsen, S. Brenner and K. D. Janda, *J. Am. Chem. Soc.* **1993**, *115*, 9812.

(495) M. C. Needels, D. G. Jones, E. H. Tate, G. L. Heinkel, L. M. Kochersperger, W. J. Dower, R. W. Barrett and M. A. Gallop, *Proc. Natl. Acad. Sci. USA* **1993**, *90*, 10700.

(496) V. Nikolaiev, A. Stierandova, V. Krchnak, B. Seligmann, K. S. Lam, S. E. Salmon and M. Lebl, *Pept. Res.* **1993**, *6*, 161.

(497) J. M. Kerr, S. C. Banville and R. N. Zuckermann, *J. Am. Chem. Soc.* **1993**, *115*, 2529.

(498) M. H. J. Ohlmeyer, R. N. Swanson, L. W. Dillard, J. C. Reader, G. Asouline, R. Kobayashi, M. Wigler and W. C. Still, *Proc. Natl. Acad. Sci. USA* **1993**, *90*, 10922.

(499) H. P. Nestler, P. A. Bartlett and W. C. Still, *J. Org. Chem.* **1994**, *59*, 4723.

(500) R. Boyce, G. Li, N. H. P., T. Suenaga and W. C. Still, *J. Am. Chem. Soc.* **1994**, *116*, 7955.

(501) W. C. Still, *Acc. Chem. Res.* **1996**, *29*, 155.

(502) Y. Cheng, T. Suenaga and W. C. Still, *J. Am. Chem. Soc.* **1996**, *118*, 1813.

(503) M. T. Burger and W. C. Still, *J. Org. Chem.* **1995**, *60*, 7382.

(504) S. S. Yoon and W. C. Still, *J. Am. Chem. Soc.* **1993**, *115*, 823.

(505) S. S. Yoon and W. C. Still, *Tetrahedron* **1995**, *51*, 567.

(506) H. Wennemers, S. S. Yoon and W. C. Still, *J. Org. Chem.* **1995**, *60*, 1108.

(507) R. Liang, L. Yan, J. Loebach, M. Ge, Y. Uozumi, K. Sekanina, N. Horan, J. Gildersleeve, C. Thompson, A. Smith, K. Biswas, W. C. Still and D. Kahne, *Science* **1996**, *274*, 1520.

(508) J. J. Baldwin, J. J. Burbaum, I. Henderson and M. H. J. Ohlmeyer, *J. Am. Chem. Soc.* **1995**, *117*, 5588.

(509) J. J. Burbaum, M. H. J. Ohlmeyer, J. C. Reader, I. Henderson, L. W. Dillard, G. Li, T. L. Randle, N. H. Sigal, D. Chelsky and J. J. Baldwin, *Proc. Natl. Acad. Sci. USA* **1995**, *92*, 6027.

(510) Z.-J. Ni, D. Maclean, C. P. Holmes, M. M. Murphy, B. Ruhland, J. W. Jacobs, E. M. Gordon and M. A. Gallop, *J. Med. Chem.* **1996**, *39*, 1601.

(511) D. Maclean, J. R. Schullek, M. M. Murphy, Z.-J. Ni, E. M. Gordon and M. A. Gallop, *Proc. Natl. Acad. Sci. USA* **1997**, *94*, 2805.

(512) E. J. Moran, S. Sarshar, J. F. Cargill, M. M. Shahbaz, A. Loi, A. M. M. Mjalli and R. W. Armstrong, *J. Am. Chem. Soc.* **1995**, *117*, 10787.

(513) K. C. Nicolaou, X.-Y. Xiao, Z. Parandoosh, A. Senyei and M. P. Nova, *Angew. Chem. Int. Ed. Engl.* **1995**, *34*, 2289.

(514) X.-Y. Xiao, Z. Parandoosh and M. P. Nova, *J. Org. Chem.* **1997**, *62*, 6092.

(515) P. Kocis, V. Krchnak and M. Lebl, *Tetrahedron Lett.* **1993**, *34*, 7251.

(516) S. E. Salmon, K. S. Lam, M. Lebl, A. Kandola, P. S. Khattri, S. Wade, M. Patek, P. Kocis, V. Krchnak, D. Thorpe and S. Felder, *Proc. Natl. Acad. Sci. USA* **1993**, *90*, 11708.

(517) M. Cardno and M. Bradley, *Tetrahedron Lett.* **1996**, *37*, 135.

(518) E. J. Ariens, A. J. Beld, J. F. Rodrigues and A. M. Simonis, in: *The Receptors: A Comprehensive Treatise*, R. D. O'Brien (Ed.), Vol. 1, p. 33, Plenum: New York (1979).

(519) P. Willett, *Perspect. Drug Discovery Des.* **1997**, *7*, 1.

(520) R. D. Brown and Y. C. Martin, *J. Chem. Inf. Comput. Sci.* **1996**, *36*, 572.

(521) R. D. Cramer, R. D. Clark, D. E. Patterson and A. M. Ferguson, *J. Med. Chem.* **1996**, *39*, 3060.

(522) R. J. Simon, R. S. Kania, R. N. Zuckermann and V. D. Huebner, *Proc. Natl. Acad. Sci. U.S.A.* **1992**, *89*, 9367.

(523) M. R. Pavia, T. K. Sawyer and W. H. Moos, *Bioorg. Med. Chem. Lett.* **1993**, *3*, 387.

(524) R. D. Brown and Y. C. Martin, *J. Chem. Inf. Comput. Sci.* **1996**, *36*, 572.

(525) D. E. Patterson, R. D. Cramer, A. M. Ferguson, R. D. Clark and L. E. Weinberger, *J. Med. Chem.* **1996**, *39*, 3049.

(526) A. R. Leach and S. R. Kilvington, *Computer-Aided Molecular Design* **1994**, *8*, 283.

(527) Tripos, 1699 South Hanley Road, St. Louis, MO 63144 (USA).

(528) D. C. I. S. Inc., 27401 Los Altos, Suite 370, MissionViejo, CA 92691 (USA).

(529) B. T. M. Simulations, 9685 Scranton Road, San Diego, CA 92121 (USA).

(530) R. H. Chemical Design Ltd., Cromwell Park, Chipping Norton, Oxon OX7 5SR (UK).

(531) R. C. Good and R. A. Lewis, *J. Med. Chem.* **1997**, *40*, 3926.

(532) R. W. Armstrong, A. P. Combs, P. A. Tempest, S. D. Brown and T. A. Keating, *Acc. Chem. Res.* **1996**, *29*, 123.

(533) M. I. Systems, 14600 Catalina Street, San Leandro, CA 94577 (USA).

(534) Beilstein Information Systems Inc., 15 Inverness Way East, P. O. Box 1154, Englewood, CO 80150 (USA).

(535) B. B. Brown, D. S. Wagner and H. M. Geysen, *Mol. Diversity* **1995**, *1*, 4.

(536) M. C. Fitzgerald, K. Harris, C. G. Shevlin and G. Siuzdak, *Bioorg. Med. Chem. Lett.* **1996**, *6*, 979.

(537) B. J. Egner, J. Langley and M. Bradley, *J. Org. Chem.* **1995**, *60*, 2652.

(538) B. J. Egner, M. Cardno and M. Bradley, *J. Chem. Soc., Chem. Commun.* **1995**, 2163.

(539) C. L. Brummel, I. N. W. Lee, Y. Zhou, S. J. Benkovic and N. Winograd, *Science* **1994**, *264*, 399.

(540) Y.-H. Chu, D. P. Kirby and B. L. Karger, *J. Am. Chem. Soc.* **1995**, *117*, 5419.

(541) R. A. Zambias, D. A. Boulton and P. R. Griffin, *Tetrahedron Lett.* **1994**, *35*, 4283.

(542) C. L. Brummel, J. C. Vickerman, S. A. Garr, M. E. Hemling, G. D. Roberts, W. Johnson, J. Weinstock, D. Gaitanopoulos, S. J. Benkovic and N. Winograd, *Anal. Chem.* **1996**, *68*, 237.

(543) N. J. Haskins, D. J. Hunter, A. J. Organ, S. S. Rahman and C. Thom, *Rapid Commun. Mass Spectrom.* **1995**, *9*, 1437.

(544) S. S. Chu and S. H. Reich, *Bioorg. Med. Chem. Lett.* **1995**, *5*, 1053.

(545) G. B. Fields and N. R. L., *Int. J. Pept. Prot. Res.* **1990**, *35*, 161.

(546) E. P. Heimer, C. D. Chang, T. L. Lambros and J. Meienhofer, *Int. J. Pept. Prot. Res.* **1981**, *18*, 237.

(547) C.-D. Chang, M. Waki, M. Ahmad, J. Meienhofer, E. O. Lundell and J. D. Huag, *Int. J. Pept. Prot. Res.* **1980**, *15*, 59.

(548) J. Meienhofer, M. Waki, E. P. Heimer, T. J. Lambros, R. C. Kakofske and C.-D. Chang, *Int. J. Pept. Prot. Res.* **1979**, *13*, 35.

(549) D. A. Campell and J. C. Bermak, *J. Am. Chem. Soc.* **1994**, *116*, 6039.

(550) S. L. Beaucage and R. P. Iyer, *Tetrahedron* **1992**, *48*, 2223.

(551) M. H. Caruthers, A. D. Barone, S. L. Beaucage, D. R. Dodds, E. F. Fisher, L. J. McBride, M. Matteucci, Z. Stabinsky and J.-Y. Tang, *Meth. Enzymol.* **1987**, *154*, 287.

(552) G.-S. Lu, S. Mojsov, J. P. Tam and R. B. Merrifield, *J. Org. Chem.* **1981**, *46*, 3433.

(553) G. C. Look, C. P. Holmes, J. P. Chinn and P. A. Gallop, *J. Org. Chem.* **1994**, *59*, 7588.

(554) E. Giralt, J. Rizo and E. Pedroso, *Tetrahedron* **1984**, *40*, 4141.

(555) W. Schoknecht, K. Albert, G. Jung and E. Bayer, *Liebigs Ann. Chem.* **1982**, 1514.

(556) E. Giralt, F. Albericio, F. Bardella, R. Eritja, M. Feliz, E. Pedroso, M. Pons and R. Rizo, in: *Innovations and Perspectives in Solid Phase Synthesis*, R. Epton (Ed.), p. 111, SPCC (UK) Ltd: Birmingham (1990).

(557) M. M. Murphy, J. R. Schullek, E. M. Gordon and M. A. Gallop, *J. Am. Chem. Soc.* **1995**, *117*, 7029.

(558) G. C. Look, M. M. Murphy, D. A. Campell and M. A. Gallop, *Tetrahedron Lett.* **1995**, *36*, 2937.

(559) R. C. Anderson, J. P. Stokes and M. Shapiro, *Tetrahedron Lett.* **1995**, *36*, 5311.

(560) R. C. Anderson, M. A. Jarema, M. J. Shapiro, J. P. Stokes and M. Ziliox, *J. Org. Chem.* **1995**, *60*, 2650.

(561) W. L. Fitch, G. Detre, C. P. Holmes, J. N. Shoolery and P. A. Kiefer, *J. Org. Chem.* **1994**, *59*, 7955.

(562) C. Dhalluin, C. Boutillon, A. Tartar and G. Lippens, *J. Am. Chem. Soc.* **1997**, *119*, 10494.

(563) J. I. Crowley and H. Rapoport, *J. Org. Chem.* **1980**, *45*, 3215.

(564) B. Yan, *J. Org. Chem.* **1995**, *60*, 5736.

(565) D. W. Vidrine, *Photoacoustic FT-IR spectroscopy of solids and liquids*, in: *FT-IR spectroscopy*, (Ed.), chapter 4, p. 125, Academic press: New York (1982).

(566) J. W. Metzger, K.-H. Wiesmuller, V. Gnau, J. Brunjes and G. Jung, *Angew. Chem. Int. Ed. Engl.* **1993**, *32*, 894.

(567) J. W. Metzger, C. Kempter, K.-H. Wiesmüller and G. Jung, *Anal. Biochem.* **1994**, *219*, 261.

(568) R. S. Youngquist, G. R. Fuentes, M. P. Lacey and T. Keough, *Rapid Commun. Mass Spectrom.* **1994**, *8*, 77.

(569) G. Jung and A. G. Beck-Sickinger, *Angew. Chem. Int. Ed. Engl.* **1992**, *31*, 367.

(570) J. A. Boutin, P. Hennig, P. H. Lambert, B. S., L. Petit, J.-P. Mathieu, B. Serkiz, J. P. Volland and J.-L. Fauchère, *Anal. Biochem.* **1996**, *234*, 126.

(571) T. Carell, E. A. Wintner, A. J. Sutherland, J. Rebek, Y. M. Dunayevskiy and P. Vouros, *Chem. Biol.* **1995**, *2*, 171.

(572) Y. Dunayevskiy, P. Vouros, T. Carell, E. A. Wintner and J. Rebek, Jr., *Anal. Chem.* **1995**, *67*, 2906.

(573) M. Stankova, O. Issakova, N. F. Sepetov, V. Krchnak, K. S. Lam and M. Lebl, *Drug Dev. Res.* **1994**, *33*, 146.

(574) M. Karas, D. Bachmann, U. Bahr and F. Hillenkamp, *Int. J. Mass Spectrom. Ion Proc.* **1987**, *78*, 53.

(575) M. Karas, U. Bahr and F. Hillenkamp, *Int. J. Mass Spectrom. Ion Proc.* **1989**, *92*, 231.

Chapter 2

Linear Assembly Strategies

2.1. Introduction

In *Chapter 1.8.4* we have analysed and compared linear and convergent approaches as the two basic strategies to assemble molecules in a rapid and efficient way. In *Table 2.1*, the most important building blocks used in linear strategies and products thereof are summarised.

Table 2.1

monomers	bond formation	polymers
amino acids	amide bond	peptides, proteins
nucleotides	phosphorester bond	oligonucleotides
mono- and disaccharides	glycosidic bond	polysaccharides
N-alkylated glycines	amide bond	peptoids

Linear strategies employ a readily available set of monomeric building blocks such as suitably protected amino acids, nucleotides, monosaccharides or *N*-alkylated glycines which can be assembled in a repetitive sequence of couplings to produce the corresponding products such as peptides, oligonucleotides, polysaccharides or peptoids. Most of these linear strategies are used by nature to produce the basic molecules of life such as proteins, DNA, RNA and polysaccharide derived natural products.

Linear strategies in combinatorial synthesis have been extensively summarised and since the present work focuses more on convergent strategies for the synthesis of small molecules we will try in this chapter to summarise only the most important aspects of this very broad field.

2.2. Peptides

2.2.1. General

The syntheses of peptide libraries that have appeared in the literature so far can be divided into mainly three categories:

- Libraries synthesised using a molecular biology approach on the surface of filamentous phage particles or plasmids. These phage display methods have found wide use in protein epitope mapping approaches of proteins and protein domains. Several reviews appeared in the literature summarising the work in this field.[1-3]

- Peptides synthesised as individuals or mixtures on solid supports such as resin beads derived from polystyrene, polyacrylamide, polyacrylamide/polystyrene co-polymers (cf. *Chapter 1.5*) and cleaved to be assayed *in solution.*

- Peptides *synthesised and assayed* as individuals or mixtures *on solid supports* such as pins,[4] resin beads,[5] cotton,[6] the silicon surface of microchips[7] or cellulose membranes.[8]

Mixtures of peptides can be obtained using two different strategies:

a)   as true mixtures where a peptide coupling step involves the coupling of a mixture (typically the 20 coding amino acids) of side-chain protected Boc- or Fmoc-amino acids (L- or D-isomers) in a predetermined molar ratio which compensates for different coupling ratios due to steric and electronic factors,

and

b)   as mixtures of resin beads which resulted from synthesis by the "portioning-mixing" method,[9] "divide, couple and recombine" process[10] or "split synthesis".[5] All these methods refer to the basic concept of "one bead - one compound"[11] as explained in *Chapter 1.8.3.*

2.2.2 Synthetic approaches to linear peptide libraries

Based on *Merrifield's* original strategy of solid-phase peptide synthesis,[12] two approaches proposed by *Geysen* in 1984[13,14] using pins and by *Houghton* in 1985[15,16] based on tea-bags allowed the simultaneous synthesis of peptide libraries. These technologies made it possible to synthesise thousands of peptides in quantities of milligrams to multigrams (in tea-bags) per year.

The **multipin synthesis** constitutes a partially separated, parallel synthesis of multiple peptides on polyacrylic acid-grafted polyethylene pins in 96-well microtitre format introduced by *Geysen*.[13,14] This technology has been extensively used in the area of protein epitope mapping. The covalently linked peptides can be screened in a direct binding assay (ELISA). Especially designed linkers allow to cleave the peptides from the pins and to carry out the assay in solution. This technique is quite limited by the small amounts of peptides that can be generated (50 nmol up to 2 µmol). The multipin synthesis has also been used for the synthesis combinatorial mixtures of peptides by randomisation of distinct positions of a given peptide sequence by coupling a mixture of the 20 common L-amino acids. The "mimotope strategy" introduced by *Geysen et al.* led to the identification of ligands (often hexapeptides) for antibodies and other receptors.[17]

Polypeptide synthesis using the **tea-bag method** consists of sealing 100 mg to several grams of resin into a polypropylene net. These bags thus represent spatially separated reaction compartments where peptides can be synthesised simultaneously capitalising on the fact that all of the washings, neutralisations and deprotection steps are identical and therefore not sequence dependent. The tea-bags are usually labelled and sorted prior to the coupling steps. Originally developed by *Houghton* using the Boc-/benzylester protective group strategy and diisopropylcarbodiimide (DIC) for coupling[15] this method was further developed by *Jung et al.*[18] using Fmoc-/*tert*-butylester protection and *O*-(benzotriazole-1-yl)tetramethyluronium tetrafluoroborate (TBTU) activation.

Biological applications of the tea-bag methodology include investigations of HIV gp 41 protein by *Van't Hoff et al.*,[19] scans of centrally truncated neuropeptide Y analogues,[20] and mapping of thymidine kinase.[21]

The combination of solid-phase peptide synthesis and photolithography (**light-directed combinatorial synthesis**) constitutes another technology for the synthesis of spatially separated multiple parallel synthesis of peptide libraries.[1,7] Using photo-cleavable protective groups such as the *N*-nitroveratryloxycarbonyl group (NVOC) allows the controlled synthesis of peptide libraries by the spatially controllable addition of specific reagents to specific locations. Thereby the solid support consists of a silicon wafer functionalised with linker groups suitable for peptide synthesis.

Multiple peptide synthesis has also been achieved by using polystyrene-grafted polyethylene films (PS-PE) which are produced using X-ray irradiation.[22] Small areas of the film (typically 1.5 x 3 cm) serve as reaction compartments where coupling can be carried out separately while washing and deblocking is carried out simultaneously.

2.3. Conformationally constrained peptides

While linear peptides have been used extensively in the very early stages of lead finding (cf. *Chapter 1.4*) they found only limited application as therapeutic agents due to their limited protease stability (and thus low plasma stability and bioavailability) and low penetration through cell membranes. To circumvent some of these inherent bioavailability problems associated with linear peptides several approaches have been published to modify and stabilise linear peptides by construction of peptide mimetics.[23-25] *Figure 2.3.1* schematically summarises these different approaches.

Figure 2.3.1

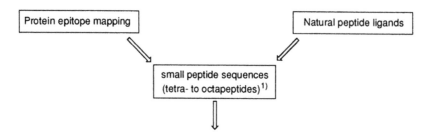

• Systematic reduction of the peptide sequence one amino acid at a time from the N- and C-termini (→ 4 to 8 residues)
• Systematic replacement of L-amino acids by D-amino acids
• Systematic replacement of side chain moieties by a methyl group ("alanine scan")
• Replacement of coding amino acids by non-natural analogues (e.g. α, α-disubstituted glycines)[2]
• Introduction of constraints by cyclisation: - cyclic peptides - depsipeptides - cyclisation via S-S bonds - templates
• Introduction of constraints by secondary structure mimetics (β-turns, loops, β-sheets and α-helix-mimetics)
• Replacements for the amide bond and introduction of transition state analogues

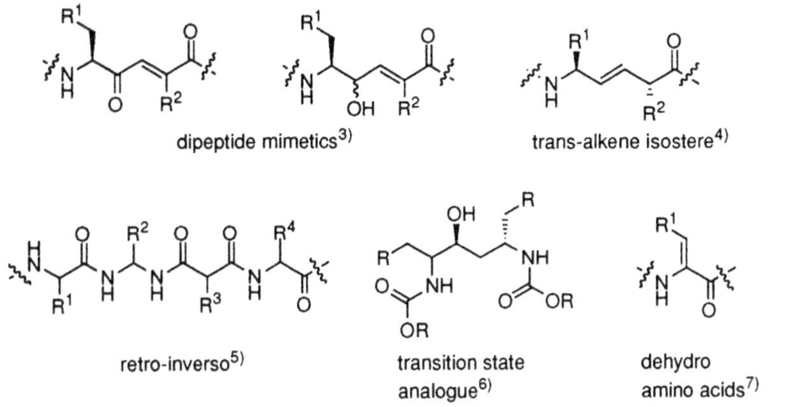

dipeptide mimetics[3)]

trans-alkene isostere[4)]

retro-inverso[5)]

transition state
analogue[6)]

dehydro
amino acids[7)]

References: 1) see review by *Hrubi et al.* [26]; 2) see review by *Spatola* [27]; 3) Refs.[28-31]; 4) Ref.[32]; 5) Ref.[33]; 6) Refs.[34,35]; 7) Ref.[36]

Some of the molecular units that have been used to replace amide bonds are depicted in *Figure 2.3.1* (lower part).

2.3.1. Solid-phase synthesis of unnatural amino acids and peptides

Among the non-standard amino acids that have been used to chemically and conformationally stabilise small peptides, α,α-disubstituted glycines have been studied[37] and depending on the nature of the α-substituents they have been shown to stabilise β-turn, 3_{10}-helical and α-helical conformations when incorporated into small peptides. Recently, a solid-phase synthesis of some of these interesting amino acids has been published by *Scott et al.*[38] as shown in *Scheme 2.3.1*.

Scheme 2.3.1

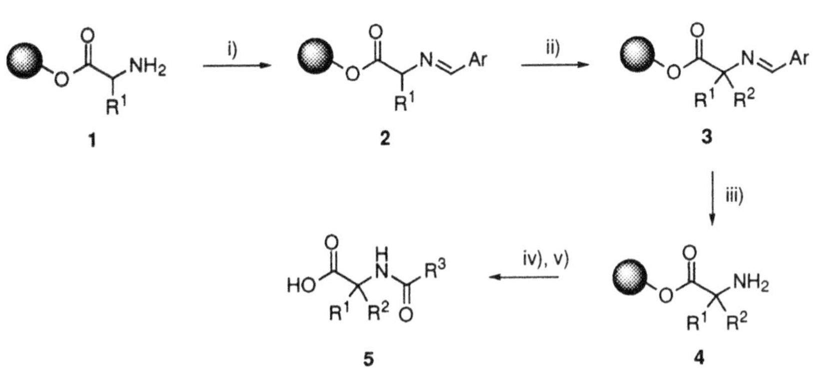

Reagents and conditions: i) Ar-CHO, ii) R^2-X, base; iii) $HONH_3{}^+Cl^-$; iv) R^4-COOH, PyBrOP; v) TFA

189

Deprotection of Fmoc-amino acid *Wang* resins under standard conditions yielded the polymer-bound amino acids **1**. Condensation with 3,4-dichlorobenzaldehyde gave aldimines **2**. Subsequent alkylation with electrophiles such as benzyl-, naphthyl- or allylbromide in the presence of 2-[(1,1-dimethylethyl)imino]- *N,N*-diethyl - 2,2,3,4,5,6- hexahydro-1,3-dimethyl-1,2,3-diazaphosphorin-2(1H)-amine (BEMP) gave the disubstituted aldimines **3**. Transketalisation with hydroxylamine hydrochloride yielded the free amine **4** which was acylated and cleaved to give the final product **5** in good yields and purities.

2.3.2. β-Turn mimetics

Turns are segments between secondary structural elements and are defined as sites in a polypeptide structure where the peptidic chain reverses its overall direction. They are composed of four (β-turn) or three (γ-turn) amino acid residues. In contrast to α-helices or β-sheets, turns are the only regular secondary structures which consist of non-repeating backbone torsional angles. Turns have been implicated as important sites for molecular recognition in biologically active peptides and globular proteins.[39,40] High-resolution crystal structures of several antibody-peptide complexes clearly showed turns as recognition motifs.[41,42] Furthermore, good correlations with turn conformations have been found in structure-activity studies of specific recognition sites in peptide hormones such as *bradykinin* and *somatostatin*.[43,44]

β–Turns have also been hypothesised to be involved in post-translational lysine hydroxylation,[45] processing of peptide hormones,[46,47] recognition of phosphotyrosine-containing peptides,[48,49] signal peptidase action,[50] receptor internalisation signals,[51,52] and in glycosylation processes.[53] These findings together with the fact that turns occur mainly at the surface of proteins make them attractive targets as peptide mimetics.[25]

Terminology of β–turns

According to *Venkatachalam*, β–turns are classified into conformational types depending on the values of four backbone torsional angles $(\Phi_1, \Psi_1, \Phi_2, \Psi_2)$ (*Table 2.2.1*).[54]

Table 2.2.1

Type[a]	Φ_1	Ψ_1	Φ_2	Ψ_2	Nomenclature[b]
I	-60	-30	-90	0	αα
I'	60	30	90	0	γγ
II	-60	120	80	0	βpγ
II'	60	-120	-80	0	εα
III	-60	-30	-60	-30	αα
III'	60	30	60	30	γγ

a) Turn nomenclature according to *Venkatalacham*[54]

b) Turn nomenclature according to *Wilmot* and *Thornton*[55]

A second criterion is the $\alpha C_i \rightarrow \alpha C_{i+3}$ distance, introduced by *Kabsch* and *Sander*, which must be shorter than 7 Å (*Figure 2.3.2*).[56] Typical for β–turn motifs is a H-bond between the carbonyl group at position *i* and the amide NH of residue *i+3*.

Figure 2.3.2

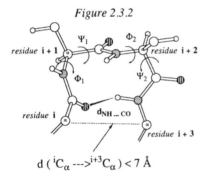

$$d\,(\,{}^iC_\alpha \,\text{--->}\,{}^{i+3}C_\alpha\,) < 7 \text{ Å}$$

To date, eight different types of β–turns have been classified and the torsional angles $(\Phi_1,\Psi_1,\Phi_2,\Psi_2)$ of the most important naturally occurring β–turns are depicted in *Table 2.2.1*.

Mimicking strategies for β–turns

Mimicking a β–turn consists in constraining correctly four torsional angles $(\Phi_1,\Phi_2,\Psi_1,\Psi_2)$ and four bonds (bonds *a-d*, cf. *Fig. 2.3.3*). Bonds *a* and *d* direct the entry and the exit of the peptide chain through the turn, respectively, whereas bonds *b* and *c* are responsible for the spatial dispositon of the amino acid side chains at position *i+1* and *i+2* of a turn. The torsional angles determine the backbone geometry of the turn and consequently the shape of the turn hydrogen

191

bond. Bonds *a* and *d*, describe the vectors of the *C*- and *N*-terminal peptide chain entering and leaving the turn. The control of these vectors is essential for cases where additional amino acid residues are placed outside of the central amide unit.

Figure 2.3.3

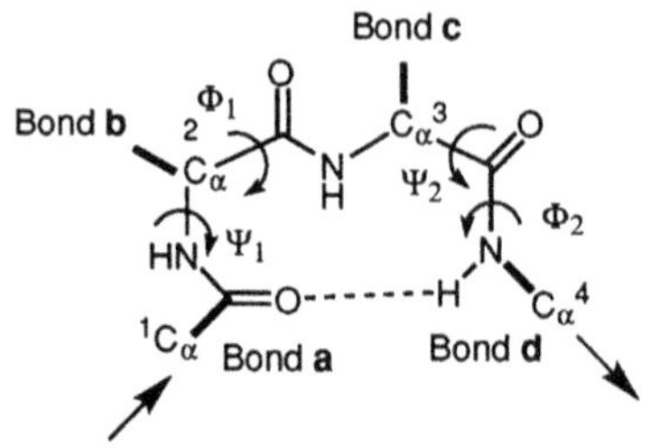

Peptidic β–turn mimetics are generally based on cyclic backbone mimetics (*e.g.* replacing the turn hydrogen bond by a covalent bond) or by introducing one or several unusual amino acids, which constrain the backbone in β–turn conformations. Synthetic approaches to non-peptidic turn mimetics can be grouped into two classes: 1) *external β–turn mimetics*, and 2) *internal β–turn mimetics*.[25]

External mimetics do not mimic *per se* the turn backbone but rather stabilise β–turn geometries of an attached peptide whereas an *internal* β–turn mimetic replaces the space in the turn which would normally be occupied by the peptide backbone. This methodology tries to correctly place key binding groups such as *b* and *c* (*Figure 2.3.3*) in 3D-space, while being less stringent with respect to the correct geometry of the turn backbone. Less effect on the backbone geometry can be expected, because only a small fragment of the peptide is replaced by the constraint. Furthermore, due to the small molecular size of the template, little steric hindrance with a receptor may be anticipated. A compilation of β-turn stabilising templates can be found in a recent review.[25]

Ellman et al.[57,58] have developed a general method for the solid-phase synthesis of β-turn mimetics **10** composed of three readily available building blocks (*Scheme 2.3.2*). α-Bromoacetic acid was coupled to the support-bound *p*-nitrophenyl alanine **6** by activation with diisopropylcarbodiimide. Subsequent treatment with the aminoalkyl thiol protected as a mixed disulfide provided the secondary amine **7**. Coupling with a Fmoc-protected amino acid by activation with HATU gave **8**. Deprotection and subsequent coupling of an α-bromo acid gave **9** which after reductive cleavage of the disulfide could be cyclised to give **10**.

Scheme 2.3.2

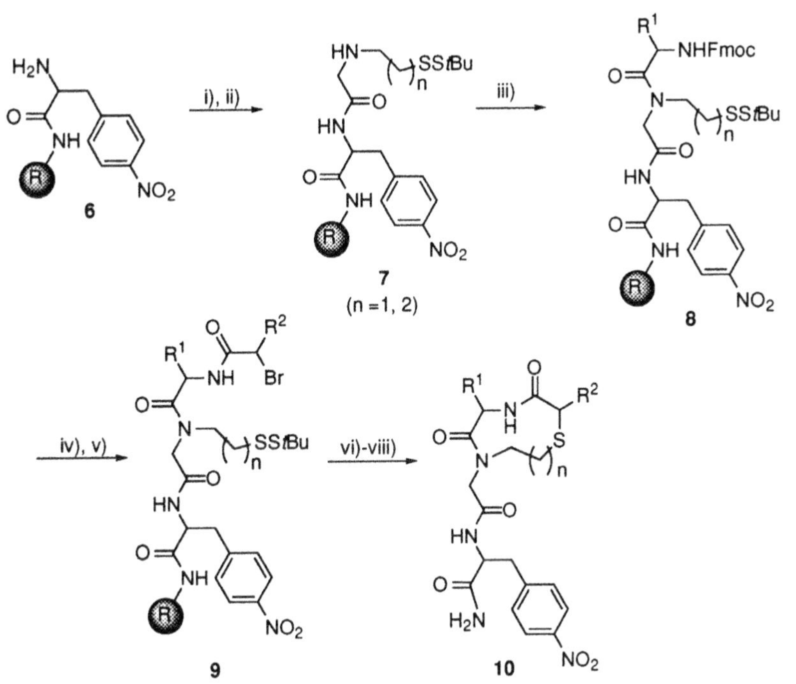

Reagents and conditions: i) bromoacetic acid, DIC; ii) $H_2N\text{-}CH_2\text{-}(CH_2)_n\text{-}SStBu$; iii) Fmoc-amino acid, HATU; iv) piperidine; v) α-halo acid, DIC; vi) PBu_3, H_2O; vii) TMG; viii) TFA

Employing this synthesis sequence turn mimetics **10** were obtained with an average purity of 75% over the 8-step process. No dimers were detected, cyclisation gave exclusively the desired macrocycles. The mimetic is constrained in a turn structure by replacing the hydrogen bond between the i and i+3-residues with a covalent backbone linkage. The flexibility of the turn mimetic and relative orientations of the side chains can be modulated by introducing different backbone linkages to provide 9- or 10-membered rings.

2.3.3. Loop mimetics

In addition to the classical secondary structure elements occurring at the surface of proteins such as γ- and β–turns, 3_{10}- and α–helices and β–sheets, loops of variable size seem to play a crucial role in many protein-protein interactions. Various amino acid derived templates have been proposed to stabilise β–turns as external templates and a compilation of loop stabilising templates is shown in *Figure 2.3.4.*[59-64]

Figure 2.3.4

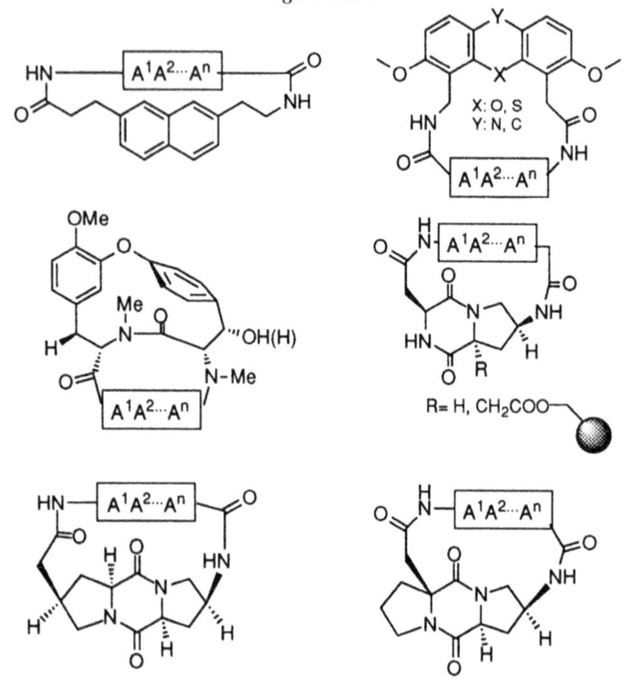

A^{1}A^2...A^n : peptide sequence containing amino acids A^1-A^n

Figure 2.3.5 shows a series of tricyclic xanthene-, phenoxazine- and phenothiazine-derived templates that have been shown to stabilise β–turn conformations in attached peptides for n=2 and 4.[24] These templates were also successfully employed for the synthesis of larger loop mimetics (n ≤ 14).

Figure 2.3.5

X= O, Y= C(CH$_3$)$_2$ xanthenes

X= O, Y= NCH$_3$ phenoxazines

X= S, Y= NCH$_3$ phenothiazines

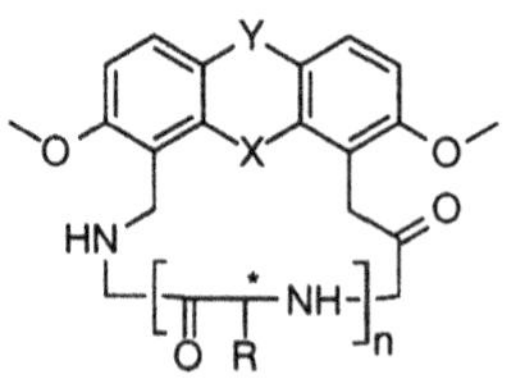

Although larger loops are far more flexible and the conformational influence enforced by the template is reduced, the templates usually increased significantly the cyclisation yields.

As a validation of this concept for loop stabilisation, cyclic peptides incorporating xanthene-derived templates corresponding to the sequence of the exposed loop structure in the snake toxin *flavoridin*[65,66] were synthesised on a solid support, and tested for binding to the GPIIb/IIIa receptor, and thus for inhibiting platelet aggregation.[24] The xanthene-derived loop incorporating the sequence Ile-Ala-Arg-Gly-Glu-Phe-Pro showed an IC_{50}-value of 15 nM and inhibited significantly platelet aggregation. In contrast, the open-chain and disulfide-bridged peptides were markedly less potent.

In order to use this template approach in view of synthesising combinatorial libraries of mimetics of exposed protein loops, *Waldmeier* and *Obrecht* [67] synthesised the unsymmetrically substituted xanthene derived templates **11** and **12** (*Figure 2.3.6*).[25]

Figure 2.3.6

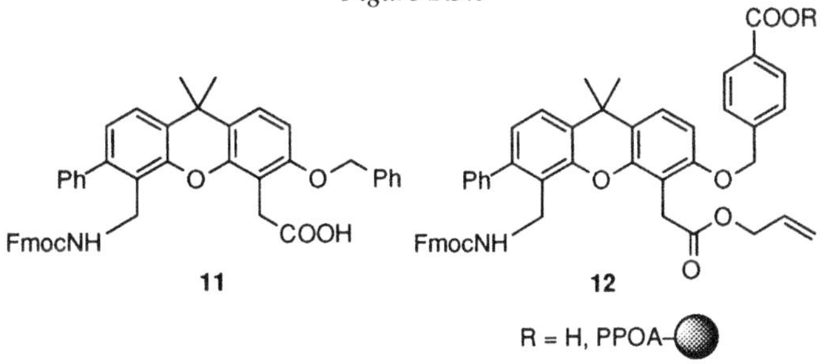

R = H, PPOA-⬤

These templates proved to be useful building blocks for the construction of β–turn stabilised peptide loop libraries using the split synthesis. It was not necessary to resort to tagging strategies for compound identification, since the cyclic molecule could be linearised by amide-cleavage, which then allowed *Edman*-degradation of the resulting open-chain peptide, as shown schematically in *Figure 2.3.7*. The selective amide cleavage results from a regioselective acid catalysed γ-lactone formation with concomitant liberation of the *N*-terminus. Since only the ring-opened products are subjected to *Edman*-sequencing, only minute amounts of the cyclic peptide have to be cleaved for this purpose.

Figure 2.3.7

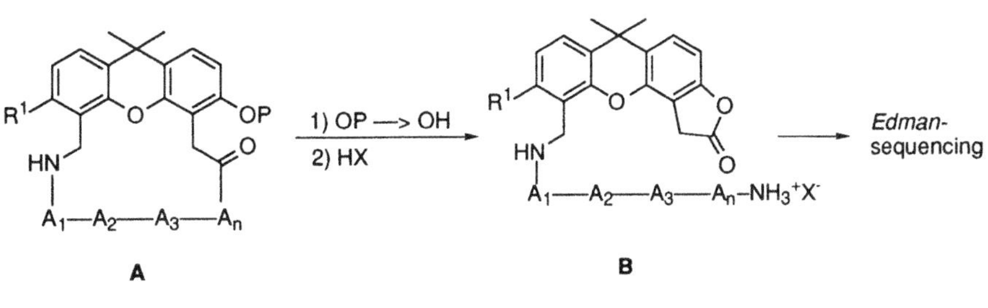

As an application of this concept, the template bridged cyclic peptides **13** and **16** containing the sequences Ala-Lys-Glu-Phe and Asp-Pro-Phe-Asp-Gly-Arg-Ala-Ile were synthesised by standard protocols,[68] ring-opened using methane sulfonic acid in MeOH at 50-60°C, and sequenced by *Edman*-degradation (*Figure 2.3.8*). Using the solid phase compatible template **12**, which was attached via the photocleavable PPOA-linker[69,70] to a polystyrene resin, template-bridged cyclic peptide libraries were synthesised and sequenced using this strategy.

Figure 2.3.8

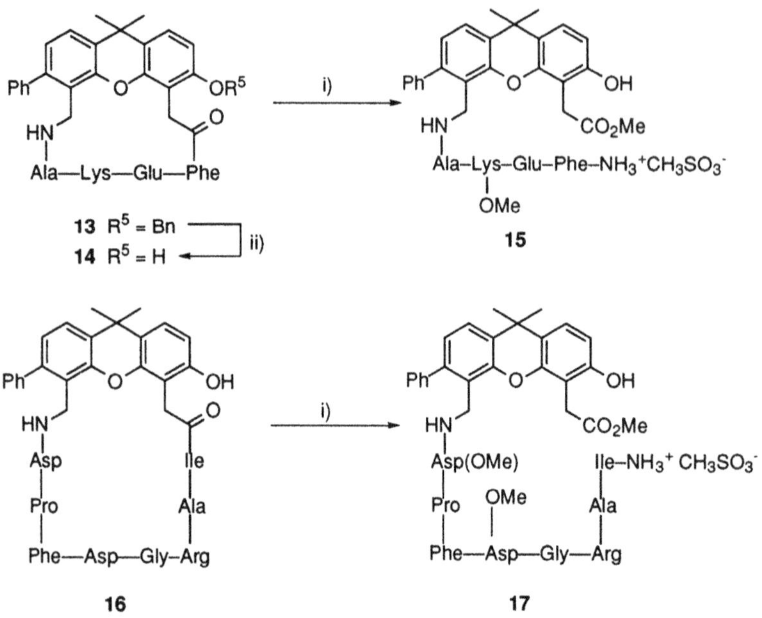

Reagents and conditions: i) MeSO₂OH, MeOH, 60°C; ii) H₂, Pd/C (95%)

The role of the xanthene-derived template can be summarised as follows:

- As previously shown, these templates stabilise efficiently β–turn type conformations in peptides and thus should also have a significant stabilising effect in larger loops.
- As a consequence of conformational stabilisation, routinely increased cyclisation yields were noticed when comparing the template-bridged loops with to the non-templated or disulfide-bridged peptides.
- As an important feature, a linker group can be attached to the tricyclic skeleton of **12** which results in the possibility to attach the template onto a solid support and synthesise template-bridged cyclic peptide libraries using the split synthesis.

- The photocleavable linker allows detachment of the cyclic peptide from the resin and to assess biological activity in homogeneous solution assays.
- Peptide ring opening via intermediate γ-lactone formation using methanesulfonic acid in MeOH liberated the *N*-terminus without concomitant cleavage of the peptide chain and allowed determination of the sequence by *Edman* degradation without having to resort to tagging strategies. The methodology should generally prove useful to efficiently construct loop-libraries corresponding to exposed loop regions of proteins and thus probe novel important insights into protein-protein interactions.

2.4. Peptoids

As already mentioned, linear small peptides show some major drawbacks[71] for the development of bioavailable, orally active drugs. In view of using the highly efficient amide coupling chemistry in combination with easily available amino acid building blocks, the so-called "peptoids" were developed[72] first at *Chiron*. *Figure 2.4.1* schematically compares a peptide with a peptoid backbone.

Figure 2.4.1

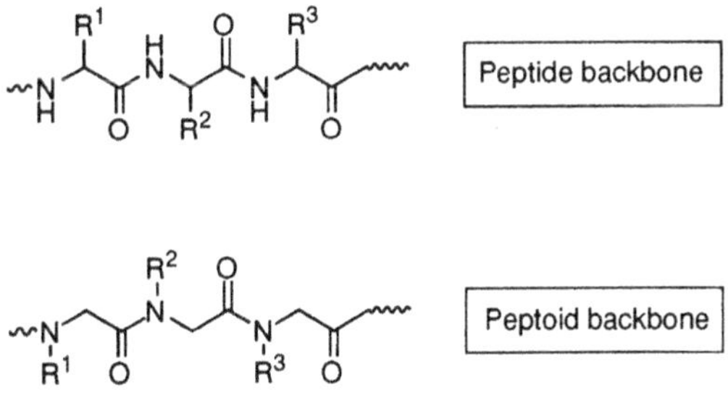

The major difference between peptides and peptoids lies in the fact that the "side chain" substituents R^1-R^3 in peptoids are attached to the amide nitrogens, thus avoiding the chiral α-centres of a peptide. There are no NH-bonds present in peptoids which has major implications on solubility, cell penetration properties and on the overall conformations which differ significantly from those of peptides.[72] In addition, compared to peptides, peptoids show an increased protease stability due to the tertiary amide bond.

Three different types of building block strategies were developed for the efficient synthesis of peptoids as schematically shown in *Scheme 2.4.1 (Approaches A-C)*.[73]

Scheme 2.4.1

Approach A: sequential coupling of N-substituted glycines

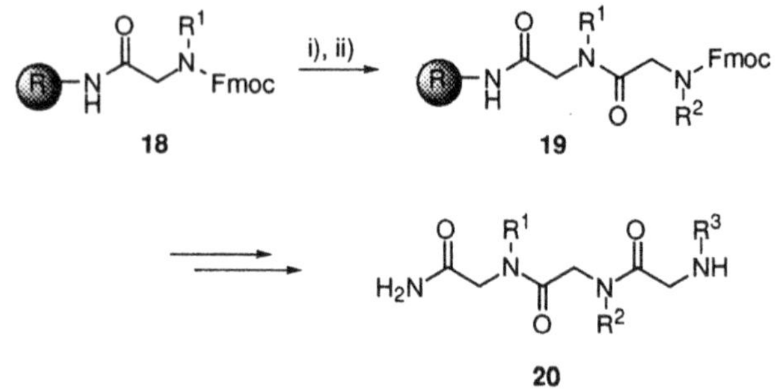

18   i), ii)   19

20

Approach B: Sequential coupling of glycine followed by reductive amination with aldehydes

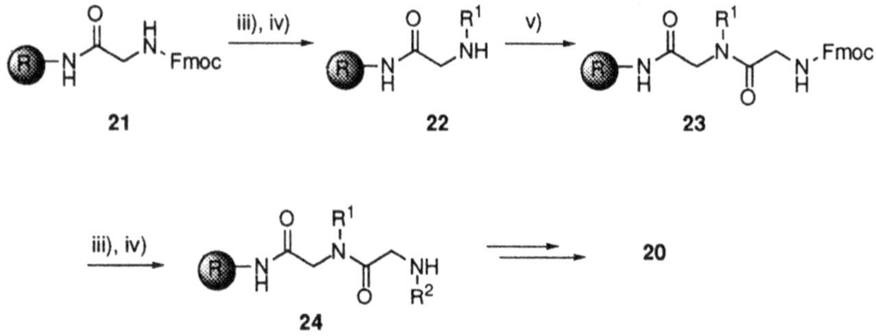

21   iii), iv)   22   v)   23

iii), iv)   24   20

Approach C: Coupling with bromoacetic acid followed by nucleophilic displacement with amines

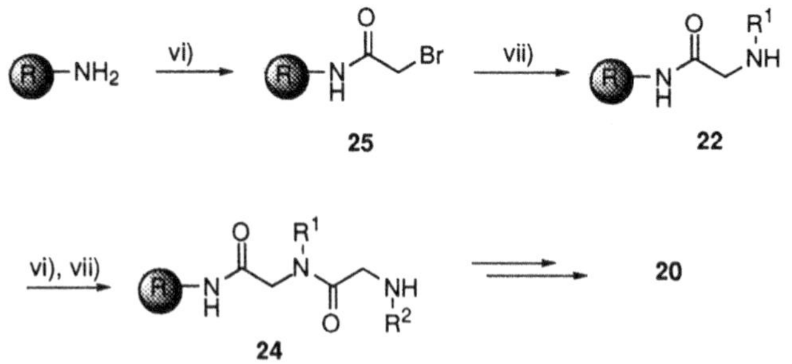

25   vi)   vii)   22

vi), vii)   24   20

Reagents and conditions: i) DBU, DMF; ii) PyBOP or PyBroP, $R^2NFmocCH_2COOH$; iii) DBU, DMF; iv) R-CHO, $NaCNBH_3$ or $Na(OAc)_3BH$, MeOH; v) Fmoc-Gly, PyBOP or PyBroP; vi) DIC, DMF, $BrCH_2COOH$; vii) $R-NH_2$, DMSO.

Approach A is based on the classical solid-phase peptide synthesis using the required *N*-substituted Fmoc-protected amino acids. Thus, deprotection of resin-bound **18** under standard conditions and coupling with Fmoc-NR^{2}CH$_2$COOH using PyBOP or PyBroP gave **19**, which was converted into the final peptoid **20**.

Approach B takes advantage of a classical peptide coupling using Fmoc-glycine as key building block. Thus, **21** is converted to **22** by Fmoc-deprotection, followed by reductive amination with an aldehyde. Repetition of this sequence via **23** and **24** yields after cleavage from the resin peptoid **20**.

Approach C is based on α-bromoacetic acid as key building block and starts from *Rink*-amine. After coupling of α-bromoacetic acid yielding **25**, followed by nucleophilic displacement with an amine, polymer-bound derivative **22** is obtained. Repetition of this cycle of events and cleavage from the resin gives rise to peptoid **20**.

Using the "split-mixed" methodology several targeted libraries were synthesised in order to find novel lead compounds. As one of the most striking examples, 18 pools of *N*-substituted glycine derived peptoids were assayed for inhibition of [^{3}H]-DAMGO (μ-specific) binding[74] to opiate receptors. The most active compound CHIR 4531 (*Figure 2.4.2*) was shown to inhibit [^{3}H]-DAMGO binding with K$_i$ = 6 ± 2 nM.

Figure 2.4.2

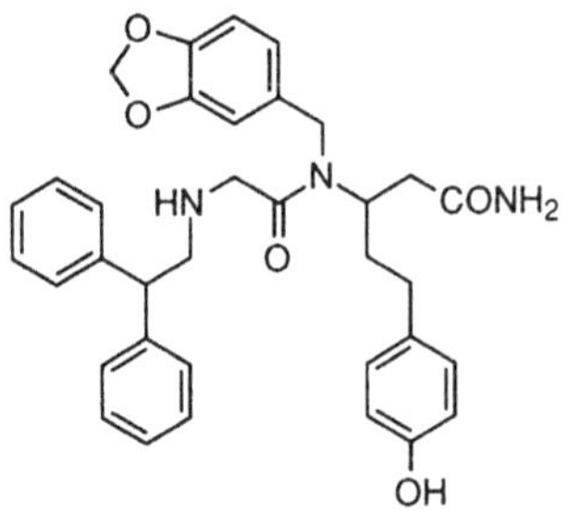

CHIR 4531

2.5. Oligonucleotides

Similarly to the polymer-supported peptide synthesis, the synthesis of oligonucleotides developed quickly into a rapidly growing field of research. Initial pioneering work has been carried out by *Letsinger* and *Khorana*[75] as indicated in *Chapter 1.2*. Many polymer supports have been reported which have been successfully used in the synthesis of oligonucleotides, including non-soluble polymers such as polyacrylamide based resins, porous glass and soluble polymers such as polyethyleneglycol derived resins (*Chapter 3.8*).

The most important strategy for the linear synthesis of oligonucleotides has been developed by *M. Caruthers*.[76] Nowadays, the phosphite triester method is widely used in the automated synthesis of oligonucleotides on silica. The principle is shown in *Scheme 2.5.1*.

Scheme 2.5.1

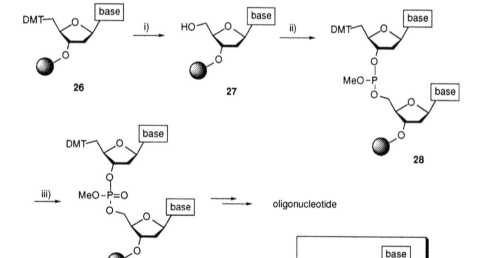

Reagents and conditions: i) dichloroacetic acid, CH₂Cl₂; ii) **30**, MeCN; iii) a) acetic anhydride, b) I₂.
(DMT: dimethoxytrityl)

The key building blocks include the appropriately protected deoxynucleoside 3'-phosphoramidites **30**. They are used in every cycle to prolong the oligonucleotide chain, followed by oxidation with iodine to obtain the desired phosphate triester **29**.

2.6. Oligosaccharides

Oligosaccharides fulfill in nature several important functions. They serve as structural materials (*e.g.* cellulose), energy storage (*e.g.* glucose and starch) and are important recognition elements on the cell surface.[77] They are usually composed of less than 20 monosaccharide units and occur as conjugates of lipids and proteins.[78]

While in the synthesis of naturally occurring oligosaccharides the use of enzymes (glycosyl transferases) proved very useful due to their substrate specificity and stereospecificity and the possibility to work without protective groups,[79] the synthesis of unnatural oligosaccharides still relies on synthetic methods.[80]

Among the basic biopolymers, the development of a polymer-supported synthesis of oligosaccharides has turned out to be the most difficult one. This is due to the requirement of stereocontrol in the formation of the glycosidic bond and the wide range of orthogonal protective groups for primary and secondary alcohols. At present, polymer-supported synthesis of oligosaccharides cannot yet compete with the classical syntheses in solution.[81] Incomplete glycosylation reactions with insufficient degree of stereocontrol, instability of the glycosyl donors under the prolonged reaction conditions required in solid-phase reactions and problems associated with incomplete protection and deprotection procedures constitute certainly some of the reasons for slower progress in this field. Soluble polymers might be the solution for the future.[82]

Two major coupling strategies were elaborated for the formation of glycosidic bonds on solid support. *Scheme 2.6.1* shows the *glycal method* developed by *Danishefsky et al.*[83,84] A polystyrene-bound glycal **31** is activated for glycosyl donation by formation of the 1,2-epoxide **32**. Addition of a glycosyl donor (**33**) generates the glycosidic bond under concomitant ring-opening and formation of a secondary alcohol in **34** which can directly be glycosylated (**36**). The other possibility is to activate the glycal as previously described and couple to **33** to give trisaccharide **37**. It is interesting to note that a silicon-based linker strategy was successfully employed.

Scheme 2.6.1

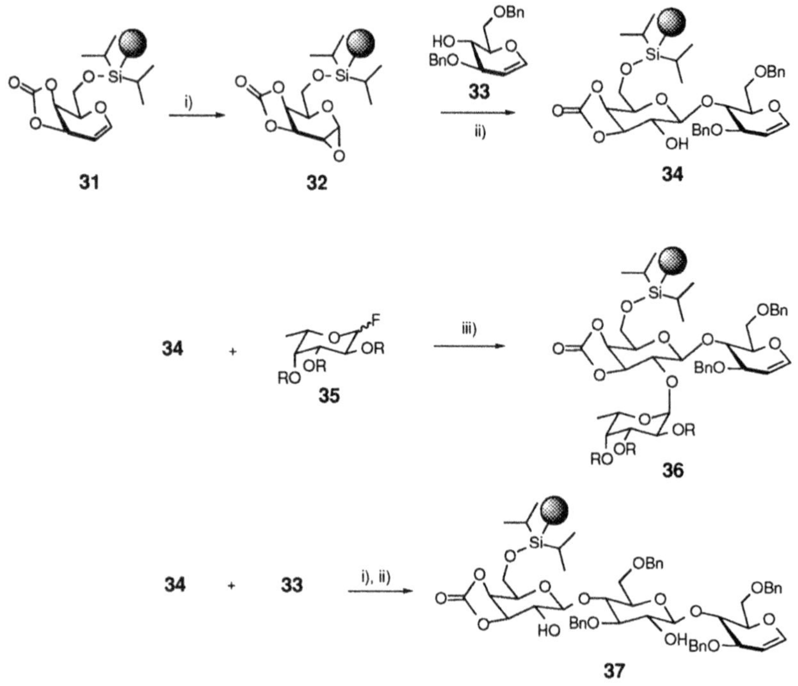

Reagents and conditions: i) 3,3-dimethyldioxirane, CH$_2$Cl$_2$, ii) ZnCl$_2$, THF; iii) Sn(OTf)$_2$, DTBP, THF.

The second strategy reported by *Kahne et al.*[85-87] is depicted in *Scheme 2.6.2*. In this sulfoxide-based method it is possible to selectively form α- or β-glycosidic bonds. Repeated glycosylation drove the reaction to completion. Oxidative cleavage with mercuric trifluoroacetate gave disaccharide **42** and **45** in 67% and 64% yield over two steps. The yields were higher on solid support as compared to those obtained in solution for this type of glycosidic linkages.

Scheme 2.6.2

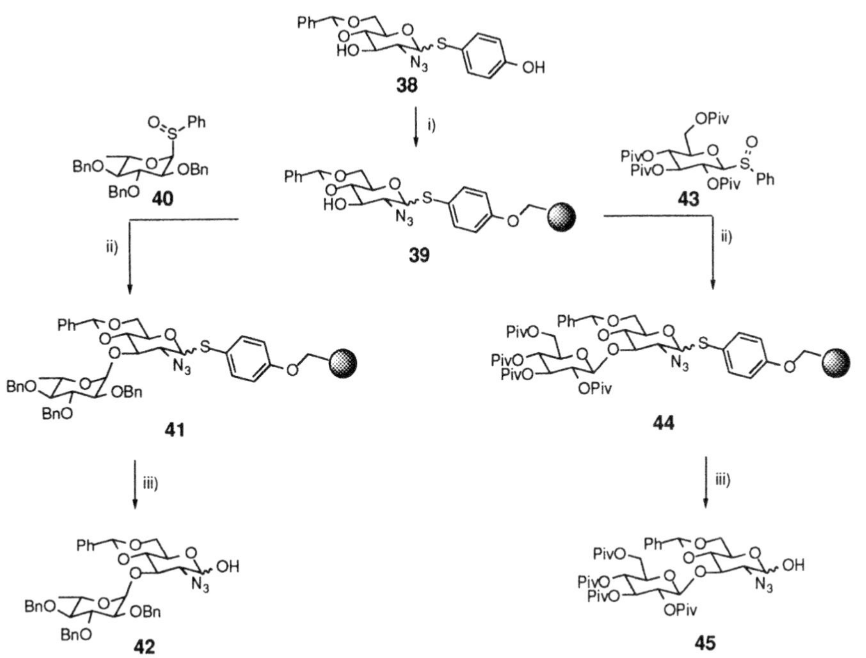

Reagents and conditions: i) Cs$_2$CO$_3$, *Merrifield* resin; ii) Tf$_2$O, 2,6-di-*tert*-butyl-4-methylpyridine, CH$_2$Cl$_2$, –60 to–30°C; iii) Hg(OCOCF$_3$)$_2$, CH$_2$Cl$_2$, H$_2$O, r.t.

Kahne, Still and coworkers reported recently the solid-phase parallel synthesis and screening of a carbohydrate library using the sulfoxide glycosylation reaction.[88] The library was constructed on TentaGel® resin and contained approximately 1300 di- and trisaccharides. A chemical tag was incorporated after every reaction step in order to monitor the history of the beads and to facilitate structure elucidation. The synthesis of the library is outlined in *Scheme 2.6.3*.

Scheme 2.6.3

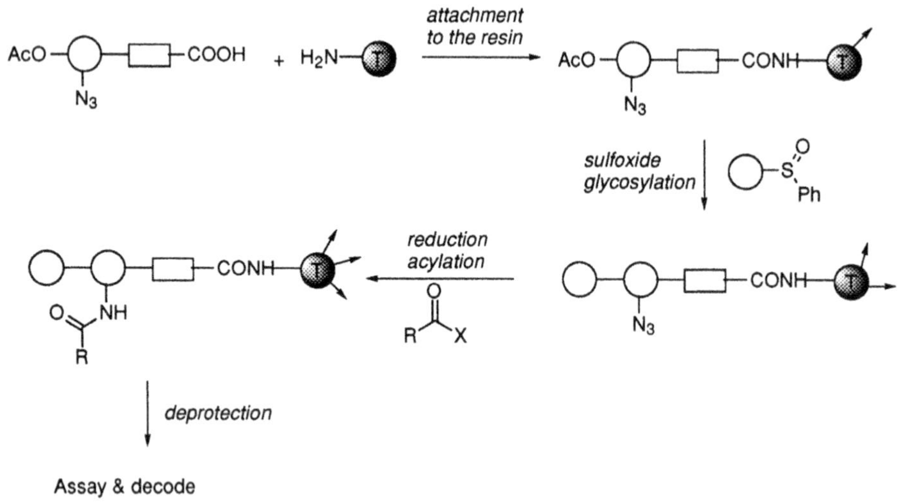

As starting building blocks were used 6 glycosyl acceptors which were attached to a TentaGel NH resin prior to coupling with 12 different gycosyl donors using the split-mixed synthesis. The amino group was derivatised with twenty different acylating agents. After deprotection, the library was assayed on bead against *Bauhinia purpurea* lectin. Two compounds could be identified which bind more tightly to the receptor than the natural ligands. Carrying out the assay on bead mimicks oligosaccharides bound to cell membranes.

Another solid-phase synthesis of oligosaccharides on soluble PEG will be discussed in *Chapter 3.8*.

2.7. Oligocarbamates and oligoureas

Besides the amide bond present in peptides, the carbamate and urea groups constitute additional chemically stable linkages for oligomeric compounds.[89] The first syntheses of oligocarbamates on solid supports were based on the sequential condensation of aminoalcohols[90] derived from α-amino acids shown in *Scheme 2.7.1*.

Scheme 2.7.1

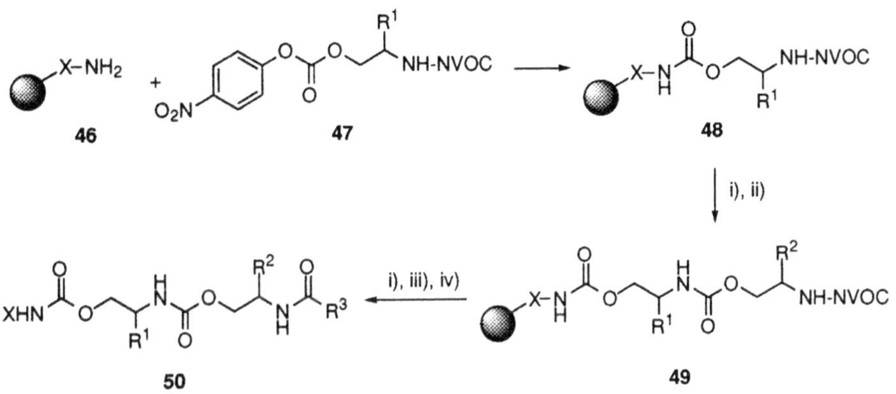

Reagents and conditions: i) hv (350nm); ii) p-NO$_2$-C$_6$H$_4$OCOCH$_2$CHR^{2}NH-NVOC; iii) R^3-COX; iv) TFA

Coupling of **46** with the corresponding p-nitrocarbonates of *N*-veratroyl (NVOC) protected aminoalcohols **47** gave polymer-bound carbamates **48** in high yields. Photochemical deprotection of the NVOC-group (350 nm), followed by a second coupling of a building block of type **47** yielded polymer-bound **49**. Deprotection of the amine and subsequent acylation followed by cleavage from the solid support gave polycarbamates of type **50**.

The urea moiety is an important structural element present in a series of enzyme inhibitors[91] such as HIV-1 protease inhibitors of type **51** as depicted in *Figure 2.7.1*.

Figure 2.7.1

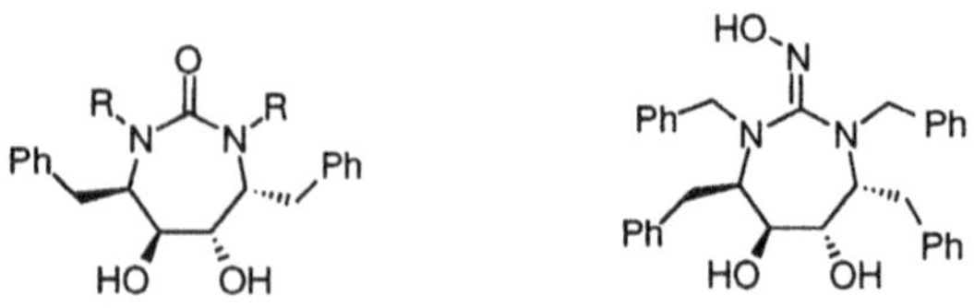

Recently, the oxamate analogue **52** (K_I = 42 nM) has been prepared and was shown to be 10-fold more active than the corresponding cyclic urea for inhibition of HIV PR.[92]

For the synthesis of oligoureas any suitably mono-protected diamine of type **53** can serve as a starting point as depicted in *Scheme 2.7.2*. Attachment of the free amine to the solid support (*e.g.* **54**), followed by deprotection and conversion into an isocyanate by treatment with triphosgene generated polymer-bound isocyanate **56** ready for treatment with a second mono-protected diamine to yield **57** after deprotection. Capping of the terminal amino group with *e.g.* an isocyanate (R''-N=C=O) and cleavage from the resin generated products of type **58**. Reaction sequences of this type have been reported by *Burgess et al.*[93]

Scheme 2.7.2

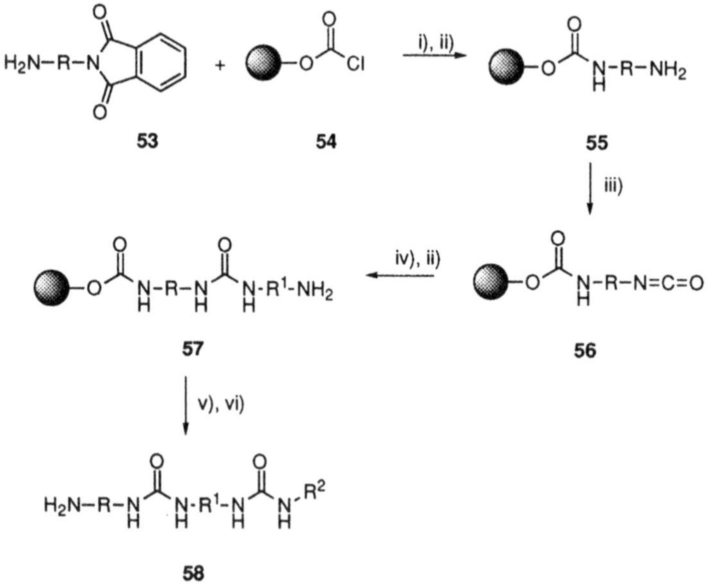

Reagents and conditions: i) CH$_2$Cl$_2$; ii) NH$_2$NH$_2$.H$_2$O, DMF 15h; iii) (Cl$_3$CO)$_2$CO, NEt$_3$, CH$_2$Cl$_2$; iv) H$_2$N-R^1-NPt; v) R^2-N=C=O; vi) base

2.8. Peptide–Nucleic acids (PNA)

Similarly to peptides, oligonucleotides are readily biologically degraded by various enzymes. Their use as orally active therapeutic agents is therefore very problematic. With the aim to overcome these problems, hybride oligomers designated as peptide–nucleic acids (PNA) have been reported for the first time in 1991 by *Nielsen et al.*[94] The synthesis is schematically depicted in *Scheme 2.8.1*.

Scheme 2.8.1

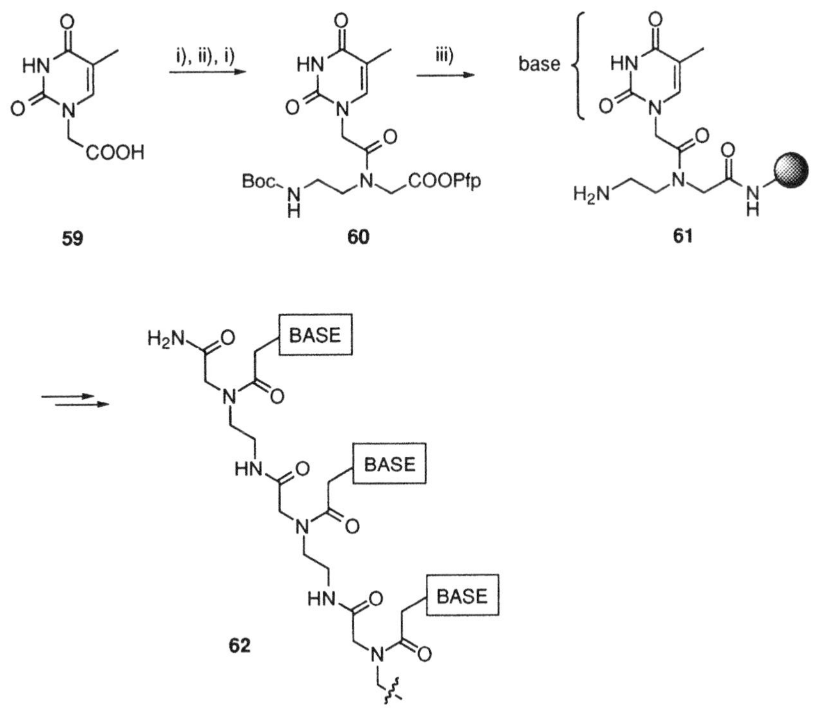

Reagents and conditions: i) DCC, pentafluorophenol; ii) BocNH(CH$_2$)$_2$NHCH$_2$COOH; iii) PS-NH$_2$

Starting from the modified bases of type **59**, activation to the corresponding pentafluorophenyl ester (PfpO), coupling to *N*-(*N*-Boc-aminoethyl)glycine and a second activation leads to the monomeric PNA-building block **60** (corresponding to thymine) which can be coupled to a polymeric support to yield **61**. Repetitive couplings of PNA-monomers and final cleavage from the resin gave rise to PNA-products of type **62** which are currently under evaluation as antisense probes for DNA therapy and diagnostic purposes.

References

(1) E. M. Gordon, R. W. Barrett, W. J. Dower, S. P. Fodor and M. A. Gallop, *J. Med. Chem.* **1994**, *37*, 1385.

(2) G. P. Smith and J. K. Scott, *Meth. Enzymol.* **1993**, *217*, 228.

(3) J. K. Scott and L. Craig, *Curr. Opin. Biotechnol.* **1994**, *5*, 40.

(4) H. M. Geysen, S. J. Rodda and T. J. Mason, *Mol. Immunol.* **1986**, *23*, 709.

(5) K. S. Lam, S. E. Salmon, E. M. Hersh, V. J. Hruby, W. M. Kamiersky and R. J. Knapp, *Nature* **1991**, *354*, 82.

(6) J. Eichler and R. A. Houghton, *Biochemistry* **1993**, *32*, 11035.

(7) S. P. A. Fodor, J. L. Read, M. C. Pirrung, L. Stryer, A. T. Lu and S. Solas, *Science* **1991**, *251*, 767.

(8) A. Kramer, R. Volkmer-Engert, R. Malin, U. Reincke and J. Schneider-Mergener, *Pept. Res.* **1993**, *6*, 314.

(9) A. Furka, F. Sebestyen, M. Asgedom and G. Dibo, *Int. J. Pept. Prot. Res.* **1991**, *37*, 487.

(10) R. A. Houghten, C. Pinilla, S. E. Blondelle, J. R. Appel, C. T. Dooley and J. H. Cuervo, *Nature* **1991**, *354*, 84.

(11) M. Lam, in: *Combinatorial Peptide and Nonpeptide Synthesis*, G. Jung (Ed.), p. 173, VCH: Weinheim (1996).

(12) R. B. Merrifield, *J. Am. Chem. Soc.* **1963**, *85*, 2149.

(13) H. M. Geysen, R. H. Meloen and S. J. Barteling, *Proc. Natl. Acad. Sci. USA* **1984**, *81*, 3998.

(14) S. Rodda, G. Tribbick and H. M. Geysen, in: *Combinatorial Peptide and Nonpeptide Libraries*, G. Jung (Ed.), p. 303, VCH: Weinheim (1996).

(15) R. A. Houghten, *Proc. Natl. Acad. Sci. USA* **1985**, *82*, 5131.

(16) C. Pinilla, J. Appel, C. Dooley, S. Blondelle, J. Eichler, B. Dörner, J. Ostresh and R. A. Houghton, in: *Combinatorial Peptide and Nonpeptide Libraries*, G. Jung (Ed.), p. 139, VCH: Weinheim (1996).

(17) H. M. Geysen and T. I. Mason, *Bioorg. Med. Chem. Lett.* **1993**, *3*, 397.

(18) A. G. Beck-Sickinger, H. Dürr and G. Jung, *Pept. Res.* **1991**, *4*, 88.

(19) W. Van't Hoff, P. C. Driedijk, M. Van den Berg, A. G. Beck-Sickinger, G. Jung and R. C. Aalbere, *Mol. Immunol.* **1991**, *28*, 1225.

(20) A. G. Beck-Sickinger, W. Gaida, G. Schnorrenberg, R. Lang and G. Jung, *Int. J. Pept. Prot. Res.* **1990**, *36*, 522.

(21) N. Zimmermann, A. G. Beck-Sickinger, G. Folckers, S. Krickl, I. Müller and G. Jung, *Eur. J. Biochem.* **1991**, *200*, 519.

(22) R. H. Berg, K. Almdal, W. Batsberg Pedersen, A. Holm, J. P. Tam and B. Merrifield, *J. Am. Chem. Soc.* **1989**, *111*, 8024.

(23) R. Hirschmann, *Angew. Chem. Int. Ed. Engl.* **1991**, *30*, 1278.

(24) K. Müller, D. Obrecht, A. Knierzinger, C. Stankovic, C. Spiegler, W. Bannwarth, A. Trzeciak, G. Englert, A. M. Labhardt and P. Schönholzer, in: *Perspectives in Medicinal Chemistry*, B. Testa, E. Kyburz, W. Fuhrer and R. Giger (Ed.), p. 513, Verlag Helv. Chim. Acta: Basel (1993).

(25) D. Obrecht, M. Altorfer and J. A. Robinson, *Novel Peptide Mimetic Building Blocks and Strategies for Efficient Lead Finding*, in: *Advances in Medicinal Chemistry*, B. E. Maryanoff and A. B. Reitz (Ed.), in press, p. JAI Press Inc.: Greenwich, Connecticut (1998).

(26) V. J. Hrubi, F. Al-Obeidi and W. Kazmierski, *Biochem. J.* **1990**, *268*, 249.

(27) A. F. Spatola, in: *Chemistry and Biochemistry of Amino Acids, Peptides and Proteins*, B. Weinstein (Ed.), Vol. 7, p. 267, Maral Dekker: New York (1983).

(28) G. J. Hanson and T. Lindbarg, *J. Org. Chem.* **1985**, *50*, 5399.

(29) D. H. Rich and M. W. Holladay, *Tetrahedron Lett.* **1983**, *24*, 4401.

(30) D. H. Rich, *J. Med. Chem.* **1985**, *28*, 263.

(31) M. Hagihara, N. J. Anthony, T. J. Stout, J. Clardy and S. L. Schreiber, *J. Am. Chem. Soc.* **1992**, *114*, 6568.

(32) A. Spaltenstein, P. A. Carpino, F. Miyake and P. B. Hopkins, *Tetrahedron Lett.* **1986**, *27*, 2095.

(33) M. Chorev and M. Goodman, *Acc. Chem. Res.* **1993**, *26*, 266.

(34) A. K. Gosh, S. P. McKee, W. J. Thompson, P. L. Parke and J. C. Zugai, *J. Org. Chem.* **1993**, *58*, 1025.

(35) J. R. Huff, *J. Med. Chem.* **1991**, *34*, 2305.

(36) C. Shin, Y. Yonegawa, T. Obara and H. Nishio, *Bull. Chem. Soc.* **1988**, *61*, 885.

(37) D. Obrecht, M. Altorfer, U. Bohdal, J. Daly, W. Huber, A. Labhardt, C. Lehmann, K. Müller, R. Ruffieux, P. Schönholzer, C. Spiegler and C. Zumbrunn, *Biopolymers* **1997**, *42*, 575.

(38) W. L. Scott, C. Zhou, F. Zhigiang and M. J. O'Donnell, *Tetrahedron Lett.* **1997**, *38*, 3695.

(39) G. Hoelzemann, *Kontakte (Merck, Darmstadt)* **1991**, 3.

(40) G. Hoelzemann, *Kontakte (Merck, Darmstadt)* **1991**, 55.

(41) I. A. Wilson and R. L. Stanfield, *Curr. Opion. Struct. Biol.* **1994**, *4*, 857.

(42) I. A. Wilson, J. B. Ghiara and R. L. Stanfield, *Res. Immunol.* **1994**, *145*, 73.

(43) D. F. Veber, *Design of a highly active cyclic hexapeptide analog of somatostatin*, in: *Peptides, Synthesis, Structure and Function. Proceedings of the Seventh American Peptide Symposium*, D. H. Rich and V. J. Gross (Ed.), p. 685, Pierce Chemical Company: Rockford, IL (1981).

(44) D. J. Kyle, S. Chakravarty, J. A. Sinsko and T. M. Storman, *J. Med. Chem.* **1994**, *37*, 1347.

(45) P. Jiang and V. S. Ananthanarayanan, *J. Biol. Chem.* **1991**, *266*, 22960.

(46) N. Brakch, M. Rholam, H. Boussetta and P. Cohen, *Biochemistry* **1993**, *32*, 4925.

(47) L. Paolillo, M. Simonetti, N. Brakch, G. D'Auria, M. Saviano, M. Dettin, M. Rholam, A. Scatturin, C. DiBello and P. Cohen, *EMBO J.* **1992**, *11*, 2399.

(48) V. Mandiyan, R. O'Brien, M. Zhou, B. Margolis, M. A. Lemmon, J. M. Sturtevant and J. Schlessinger, *J. Biol. Chem.* **1996**, *271*, 4770.

(49) T. Trüb, W. E. Choi, G. Wolf, E. Ottinger, Y. Chen, M. Weiss and S. E. Shoelson, *J. Biol. Chem.* **1995**, *270*, 18205.

(50) G. A. Barkocy-Gallagher, J. G. Cannon and P. J. Bassford, *J. Biol. Chem.* **1994**, *269*, 13609.

(51) A. Bansal and L. Gierasch, *Cell* **1991**, *67*, 1195.

(52) W. Eberle, C. Sander, W. Klaus, B. Schmidt, K. von Figura and C. Peters, *Cell* **1991**, *67*, 1203.

(53) B. Imperiali, J. R. Spencer and M. D. Struthers, *J. Am. Chem. Soc.* **1994**, *116*, 8424.

(54) C. M. Venkatachalam, *Biopolymers* **1968**, *6*, 1425.

(55) C. M. Wilmot and J. M. Thornton, *J. Mol. Biol.* **1988**, *203*, 221.

(56) W. Kabsch and C. Sander, *Biopolymers* **1983**, *22*, 2577.

(57) A. A. Virgilio and J. A. Ellman, *J. Am. Chem. Soc.* **1994**, *116*, 11580.

(58) J. A. Ellman, *Acc. Chem. Res.* **1996**, *29*, 132.

(59) R. Sarabu, K. Lovey, V. Madison, D. C. Fry, D. N. Greeley, C. M. Cook and G. L. Olson, *Tetrahedron* **1993**, *49*, 3629.

(60) D. L. Boger, M. A. Patane and J. Zhou, *J. Am. Chem. Soc.* **1995**, *117*, 7357.

(61) R. Beeli, M. Steger, A. Linden and J. A. Robinson, *Helv. Chim. Acta.* **1996**, *79*, 2235.

(62) C. Bisang, C. Weber and J. A. Robinson, *Helv. Chim. Acta* **1996**, *79*, 1825.

(63) F. Emery, C. Bisang, M. Favre, L. Jiang and J. A. Robinson, *J. Chem. Soc., Chem. Commun.* **1996**, 2155.

(64) M. Pfeifer, A. Linden and J. A. Robinson, *Helv. Chim. Acta.* **1997**, *80*, 1513.

(65) W. Klaus, C. Broger, P. Gerber and H. Senn, *J. Mol. Biol.* **1993**, *232*, 897.

(66) H. Senn and W. Klaus, *J. Mol. Biol.* **1993**, *232*, 907.

(67) P. Waldmeier, Ph.D thesis, University of Zürich (1997).

(68) W. Bannwarth, F. Gerber, A. Grieder, A. Knierzinger, K. Müller, D. Obrecht and A. Trzceciak, Can. Pat. Appl. CA 2101599, 1995.

(69) N. A. Abraham, *Tetrahedron Lett.* **1991**, *32*, 577.

(70) D. Bellof and M. Mutter, *Chimia* **1985**, *39*, 10.

(71) J. J. Plattner and D. W. Norbeck, *Obstacles to drug development from peptide leads*, in: *Drug Discovery Technologies*, C. R. Clark and W. H. Moos (Ed.), p. 92, Ellis Horwood Ltd.: Chichester, U.K. (1990).

(72) L. S. Richter, D. C. Spellmeyer, E. J. Martin, G. M. Figliozzi and R. N. Zuckermann, *Automated synthesis of non-natural oligomer libraries: the peptoid concept*, in: *Combinatorial Peptide and Nonpeptide Libraries*, G. Jung (Ed.), p. 387, VCH: Weinheim (1996).

(73) R. J. Simon, R. S. Karua, R. N. Zuckermann, V. D. Huebner, D. A. Jewell, S. Banville, S. Ng, L. Wang, S. Rosenberg, C. K. Marlowe, D. C. Spellmeyer, R. Tan, A. D. Frenkel, D. V. Santi, F. E. Cohen and P. A. Bartlett, *Proc. Natl. Acad. Sci. USA* **1992**, *89*, 9367.

(74) M. G. C. Gillan and H. C. Kosterlitz, *Br. J. Pharm.* **1982**, *77*, 461.

(75) R. L. Letsinger and V. Mahadevan, *J. Am. Chem. Soc.* **1965**, *87*, 3526.

(76) M. H. Caruthers, *Science* **1985**, *230*, 281.

(77) C. M. Taylor, *Strategies for combinatorial libraries of oligosaccharides*, in: *Combinatorial Chemistry*, S. R. Wilson and A. W. Czarnik (Ed.), p. 207, John Wiley & Sons, Inc.: New York (1997).

(78) H. Paulsen, *Angew. Chem. Int. Ed. Engl.* **1990**, *29*, 823.

(79) E. J. Toone, E. S. Simon, M. D. Bednarsky and G. M. Whitesides, *Tetrahedron* **1989**, *45*, 5365.

(80) K. Toshima and K. Tatsuta, *Chem. Rev.* **1993**, *93*, 1503.

(81) Y. Wang, H. Zhang and W. Voelter, *Chem. Lett.* **1995**, 273.

(82) D. J. Gravert and K. D. Janda, *Chem. Rev.* **1997**, *97*, 489.

(83) J. T. Randolph, K. F. McClure and S. J. Danishefsky, *J. Am. Chem. Soc.* **1995**, *117*, 5712.

(84) S. J. Danishefsky, K. F. McClure, J. T. Randolph and R. B. Ruggeri, *Science* **1993**, *260*, 1307.

(85) L. Yan, C. M. Taylor, R. Goodnow Jr. and D. Kahne, *J. Am. Chem. Soc.* **1994**, *116*, 6953.

(86) S.-H. Kim, D. Augeri, D. Yang and D. Kahne, *J. Am. Chem. Soc.* **1994**, *116*, 1766.

(87) D. Kahne, S. Walker, Y. Cheng and D. Van Engen, *J. Am. Chem. Soc.* **1989**, *111*, 6881.

(88) R. Liang, L. Yan, J. Loebach, M. Ge, Y. Uozumi, K. Sekanina, N. Horan, J. Gildersleeve, C. Thompson, A. Smith, K. Biswas, W. C. Still and D. Kahne, *Science* **1996**, *274*, 1520.

(89) J. S. Früchtel and G. Jung, *Angew. Chem. Int. Ed. Engl.* **1996**, *35*, 17.

(90) C. Y. Cho, E. J. Moran, S. R. Cherry, J. C. Stephans, S. P. A. Fodor, C. L. Adams, A. Sundaram, J. W. Jacob and P. G. Schultz, *Science* **1993**, *261*, 1303.

(91) P. Y. S. Lam, P. K. Jadhav, C. J. Eyermann, C. N. Hodge, Y. Ru, L. T. Bacheler, J. L. Meck, M. J. Otto, M. M. Rayner, Y. N. Wong, C.-H. Chang, P. C. Weber, D. A. Jackson, T. R. Sharpe and S. Erickson-Viitanen, *Science* **1994**, *263*, 380.

(92) P. K. Jadhav, W. F.J. and M. Hon-Wah, *Bioorg. Med. Chem. Lett.* **1997**, *7*, 2145.

(93) K. Burgess, D. S. Linthicum and H. Shin, *Angew. Chem. Int. Ed. Engl.* **1995**, *34*, 907.

(94) P. E. Nielsen, M. Egholm, R. H. Berg and O. Buchhardt, *Science* **1991**, *254*, 1497.

Chapter 3

Library Synthesis in Solution Using Liquid-Liquid and Liquid-Solid Extractions and Polymer-Bound Reagents

3.1. General

As mentioned earlier in *Chapter 1.8*, solution-phase chemistry has the advantage over solid-supported chemistry in that less effort has to be put into method development work and that essentially the whole plethora of organic reactions is available. Strategies that can be employed in order to reduce the tedious purification procedures associated with solution-phase chemistry will be discussed in the following paragraphs. Lets consider the simple case of an amide library (R^1CONHR^2) generated from acid R^1COOH using an intermediate activation followed by treatment with amines R^2NH_2 as shown in *Figure 3.1.1*.

Figure 3.1.1

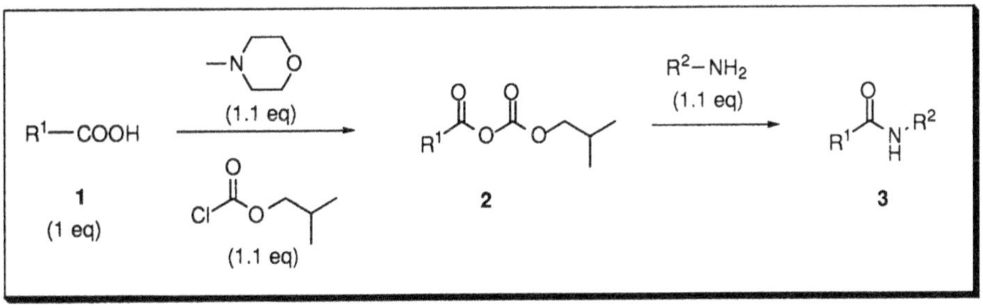

By-products: N-methylmorpholine hydrochloride, 2-methylpropanol, excess amine, excess base and excess chloroformate

Although this reaction is known to proceed in high yields it produces the by-products mentioned in the above *Figure*. In the following paragraphs a number of approaches to reduce the work-up efforts will be discussed:

- Multicomponent one-pot reactions
- Multigeneration reactions
- Liquid-liquid phase extractions (LPE)
- Liquid-solid phase extractions (SPE) using polymer-bound scavengers
- Solution reactions using polymer-bound reagents or catalysts

Multicomponent one-pot reactions producing no by-products

The first strategy to circumvent laborious extraction and purification procedures is to perform the amide formation reaction in a multicomponent format, which allows the synthesis of the amide library without vessel transfer already in the screening format (usually 96-well plates). Thus, using the *Ugi* four component reaction (entry a) or 3-amino-2*H*-azirines[1] (entry b) the peptide coupling can be carried out without activation of the acid derivative (*Figure 3.1.2*). Virtually all the atoms of all the reagents are present in the product.

Figure 3.1.2

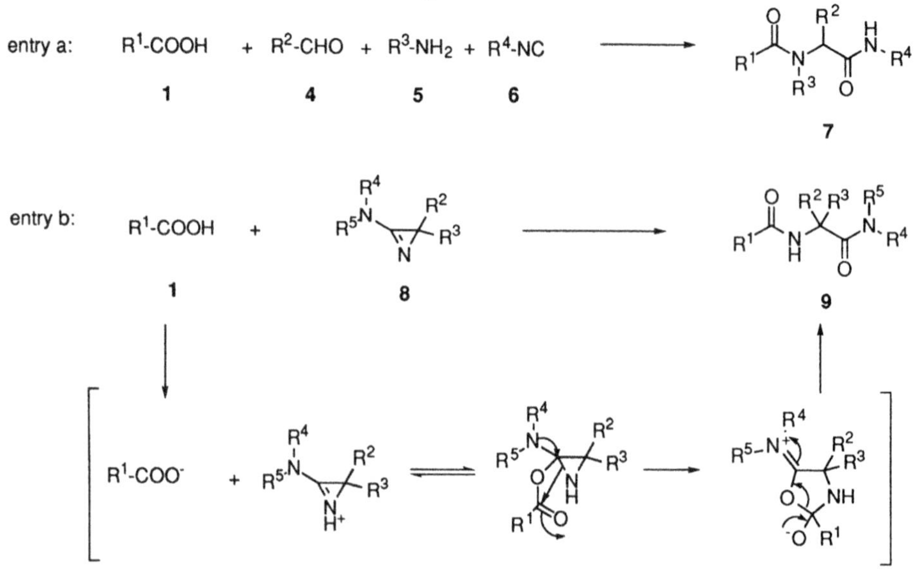

While the azirine reaction proceeds in essentially quantitative yield producing no by-products, the *Ugi* reaction usually yields products of type **7** in 60-90% yield along with some starting materials **1**, **4**, **5** and **6** which can be trapped using a polymer-supported quenching (PSQ) method of purification as shown later.

Multigeneration reactions

The concept of high yielding multigeneration reactions in which an essentially quantitative transformation generates a novel functional group ready for being engaged in the next step has been developed by *Boger et al.*[2] This strategy offers the benefit that the amount of by-products in the reaction mixture can be minimised and separated by simple extractions. As a consequence, the last generation products are not contaminated with products from previous generations (*Figure 3.1.3*).

Figure 3.1.3

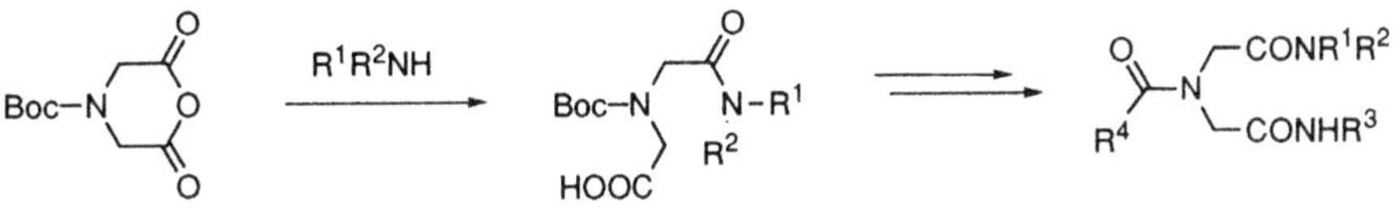

Liquid-liquid phase extractions

Excesses of reagents and by-products are removed by simple acid-base aqueous extractive work-up procedures which ultimately can be automated. Looking at our model case shown in *Figure 3.1.1* it is obvious that aqueous extraction is efficient for the removal of several by-products and excesses of reagents.

Basic extraction: removal of remaining acid and saponification of excess chloroformate.

Acidic extraction: elimination of basic components (*N*-methylmorpholine and excess amine).

In the *organic phase* remains: product **3** and 2-methylpropanol (which can be evaporated).

Liquid-solid phase extractions using polymer-bound scavengers

This strategy, although quite expensive, offers the advantage that excesses of reagents and by-products can be removed by simple filtration. Refering to our model reaction, the following strategies can be considered (*Figure 3.1.4*).

Figure 3.1.4

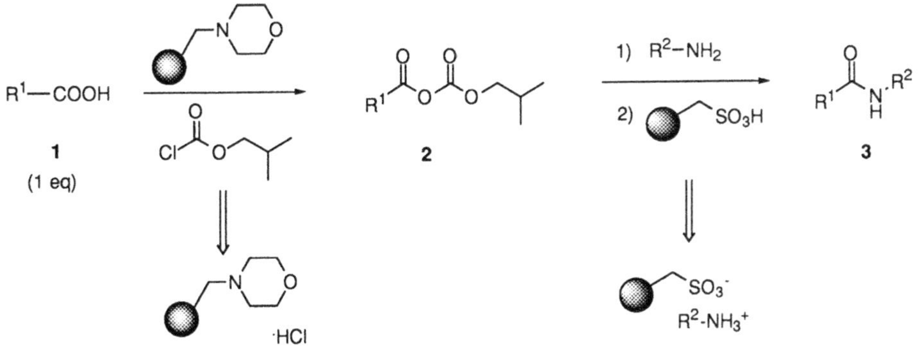

The organic base is covalently linked to an insoluble support, removing in the first step of the process the hydrochloride salt generated in the course of the reaction without aqueous work-up.

After amide coupling, the excess of amine can be easily sequestered using an acidic ion exchange resin. This protocol allows removal of by-products and excess reagents by simple filtrations generating products of high purity.

Solution reactions using polymer-bound reagents or catalysts

For this purpose, our model reaction can be modified as shown in *Figure 3.1.5* using a polymer-bound chloroformate linking the mixed anhydride intermediate to a solid support and, thus, allowing easy removal of excess base by simple washing. After coupling with amines R^2-NH_2, only amide **3** and excess amine remain in solution which can be easily separated as previously described.

Figure 3.1.5

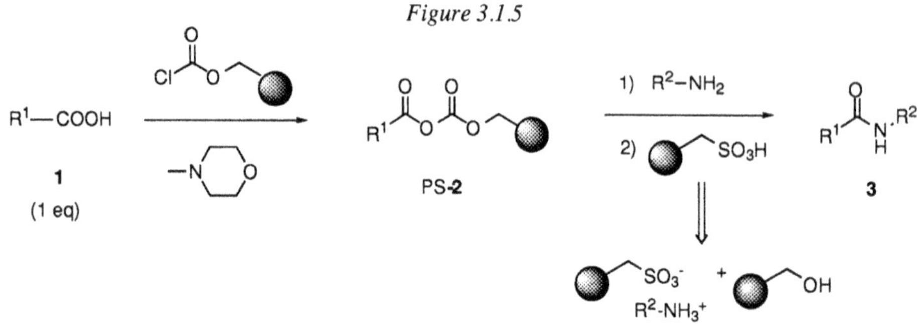

3.2. Applications of solution-phase multicomponent and multigeneration reactions

3.2.1. Optimisation of the biological activity using a genetic algorithm

Genetic algorithms constitute an interesting approach for rational optimisation of multiparameter systems. Applied to combinatorial chemistry this means that starting from a given set of variables a specific parameter is optimised, in this case biological activity, using the genetic operations replication, mutation and crossover. As exemplified in *Scheme 3.2.1*, *Weber et al.*[3] used a genetic algorithm to optimise the biological activity of thrombin inhibitors.

Scheme 3.2.1

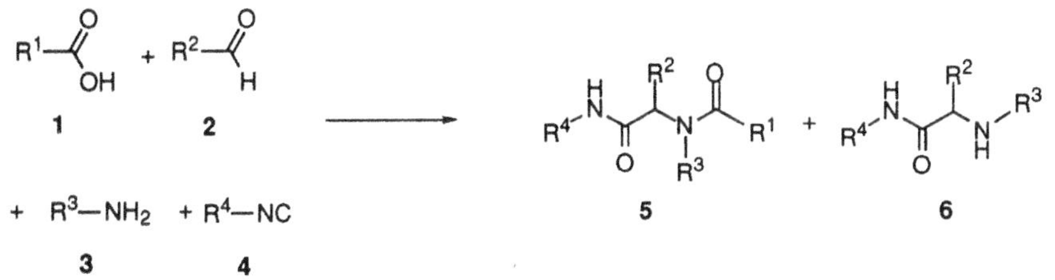

Based on the building blocks used for the most active products of the first generation and applying the rules of the genetic algorithm (mutation and crossover) a new set of inputs was determined and used for the synthesis of the next generation. After 16 generations, the biological activity of the most active compounds laid in the submicromolar range. The process was continued for twenty generations corresponding to 400 synthesised products (0.25% of the possible 160'000). The two most active compounds out of this library are shown in *Figure 3.2.1*.

Figure 3.2.1

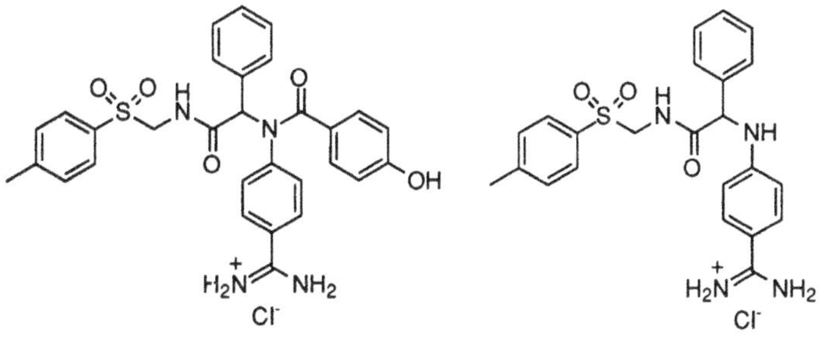

3.2.2. Synthesis of heterocycles using amino-azirine building blocks

A nice example of a high-yielding two component reaction incorporating all atoms in the products constitutes the ring-expansion reaction of 3-amino-2*H*-azirines 7 with NH-acidic heterocycles[4-7] as shown in the following examples (*Scheme 3.2.2*).

Scheme 3.2.2

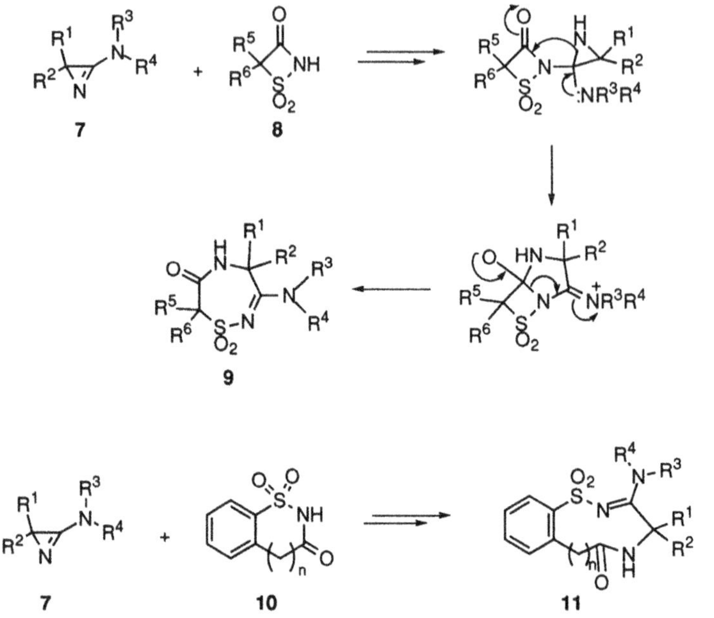

Reaction with 4-membered cyclic sulfonamides 8 afforded 7-membered heterocycles 9 in high yield (60-90%), and reaction with 6-membered sulfonamide 10 (n=1) proceeded in 85-99% yield incorporating all atoms of both building blocks into the products. No by-products were thus formed.

3.2.3. The use of 1-isocyanocyclohexene in *Ugi* four component condensations

The lack of commercially available isocyanides is one of the major restrictions of the *Ugi* four component reaction. A solution to this problem has been presented by *Armstrong et al.*[8] The use of 1-isocyanocyclohexene permitted facile conversion of the *Ugi* products into several new core structures as shown in *Scheme 3.2.3*.

Scheme 3.2.3

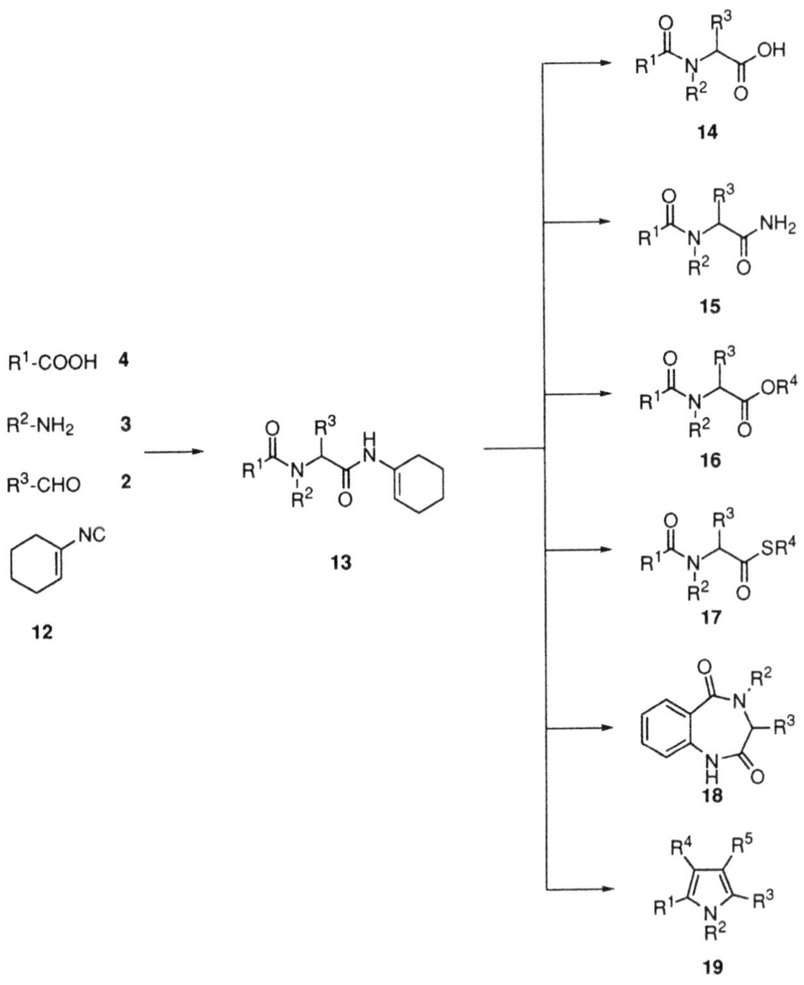

Acidic hydrolysis of the reactive enamide led to the corresponding carboxylic acids **14** whereas alcoholysis gave esters **16** and aminolysis amides **15**. The mechanism of the hydrolysis was shown to proceed via münchnone derivatives **20** which, instead of being opened with a nucleophile, reacted as a 1,3-dipole in [3+2] cycloaddition reactions with propiolic esters or acetylene dicarboxylic esters to give after elimination of carbon dioxide protected pyrroles **19** (*Scheme 3.2.4*)

Scheme 3.2.4

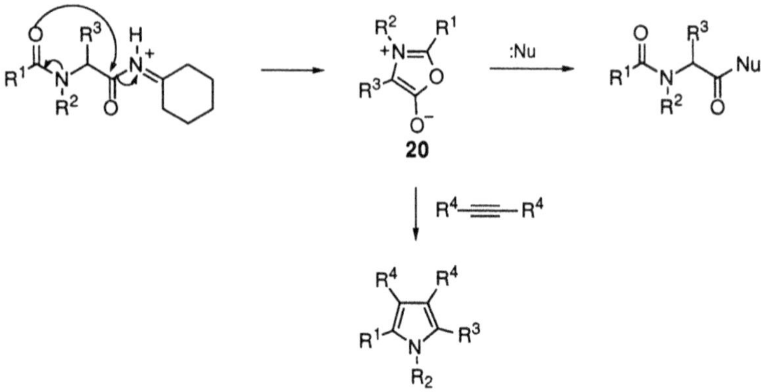

Cyclisation was achieved by intramolecular substitution as exemplified by the synthesis of 1,4-benzodiazepine-2,5-diones **18**.

3.2.4. Multigeneration synthesis of a thiazole library

Maehr et al.[9] reported an elegant multigeneration approach towards the synthesis of a thiazole library in solution using the *Hantzsch* thiazole synthesis (*Scheme 3.2.5*). Starting from (2-aminothioxoethyl)phosphonic acid diisopropyl ester **21** and 10 different α-halo ketones **22**, a first set of thiazoles of type **23** was generated. Subsequent *Wittig-Horner* reaction with 7 nitrobenzaldehydes led to the second generation of thiazoles **24**, which after reduction of the nitro group gave the third generation of products **25**. Acylation with 10 different cyclic anhydrides furnished the fourth generation of thiazoles **26**.

Scheme 3.2.5

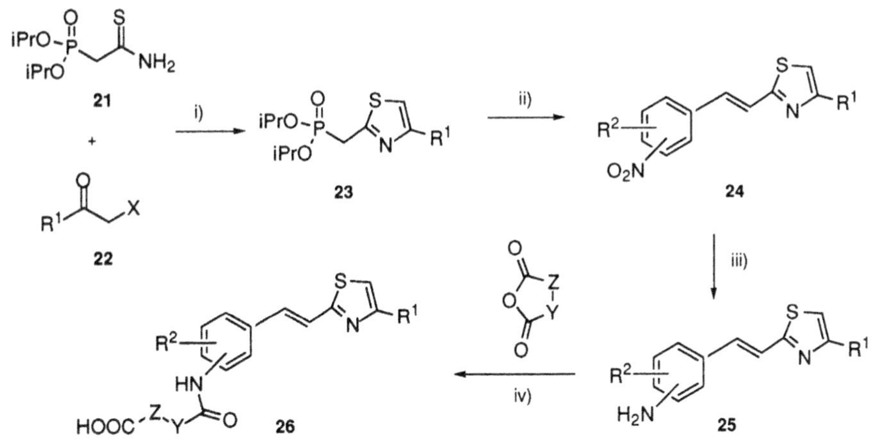

Reagents and conditions: i) MeOH, Δ; ii) nitrobenzaldehyde, K₂CO₃;
iii) SnCl₂, EtOH, 55°C; iv) toluene, 90°C.

3.3. Applications of liquid-liquid phase extractions

3.3.1. Multigeneration synthesis using an "universal template"

Boger et al.[2,10,11] described a simple, versatile and general approach to a solution phase, parallel synthesis of conformationally flexible products of type **30** (*Scheme 3.3.1*) on a multi-milligram scale.

Scheme 3.3.1

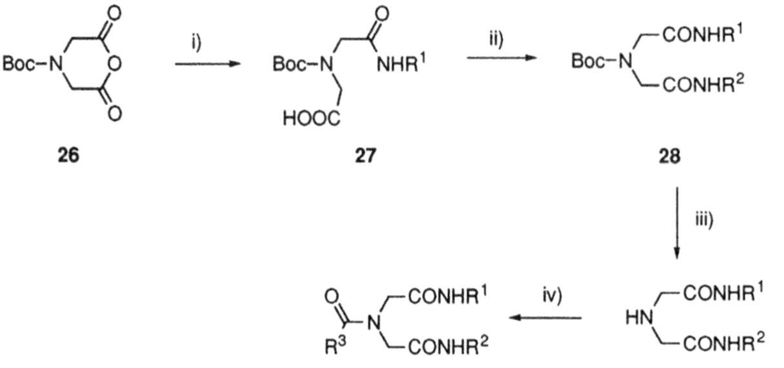

Reagents and conditions: i) H_2NR^1; ii) PyBOP, DIEA, H_2NR^2; iii) HCl, dioxane; iv) R^3COOH, PyBOP, DIEA

Thus, a 1158 compound library with a high degree of diversity was synthesised using the following procedure: Nucleophilic ring-opening of symmetrical anhydride **26** with different amines gave acids **27**, which were purified by acid/base extraction and further coupled to a second amine to give **28**. Liquid/liquid extraction allowed the removal of coupling reagents and excess amine. After deprotection of the secondary amine and acylation under the same standard conditions and purification by acid/base extraction, the products **30** were isolated in overall yields ranging from 10-71% and in purities >90%.

3.3.2. Non-aqueous liquid-liquid extraction using fluorinated solvents

Fluorocarbon-derived solvents are usually immiscible in organic solutions and can therefore be used in standard liquid/liquid extractions. *Curran* and *Wipf* developed a new strategy in which a fluorinated substrate was linked to one of the starting materials thus rendering it soluble in fluorinated solvents.[12,13] The principle is depicted in *Figure 3.3.1*.

Figure 3.3.1

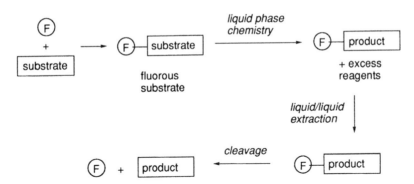

After every reaction step, the product can be extracted from the 'organic' phase with fluorinated solvents. Excess reagents remain in the 'organic' phase and can be easily removed. The strategy has also been successfully used in a *Ugi* multicomponent-type reaction condensing derivatised benzoic acid **31** with amines, aldehydes and isonitriles as shown in *Scheme 3.3.2*. The reaction was carried out in TFE and products **32** were isolated by liquid-liquid extraction. After cleavage of the fluorinated chain with TBAF followed by another extraction, the desired products **33** were isolated in good to excellent yields and with purities >80% (determined by GC).

Scheme 3.3.2

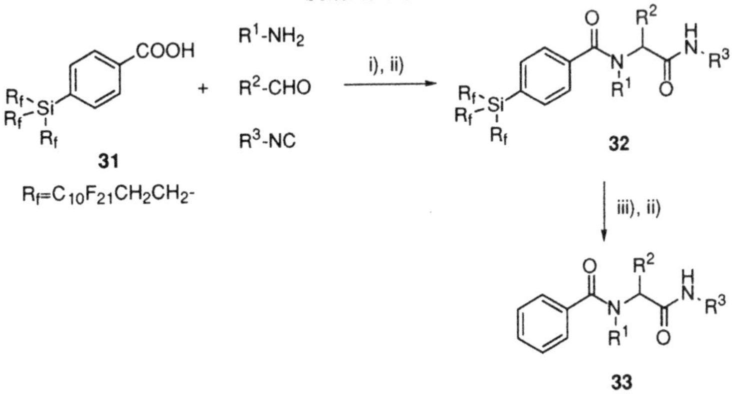

Reagents and conditions: i) TFE, 90°C, 48h; ii) liquid-liquid extraction; iii) TBAF, THF, 25°C

An analogous version of the *Biginelli* reaction was also developed[12,13] following the same strategy. In addition, fluorinated stannane reagents have been used in *Stille*-type couplings. The usually tedious separation of stannanes from the products could be achieved by simple extraction with trifluorobenzene.[14]

3.4. Applications of liquid-solid phase extractions and polymer-bound scavengers

3.4.1. Solid-phase extraction using ion exchange resins

Several groups reported the use of ion exchange resins for the parallel purification of small-molecular-weight compound libraries. Reaction conditions for the synthesis of the library are chosen such that the product of interest has distinctly different physical properties than the by-products or the excess reagents.

Suto et al.[15] described the purification of an amide library with basic ion exchange resins. The reaction was performed using an excess of acid chloride as acylating agent (see Scheme 3.4.1). After completion of the reaction, addition of a small amount of water led to the hydrolysis of the excess acid chloride. Carboxylic acids as well as hydrochloric acid formed during the reaction were then adsorbed on the resin. The desired amides remained in solution and were isolated in excellent yields and HPLC purities >98%.

Scheme 3.4.1

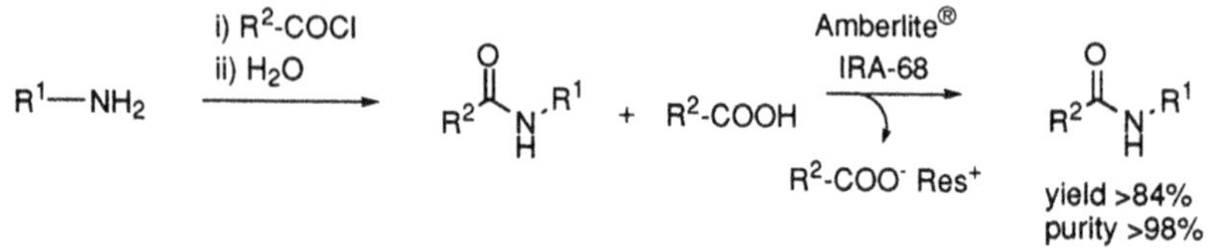

R^1—NH$_2$ $\xrightarrow[\text{ii) H}_2\text{O}]{\text{i) R}^2\text{-COCl}}$ R^2—C(O)—N(H)—R^1 + R^2-COOH $\xrightarrow{\substack{\text{Amberlite}^\circledR \\ \text{IRA-68}}}$ R^2-COO$^-$ Res$^+$ R^2—C(O)—N(H)—R^1 yield >84% purity >98%

Similarly, excess amines could be extracted with acidic ion exchange resins as exemplified by the same group in the synthesis of ureas starting from isocyanates and amines.[15]

A different approach has been chosen by Siegel et al. at Eli Lilly. After a reductive amination reaction (depicted in Scheme 3.4.2) using one equivalent of amine and three equivalents of aldehyde the reaction mixture was loaded on an ion exchange column. The desired resin-bound amines could be separated from excess aldehyde by filtration and the products were subsequently washed from the ion exchange resin by treatment with methanolic ammonia.

Scheme 3.4.2

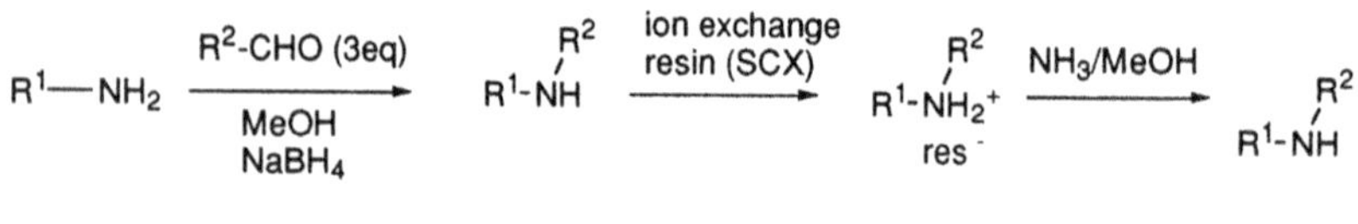

R^1—NH$_2$ $\xrightarrow[\substack{\text{MeOH} \\ \text{NaBH}_4}]{\text{R}^2\text{-CHO (3eq)}}$ R^1-N(H)-R^2 $\xrightarrow{\substack{\text{ion exchange} \\ \text{resin (SCX)}}}$ R^1-N(H$_2^+$)-R^2 res$^-$ $\xrightarrow{\text{NH}_3/\text{MeOH}}$ R^1-N(H)-R^2

An alternative cleavage procedure from ArgoGel-*Wang* resin has been explored by *Porco Jr. et al.*[16] as the standard cleavage conditions (TFA) provided products only in insufficient purity. Oxidative cleavage using 2,3-dichloro-5,6-dicyanobenzoquinone (DDQ) proceeded very cleanly but isolation of the products remained tedious. A mixed bed ion exchange scavenger was developed which reduced excess DDQ to DDQH and subsequently scavenged DDQH from the reaction solution. Amberlyst® A-26 (OH-form) was loaded with ascorbic acid for the reduction and Amberlyst® A-26 (HCO$_3^-$) was used for the subsequent trapping of DDQH.

3.4.2. Solid-phase extraction using covalent scavengers

Reagents of by-products can also be trapped by covalent binding to a polymer. *Kaldor et al.*[17,18] reported the use of aminomethyl polystyrene for the extraction of isocyanates, acid chlorides or sulfonyl chlorides from reaction mixtures (*Scheme 3.4.3*, entry a). Similarly, amines were trapped with polymer-supported isocyanates, acid chlorides or aldehydes (*Scheme 3.4.3*, entry b).

Scheme 3.4.3

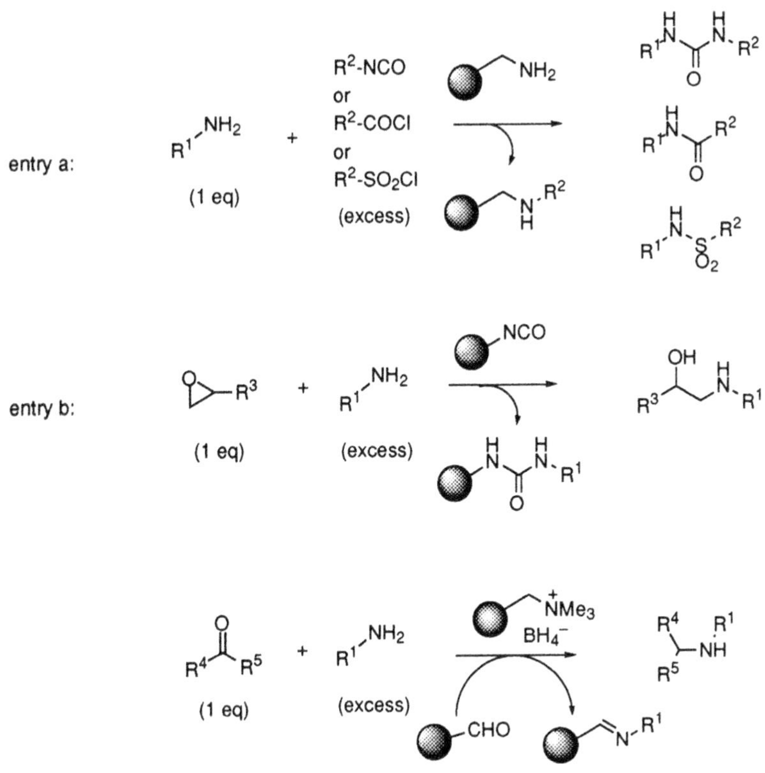

225

3.4.3. Synthesis of a thiazole library using liquid- and solid-phase extractions

As an example, we present a solution-phase multigeneration strategy towards highly substituted thiazoles based on the *Hantzsch* condensation of thioureas **33** with 2-bromomethyl ketones **34** to give 2-aminothiazoles **35** in high yields, as shown in *Scheme 3.4.4*, using liquid-liquid phase extraction (LPE) and liquid-solid phase extraction (SPE).

Scheme 3.4.4

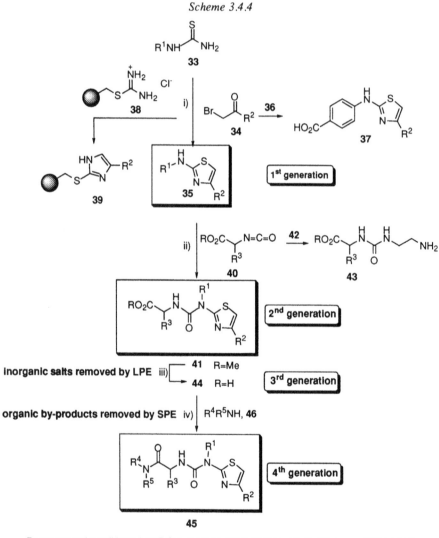

Reagents and conditions: i) a) EtOH, dioxane, 60°C, b) **36** or **38**; ii) a) toluene, 70°C, b) **42**;

iii) LiOH, dioxane, H$_2$O, r.t.; iv) EDCI, cat. DMAP, CH$_2$Cl$_2$, r.t.

(**36**=*N*-(4-carboxyphenyl)thiourea; **42**=1,2-diaminoethane)

226

The excess of **34** was trapped either with *N*-(4-carboxyphenyl)thiourea (**36**) yielding **37** and removed by LPE,[19] or with polymer-bound thiouronium salt **38** to yield polymer-bound imidazole **39** (in analogy to a liquid-phase strategy developed by *Dodson et al.*[20]) and removed by SPE. Subsequent treatment of thiazoles **35** with a series of amino acid-derived isocyanates **40** gave the second generation of thiazoles **41** in essentially quantitative yields. Excess of **40** was trapped with 1,2-diaminoethane (**42**) to yield the highly polar ureas **43** which were easily removed by SPE. Saponification gave the third generation of thiazole acids **44**, which could be efficiently transformed into the corresponding amides **45** by EDCI (1-ethyl-3-[3-(dimethylamino)propyl] carbodiimide hydrochloride) activation and coupling with amine **46**. Again, all the excesses of reagents could be removed easily either by SPE or LPE. This aminothiazole synthesis, comprising four generations of products **35**, **41**, **44** and **45**, could be optimised, automated and performed on a robotic system.

3.4.4. Synthesis of a pyrazole library using purification by solid phase extraction

Another example of a multistep synthesis using only PSQ purification procedures was described by *Hodges et al.*[21] In the first step, 1,3-diketone **47** was reacted with an excess of hydrazine **48** in the presence of morpholinomethyl-polystyrene (**49**). Removal of excess **48** by quenching with polymer-bound isocyanate **50** afforded pyrazole **51** with a purity of 97% (HPLC) but in quite moderate yield (48%). In the second step, **51** was converted into a mixed anhydride (by treatment with isobutyl chloroformate in the presence of **49**) and subsequently transformed *in situ* into **53** by reaction with an amine. Addition of the polymer-bound reagents **50** and **52** removed excesses of chloroformate and amine by filtration and **53** was isolated in 75% yield with a purity of 97% (HPLC).

Scheme 3.4.5

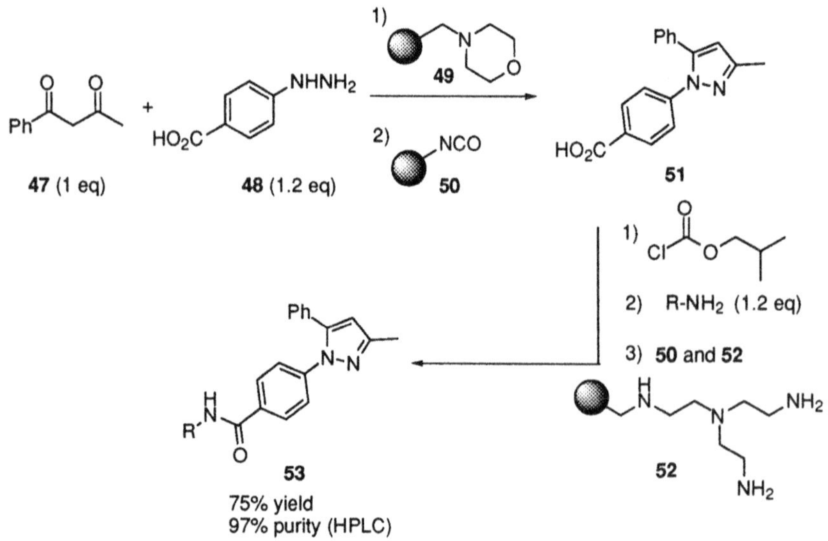

3.5. Applications of polymer-bound reagents

3.5.1. Polymer-supported *Burgess* reagent

The *Burgess* reagent[22] (**54**, *Figure 3.5.1*) is a versatile and mild reagent for the cyclodehydration of β-hydroxyamides or thioamides to produce oxazolines or thiazolines.

Figure 3.5.1

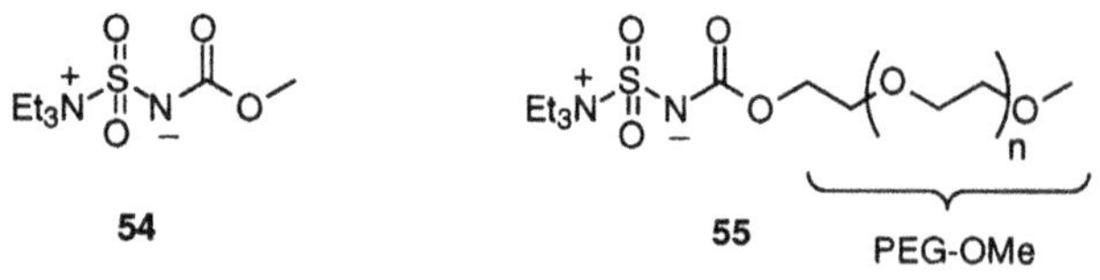

The problem related with this commercially available compound is its sensitivity towards oxidation as well as moisture. *Wipf et al.*[23] reported the synthesis of a PEG-grafted *Burgess* reagent (**55**) which proved to be highly effective for cyclodehydration reactions as shown in *Scheme 3.5.1*.

Scheme 3.5.1

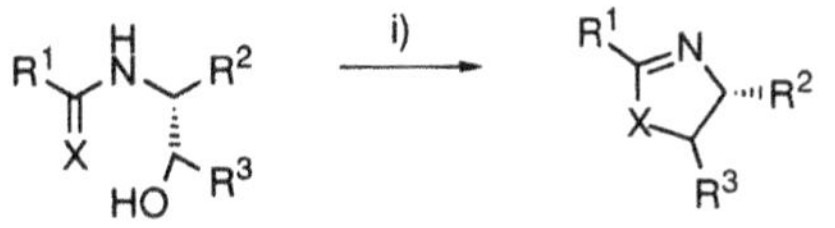

(X = O, S)

Reagents and conditions: i) **55**, dioxane/THF (1:1), 85°C, 3h

Cyclodehydrations with PEG-bound *Burgess* reagents proceeded in high yields (76-98%), with little (<2%) epimerisation and with fewer side-products. Furthermore, **55** is easily prepared on a large scale and shows an increased stability compared to the standard reagent **54**.

3.5.2. Polymer-supported silyl enol ethers

A library of γ-amino alcohols has been synthesised using polymer-supported silyl enol ethers. The strategy followed by *Kobayashi et al.*[24] is outlined in *Scheme 3.5.2*.

Scheme 3.5.2

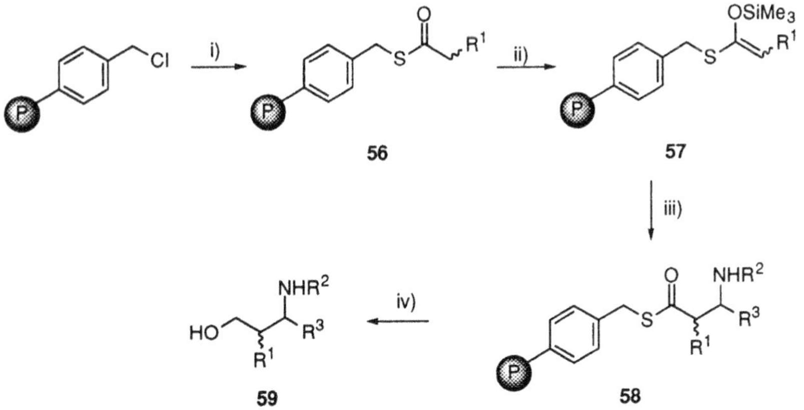

Reagents and conditions: i) R^1CH_2COSK, DMF; ii) TMSOTf, NEt$_3$, CH$_2$Cl$_2$;
iii) imine, Sc(OTf)$_3$ (10 mol%), CH$_2$Cl$_2$; iv) LiBH$_4$, Et$_2$O, r.t.

Starting from commercially available *Merrifield* resin, silyl enol ether **57** was generated in two steps. *Lewis* acid catalysed addition of **57** to a variety of imines gave amino thioesters **58** which upon reduction with LiBH$_4$ afforded amino alcohols **59** in moderate to good yields (50-80%). After reduction, the resin was recovered in its thiol form and could be further acylated and re-used.

229

3.5.3. Polymer-supported oxazolidinones

Enantiomerically pure oxazolines and oxazolidinones have found widespread application in organic synthesis as chiral auxiliaries. They have been mainly used for the synthesis of enantiomerically pure amino acids[25] but also as chiral auxiliaries to produce non-racemic enolates as pioneered by *Evans*.[26] The reaction types proceeding with high stereocontrol include enolate alkylation, enolate oxidation, enolate halogenation, enolate amination, enolate acylation, aldol reaction and *Diels-Alder* reactions.

Scheme 3.5.3

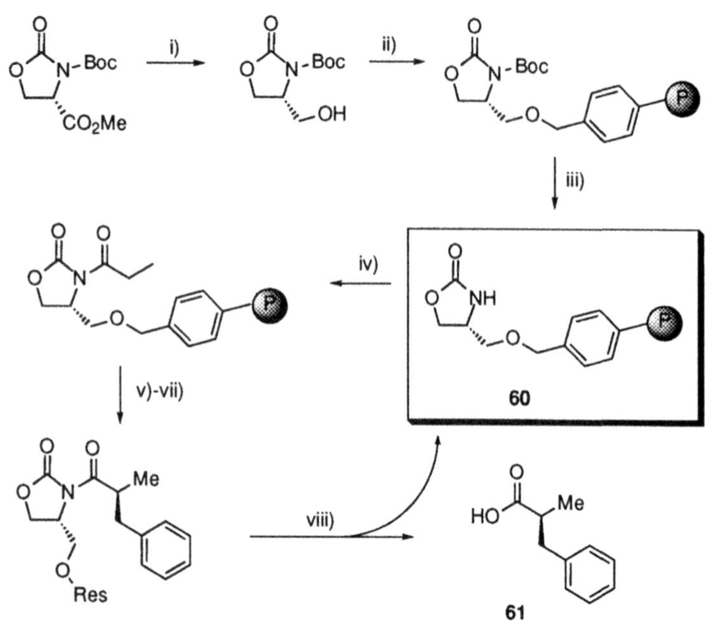

Reagents and conditions: i) NaBH₄, EtOH, 0°C; ii) a) KH, DMF, 0°C, b) *Merrifield* resin, 18-Crown-6, 80°C; iii) HCl (dil.), CH₂Cl₂; iv) propionic anhydride, NEt₃, DMAP, THF, reflux; v) LDA (2 eq), THF, 0°C; vi) BnBr (2 eq); vii) NH₄Cl (sat.); viii) LiOH, THF, H₂O.

Shuttleworth and *Allin*[27] reported the use of polystyrene-supported oxazolidinone **60** for the synthesis of chiral carboxylic acids **61** (*Scheme 3.5.3*). Preparation of **60** was accomplished in five steps. Acylation, enolate alkylation and subsequent hydrolysis afforded carboxylic acid **61** in 42% yield and with an enantiomeric excess of 96%. The chiral auxiliary **60** was recovered and could be re-used.

Following the same procedure, *Burgess et al.*[28] reported the synthesis of chiral primary alcohols. Enantiomeric excesses of the products cleaved by reduction with LiBH₄ ranged from 71-90 %.

3.5.4. Polymer-supported triphenylphosphines

Triphenylphosphine is a versatile reagent in a wide range of organic reactions such as *Wittig* reactions, *Mitsunobu* reactions,[29,30] *Staudinger* reactions, reductive cleavage of disulfides, and aza-*Wittig* reactions. The problem related with the use of triphenylphosphine is, however, the separation of reaction products from the simultaneously formed triphenylphosphine oxide. The use of polymer-supported triphenylphosphine[31] allows straightforward removal of either excess triphenylphosphine or triphenylphosphine oxide by filtration as illustrated by the following examples.

Wittig reactions[32]

Transformation of the polymer-supported triphenylphosphine into phosphonium salt **62** followed by reduction and acylation yielded the corresponding phosphonium salt **63** which was converted under different reaction conditions[33] into products **64-66** as shown in *Scheme 3.5.4*.

Scheme 3.5.4

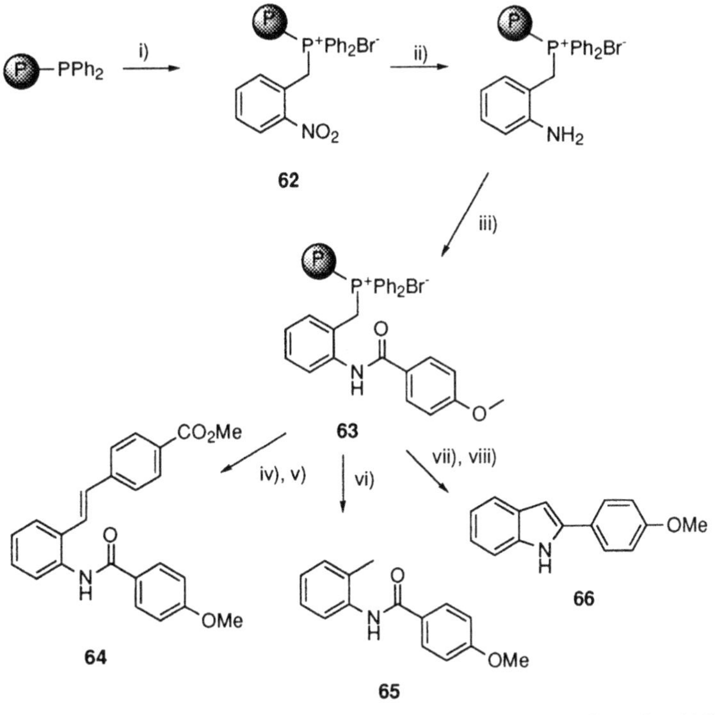

Reagents and conditions: i) 2-nitrobenzyl bromide, DMF, 70°C; ii) a) $Na_2S_2O_4$, EtOH, reflux, b) HBr, MeOH, dioxane; iii) PMB chloride, py, CH_2Cl_2; iv) methyl 4-formylbenzoate, NaOMe, MeOH, reflux; v) aminomethyl resin, AcOH, MeOH/dioxane; vi) NaOMe, MeOH, reflux; vii) toluene, DMF, distillation; viii) KOtBu, reflux.

Thus, conditions developed for the *Wittig* cleavage reaction were found to be suitable for a wide range of substituted benzyl phosphonium salts and aromatic aldehydes. Removal of excess aldehyde by quenching with an aminomethyl derived resin afforded styrene **64** as a 3:1 *E/Z* mixture in a 82% overall yield. On the other hand, hydrolysis of **63** with sodium methoxide gave **65** in 81% yield. Application of an intramolecular *Wittig* reaction[34] under strictly anhydrous conditions gave indole **66** in 78% yield.

The transformations and cleavage reactions of polymer-supported phosphonium salts of type **63** can be considered as nice examples of traceless linker strategies.

Mitsunobu reactions

In recent years, the *Mitsunobu* reaction[29] has been studied extensively in solution and has found many useful applications. In an effort to overcome many of the problems associated with solution synthesis, the use of highly loaded triphenylphosphine polystyrene-derived resin[31] represents a simple solution to the problem of purification. Thus, reaction of a series of alcohols of type **67** with carboxylic acids in the presence of diethyl azodicarboxylate (DEAD) and PS-PPh$_2$ in THF between 0°C and room temperature resulted in the clean formation of the expected esters of type **68**.[35] Filtration through a plug of Al$_2$O$_3$ gave the essentially pure products in high yields (70-90%).

Scheme 3.5.5

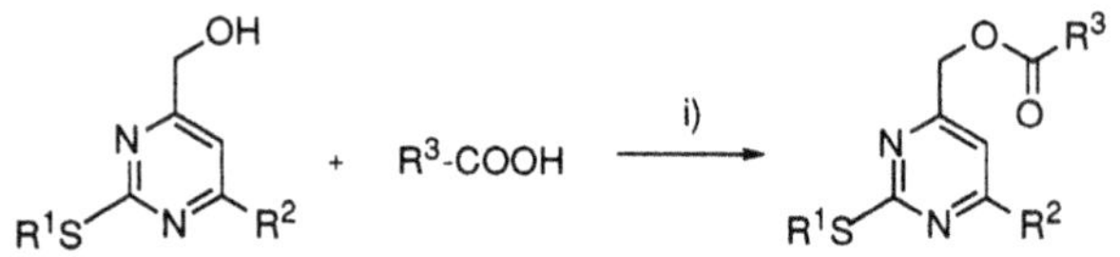

67   **68**

70-90% yield

Reagents and conditions: i) PS-PPh$_2$, DEAD, THF, 0°C-r.t.

Staudinger reaction of azido nucleosides for the synthesis of amino nucleosides

Holletz et al.[36] reported the use of polymer-supported triphenylphosphine for the reduction of azido groups in the sugar moiety of nucleosides as shown in *Scheme 3.5.6*.

Scheme 3.5.6

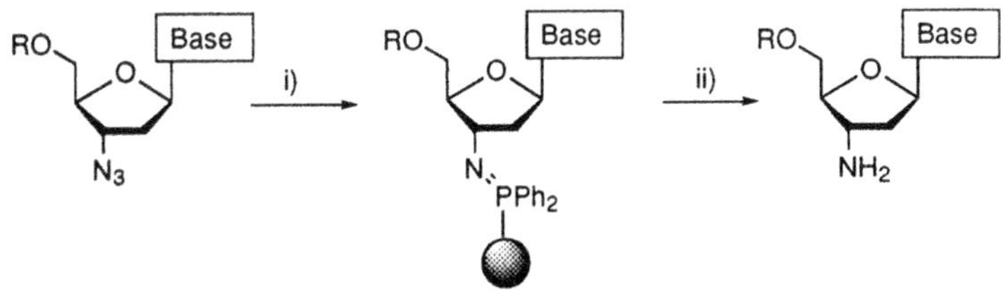

Reagents and conditions: i) PS-PPh$_2$, dioxane or py, 3h, r.t., 100%; ii) dioxane / NH$_3$, r.t., 2-12h, 89-100%

In a *Staudinger*-type reaction,[30] the azido group was converted into the corresponding phosphoryl imine which after hydrolysis gave the desired amino nucleosides. The starting azido nucleosides were used either with protected or unprotected bases with good results. Filtration of the crude mixture led to pure products in in excellent yields (89-100%).

Reductive cleavage of disulfides

The reductive cleavage of disulfides has been accomplished in a number of ways including reductions with NaBH$_4$, LiAlH$_4$ or Zn/AcOH. The use of triphenylphosphine in aqueous organic solvents appeared as a particularly attractive approach for disulfide cleavage because of its mild conditions and high yields. Although the reaction is quite clean and affords little by-products it still necessitates the separation of the thiol from the triphenylphosphine oxide. Once again, the use of a polymer-bound triphenylphosphine solved the problem of purification. *Amos et al.*[37] reported the reductive cleavage of a range of disulfides **(69)** using resin-bound triphenylphosphine in a mixture of THF and 0.1N HCl (*Scheme 3.5.7*). Although the reaction rates depended greatly on the substitution pattern of the disulfide, the yields were generally very high (76-99%). As thiols are sensitive to oxidation, a one-pot reductive cleavage/alkylation procedure was also elaborated. Thus, after reductive cleavage, the intermediate thiols were directly alkylated or acylated affording the corresponding sulfides **(71)** in high yields[38] (89-95%).

Scheme 3.5.7

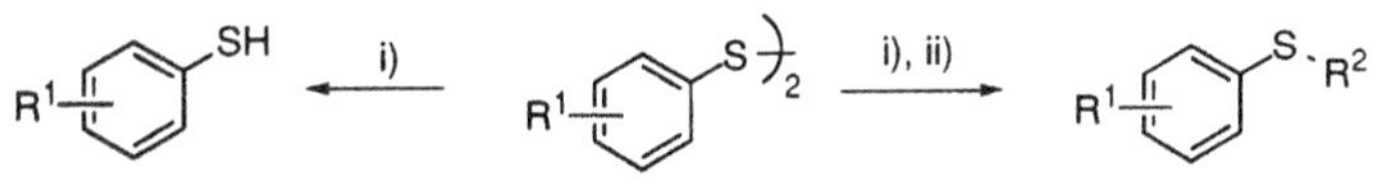

Reagents and conditions: i) PS-PPh$_2$, THF/0.1 N HCl (12:1); ii) R^2-X, NEt$_3$

3.5.5. Simultaneous multistep synthesis using several polymer-supported reagents

A still not fully explored advantage of polymer-supported reagents constitutes the simultaneous use of several polymer-bound reagents in the same reaction compartment, assuming that the reactive groups attached to the polymers would only react with the molecules that remain in solution. This concept allows the combination of two or more otherwise mutually incompatible reagents to be used in the same reaction flask. *Parlow*[39] at Monsanto reported a three step solution phase synthesis of pyrazoles using exclusively polymer-supported reagents and carried out sequentially or in a one-pot procedure as depicted in *Scheme 3.5.8*. First, benzylic alcohol **72** was oxidised to acetophenone **73** with polymer-bound pyridinium dichromate, followed by perbromide oxidation to yield bromomethyl ketone **74** and substitution reaction to give **75**. The stepwise reaction gave an overall yield of 42% whereas the one-pot procedure gave 48% yield.

Scheme 3.5.8

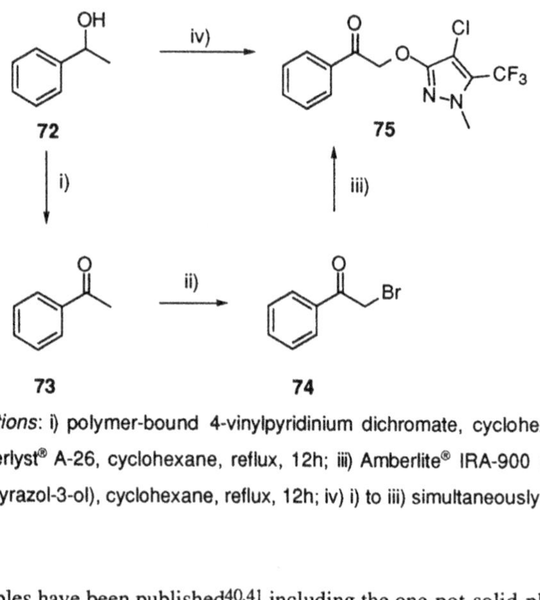

Reagents and conditions: i) polymer-bound 4-vinylpyridinium dichromate, cyclohexane, reflux, 12h; ii) perbromide on Amberlyst® A-26, cyclohexane, reflux, 12h; iii) Amberlite® IRA-900 (4-chloro-1-methyl-5-(trifluoromethyl)-1H-pyrazol-3-ol), cyclohexane, reflux, 12h; iv) i) to iii) simultaneously.

Several other examples have been published[40,41] including the one-pot solid-phase cleavage of α-diols to primary alcohols[42] as shown in *Scheme 3.5.9*. In this successful application, a 1:1 mixture of periodate- and borohydride-resins were filled into a column and an aqueous solution of nucleoside was slowly pumped through. The intermediately formed, unstable dialdehydes were reduced *in situ* to the corresponding diols. Yields were good for the base being adenine (70%), cytosine (79%) and uracil (73%) and moderate for guanidine (40%).

Scheme 3.5.9

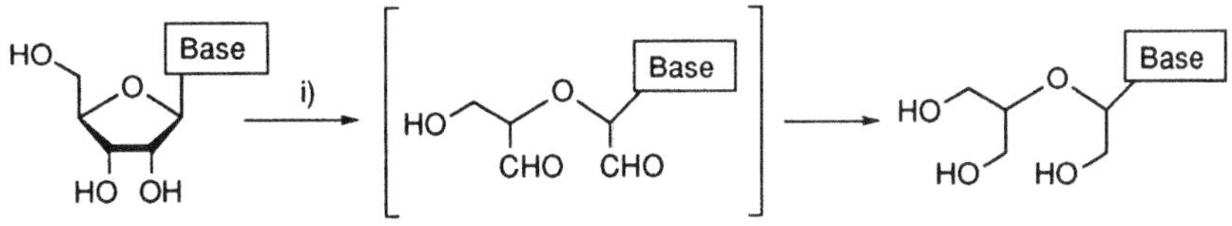

Reagents and conditions: i) periodate and borohydride on Amberlyst® A-27 (1:1), H_2O, r.t.

3.6. Applications of polymer-bound catalysts

3.6.1. Catalysts for oxidations

Sharpless **asymmetric dihydroxylation**

The *Sharpless* catalytic asymmetric dihydroxylation of olefins,[43] using catalytic amounts of osmium tetroxide in the presence of chinchona alkaloid derivatives, allowed access to a variety of enantiomerically pure 1,2-diols (*Scheme 3.6.1*).

Scheme 3.6.1

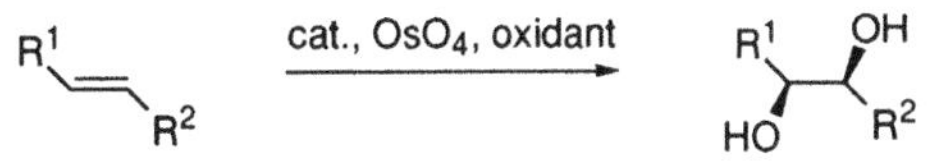

Utilisation of a polymer-bound catalyst in this reaction would allow to reduce the cost of preparation by facile recycling of the catalyst. Initial investigations in this area have been reported by *Sharpless*[44] who was able to oxidise trans-stilbene in 81-87% yield and with 85-93% enantiomeric excess using the polymer-supported catalyst **76** (*Figure 3.6.1*), OsO_4 and *N*-methylmorpholine *N*-oxide (NMO). The yields were high but enantiomeric excesses were inferior to the corresponding homogeneous reactions.

Figure 3.6.1

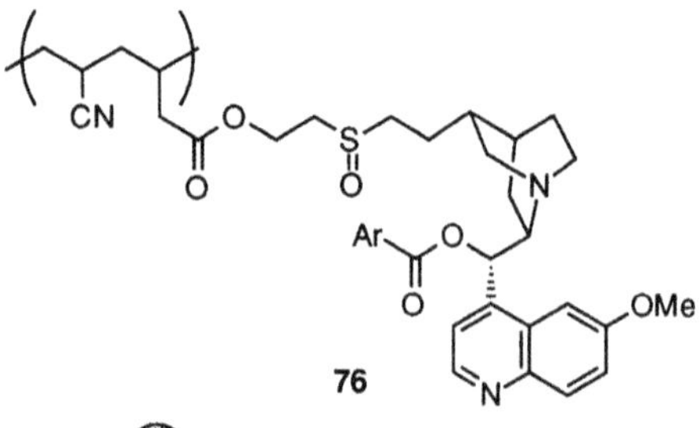

76

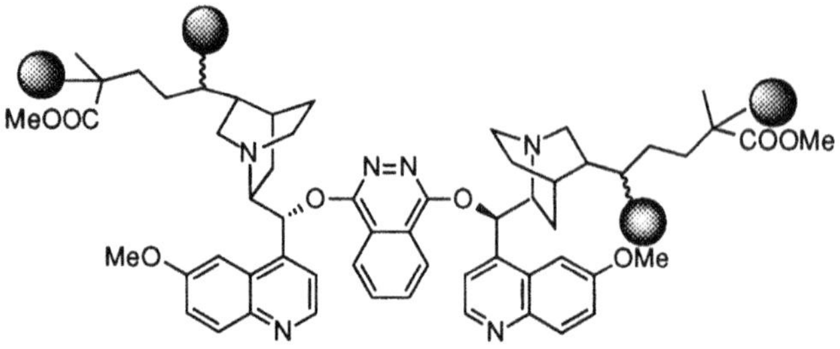

77

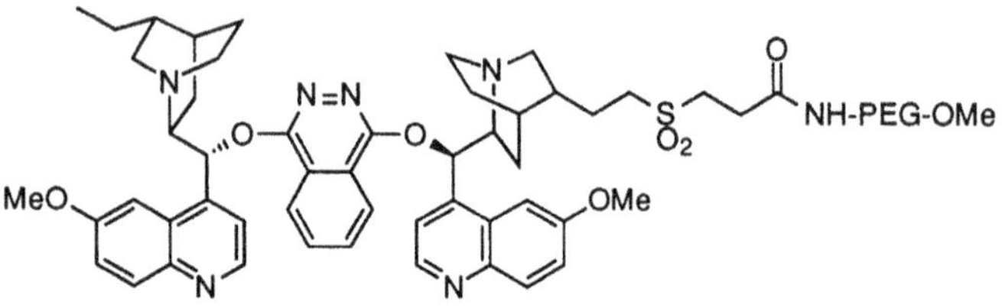

78

Several other catalysts have been studied[45] showing similar results. The only polymer-bound catalyst exhibiting excellent e.e. values, a co-polymer of methyl acrylate and a 1,4-bis(9-O-quininyl)phthalazine ((QN)$_2$-PHAL) derivative, was developed by *Song et al.*[46] Reaction of trans-stilbene or methyl cinnamate using **77** (0.02 eq), OsO$_4$ (0.01 eq) and K$_3$Fe(CN)$_6$-K$_2$CO$_3$ as co-oxidant gave dihydroxylated products in 82-93% yield and with e.e. values >99%.

In a similar approach, a (DHQD)$_2$PHAL ligand was bound to a MeO-PEG-NH$_2$ resin to give the soluble catalyst **78**.[47] Dihydroxylation of a variety of substrates using **78**, OsO$_4$ and K$_3$Fe(CN)$_6$ proceeded in good to excellent yields (83-95%) and high enantiomeric excesses (>97%). The polymer-bound catalyst is completely soluble in *tert*-butanol/water or acetone/water and accelerates the reaction similar to the free ligand. In addition, precipitation with diethyl ether allows virtually quantitative recovery of the catalyst.

Sharpless epoxidation of allylic alcohols

The asymmetric epoxidation of allylic alcohols developed by *Sharpless*[48] is a key reaction in asymmetric organic synthesis (*Scheme 3.6.2*). Enantiomeric purities of the obtained products are usually very high (>98% e.e.).

Scheme 3.6.2

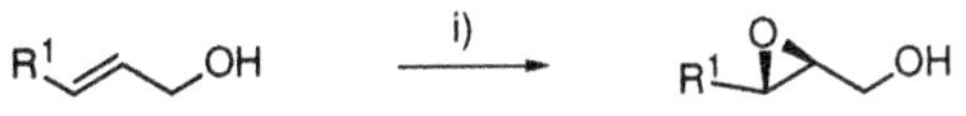

Reagents and conditions: i) L-(+)-diethyl tartrate, Ti(O-iPr)₄, tBuOOH, molecular sieves 4Å, CH₂Cl₂, -20°C

Thus, polymeric tartrate esters **79** (*Scheme 3.6.3*) were synthesised and tested as catalysts in asymmetric *Sharpless*-epoxidations.

Scheme 3.6.3

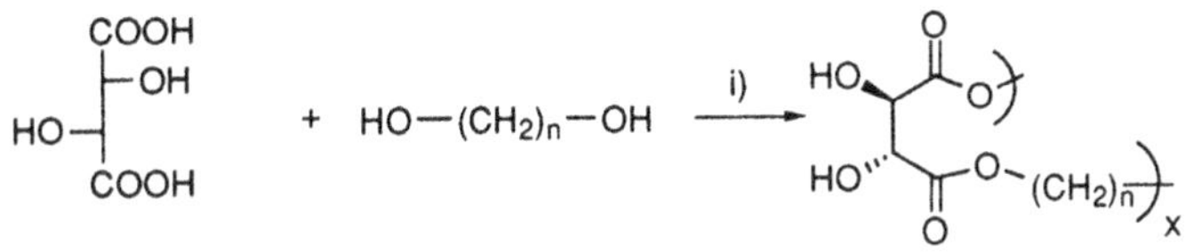

| L-(+)-tartaric acid | n = 2, 6, 8, 12 | **79** |

Reagents and conditions: i) p-toluene sulfonic acid, ca. 120°C, 3d

Unfortunately, epoxidations of trans-hex-2-en-1-ol using the polymeric catalysts **79** gave unsatisfactory results. Yields of isolated products varied between 44 and 80%, whereas enantiomeric excesses ranged from 8-79% e.e. depending on the amount of catalyst and titanium tetraisopropoxide used. A homogeneous experiment using L-(+)-diethyl tartrate gave the desired product with 91% e.e. It has to be noted that the polymer-supported version of the *Sharpless* epoxidation was clearly less effective than the corresponding reaction in solution.[49]

Hydroxylation of benzene using a polymer-supported VO²⁺ catalyst

The direct formation of phenols from benzene has been studied with several transition metal complexes. However, the complexes are usually used in stoichiometric amounts and cannot be recycled.

237

Based on this observation, *Kumar et al.*[50] have developed a polystyrene-based vanadyl complex
80 (*Scheme 3.6.4*), generated from diethanolamine, L-tyrosine, salicyl aldehyde and $VO(acac)_2$,
to perform hydroxylations of benzene using 1 mol% of **80** and 1 equivalent of H_2O_2. After stirring
at 65°C for 6h, a conversion of 30% was detected. The phenol could be easily separated from
unreacted benzene by distillation.

Scheme 3.6.4

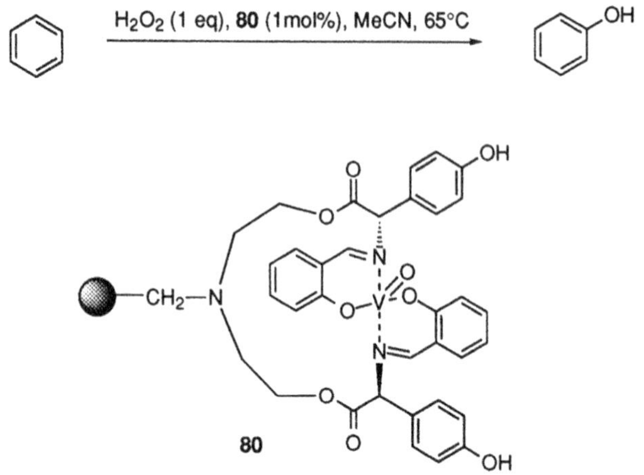

Although the reaction proceeded with only 30% conversion it offers some potential for further
investigations. The catalyst could be recovered by simple filtration and re-used another ten times
before deterioration was observed.

3.6.2. Catalysts for reductions

The use of oxazaborolidine catalysts in asymmetric reductions of ketones has been extensively
studied by *Corey et al.*[51] Recently, a polymer-bound oxazaborolidine[45] (*Scheme 3.6.5*) was
developed and successfully used in the asymmetric reduction of acetophenone. 1-Phenylethanol
was obtained in 93% yield and 98% enantiomeric excess.

Scheme 3.6.5

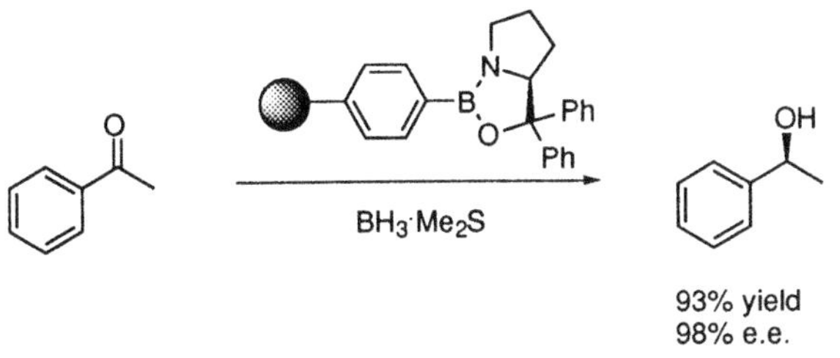

93% yield
98% e.e.

3.6.3. Catalysts for C-C bond formation

Lanthanide(III) catalysts supported on ion exchange resins

Lanthanide-mediated reactions have found extensive use in organic synthesis due to their high selectivities. Lanthanide triflates constitute recyclable *Lewis* acids tolerating aqueous reaction conditions and catalysing a variety of important organic reactions under mild conditions.

Wang et al. studied the preparation and synthetic use of lanthanide(III) catalysts supported on ion exchange resins.[52] A number of commercially available ion exchange resins were loaded with lanthanide(III) ions and their capacity of catalysing *Mukaiyama* aldol reactions has been tested. The reaction of benzaldehyde with a silyl enol ether in dichloromethane in the presence of Yb(III)-loaded resin shown in *Scheme 3.6.6* served as a model reaction for the classification of the different resins. In summary, it was found that strongly acidic ion exchange resins with large pores and surface area such as Amberlyst® XN-1010 or Amberlyst® 15 gave best results.

Scheme 3.6.6

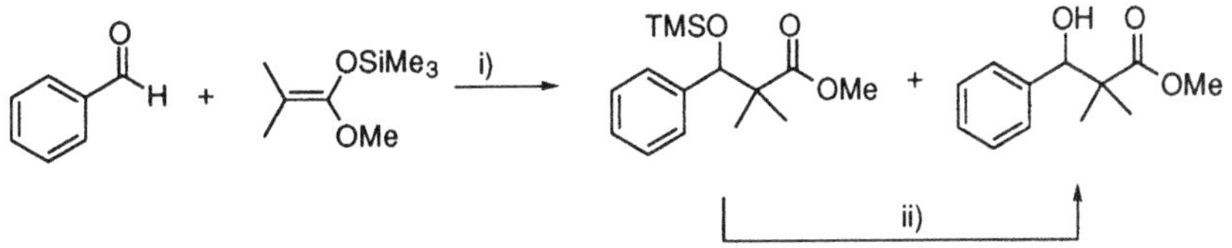

Reagents and conditions: i) Lanthanide(III) resin, CH$_2$Cl$_2$; ii) TFA, CH$_2$Cl$_2$

The reaction afforded a mixture of free hydroxyl aldol derivative and silylated product, easily desilylated with TFA in dichloromethane, with overall yields ranging from 71-83% independently of the metal ion. Yb(III) on Amberlyst® XN-1010 has been used ten times consecutively in this type of aldol reaction without any loss of catalytic activity.

Lanthanide(III) on ion exchange resins catalyse *Mukaiyama* aldol reactions in aqueous media, acetalisations, additions of silyl enol ethers to imines, aza-*Diels-Alder* reactions and the ring-opening of epoxides with alcohols as depicted in *Scheme 3.6.7*.

Scheme 3.6.7

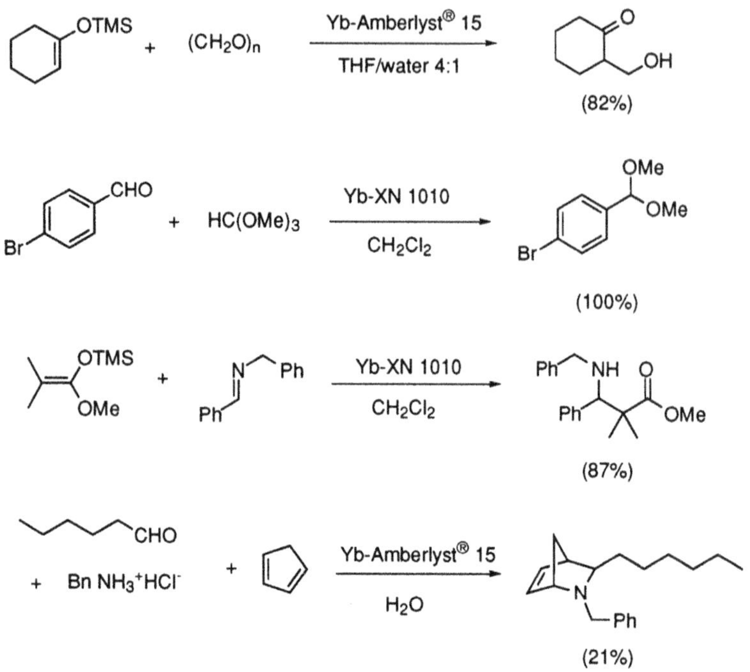

Chiral solid-supported amino alcohols as ligands for *Lewis* acid catalysts

Many supported or heterogeneous catalysts used for *Diels-Alder* reactions are known to give better results than their non-supported analogues. Nevertheless, chiral catalysts for asymmetric *Diels-Alder* reactions are scarce. *Mayoral, Luis* and coworkers[53] studied the use of a variety of chiral polymer-bound amino alcohols as catalysts in cycloaddition reactions. Reaction of cyclopentadiene with methacrolein in the presence of (*S*)-prolinol-derived resin **81** proceeded with excellent yield (98%) but poor enantioselectivity (14% e.e.) as shown in *Scheme 3.6.8*. Once again, extrapolation from solution phase chemistry to a solid-supported reaction proved difficult.

Scheme 3.6.8

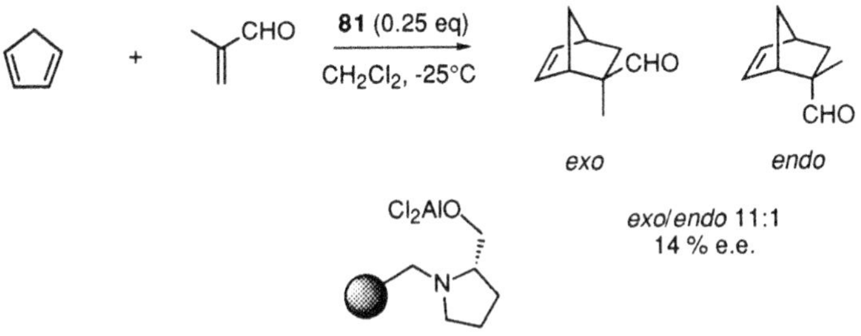

exo

endo

exo/endo 11:1
14 % e.e.

81

Enantioselective addition of diethylzinc to aldehydes

Soai et al.[54] reported the use of immobilised *N*-butylnorephedrine (**82**) as a catalyst in enantioselective addition of diethylzinc to various aldehydes producing secondary alcohols **83**.

Scheme 3.6.9

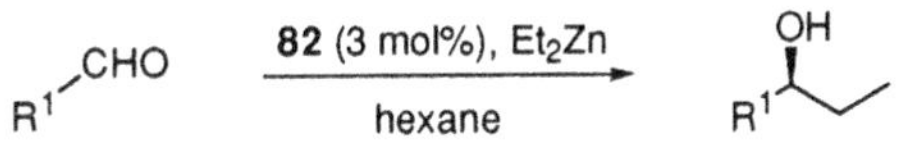

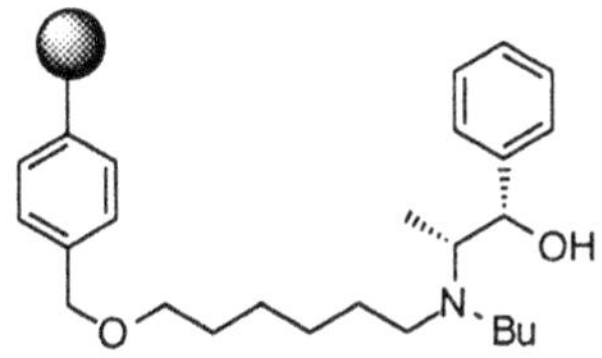

82

It is interesting to note that the authors reported best results when the reaction was performed in hexane affording **73** in good yields (53-91%) and with enantiomeric excesses between 51 and 82% (*Scheme 3.6.9*). Although the resin did not swell in this solvent it apparently still favoured interactions between the substrates and the catalyst.

During the last few years, several resin-bound catalysts have been developed for the enantioselective addition of diethylzinc to aldehydes. The related results have been summarised by *Shuttleworth* in a recent review.[45]

Immobilised Ti-TADDOL complexes

$\alpha,\alpha,\alpha',\alpha'$-Tetraaryl-1,3-dioxolane-4,5-dimethanol (TADDOL) derivatives of the general structure **84**, have been widely used as chiral ligands for the preparation of homogeneous enantioselective catalysts.[55]

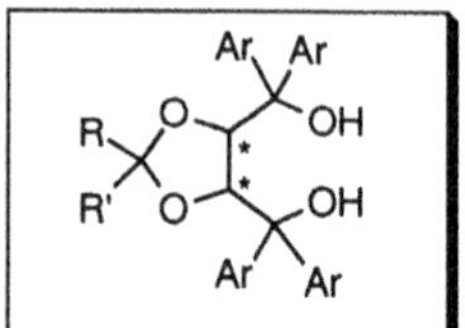

84

Recently *Mayoral* and coworkers[56] have reported the grafting of a TADDOL derivative onto several types of polystyrene resins. The immobilised TADDOLs were subsequently transformed into their chiral titanium complexes as shown in *Scheme 3.6.10*.

Scheme 3.6.10

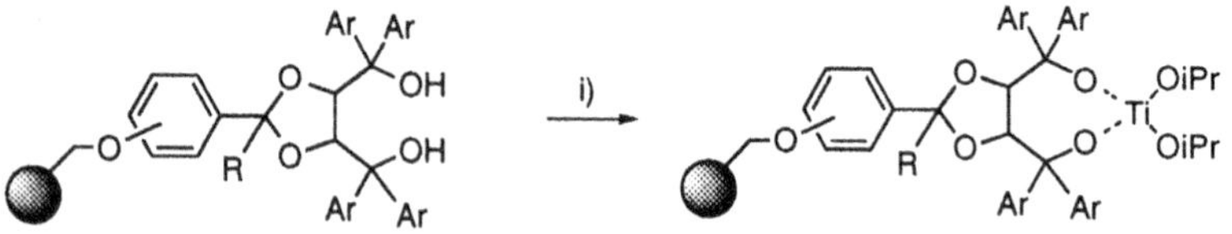

85   **86**

Reagents and conditions: i) TiCl$_2$(OiPr)$_2$, toluene/CCl$_4$, reflux, 24h

Diels-Alder cycloaddition of cyclopentadiene with 3-crotonoyl-1,3-oxazolidin-2-one **87** using immobilised Ti-TADDOLates has been studied in order to determine their catalytic activity (see *Scheme 3.6.11*).

Scheme 3.6.11

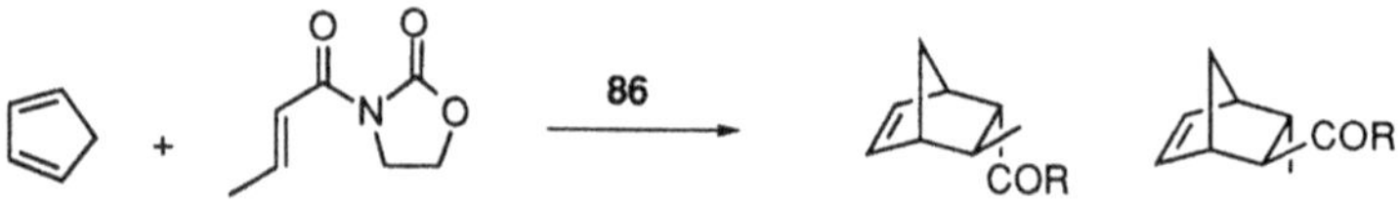

87   **88**   **89**

Cycloaddition reactions proceeded at room temperature with a 2.5-fold excess of cyclopentadiene. Conversions depended very much on the resin type and ranged from 30-100%. *Endo/exo* selectivity amounted to about 4:1 (**88**:**89**) which is similar to the one reported for the homogeneous phase reaction showing that the polymer-supported TADDOLs **86** indeed possessed catalytic activity. The observed enantioselectivity, however, was very low indicating a strong influence of the polymeric support. In one example, an enantiomeric excess of 25% was obtained. The analogous homogeneous reaction of structurally similar TADDOLs gave 33 and 38% e.e., respectively.

Very recently, *Seebach* and coworkers reported a different approach for the immobilisation of TADDOL derivatives.[57] Dendritically substituted TADDOLs (**90**) were co-polymerised with styrene in a suspension polymerisation procedure and subsequently transformed into their titanium complexes.

9 0

These polymer-supported catalysts were tested in the addition of diethyl zinc to benzaldehyde (*Scheme 3.6.9*). Additions catalysed with polymer-bound TADDOLates were quantitative and proceeded faster than the analogous homogeneous reaction. Comparable enantioselectivities were obtained under homogeneous (98% e.e.) and heterogeneous conditions (96% e.e.).

A polymer-bound catalyst of type **86** gave also quantitative conversion and 96% e.e. whereas the reaction rate was much smaller. The dendritically substituted TADDOLate seems more accessible for the substrates than the "traditionally" linked derivatives.

3.6.4. Miscellaneous

Polymer-bound palladium-phosphine complexes for allylic substitution reactions

The synthesis of a triphenylphosphine functionalised TentaGel® was reported by *Uozumi, Hayashi* and coworkers.[58] The resin was loaded with Pd^{2+} and used as a resin-bound catalyst in allylic substitution reactions as shown in *Scheme 3.6.12*.

Scheme 3.6.12

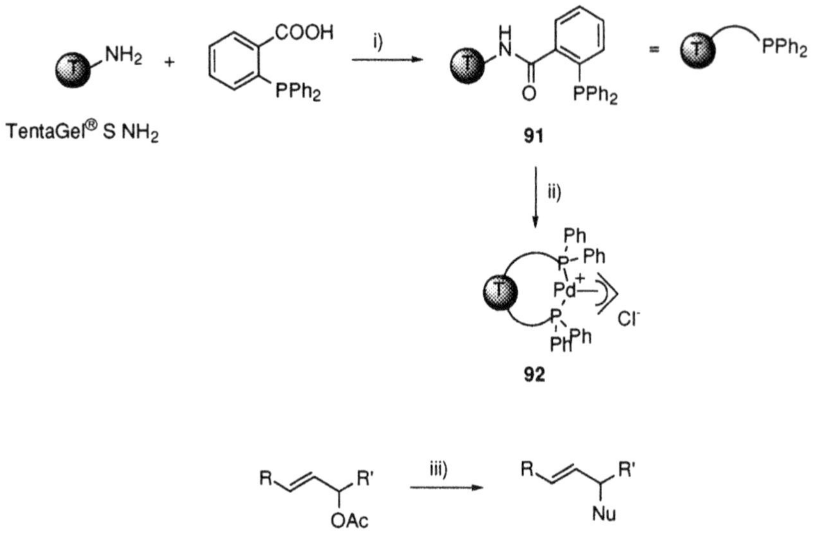

Reagents and conditions: i) EDCI (3 eq), HOBt (4 eq), DMF, r.t., 4h; ii) [PdCl(p-C$_3$H$_5$)]$_2$, CH$_2$Cl$_2$, r.t.; iii) Nu (1.5 eq), base (4.5 eq), **82** (2 mol%), H$_2$O, r.t., 24h

The range of nucleophiles usable in this substitution reaction was very broad. Carbon nucleophiles such as acetoacetate, diethyl malonate, 3-methyl-2,4-pentanedione gave alkylated products in 98%, 94% and 86% yield, respectively. Nitrogen containing nucleophiles (hydrochloride salts of amino acids) gave *N*-allylated products in 90-98% yield. Sodium phenylsulfinate and sodium azide also reacted in high yields (86% and 79%, respectively). The catalyst could be readily recovered and re-used. No loss of activity was observed over 7 continuous runs.

Synthesis of a quinoline library using a polymer-supported scandium catalyst

The three component condensation employing aldehydes, anilines and electron-rich olefins leading to substituted tetrahydroquinolines **93** constitutes an attractive approach for combinatorial and parallel synthesis (*Scheme 3.6.13*).

Scheme 3.6.13

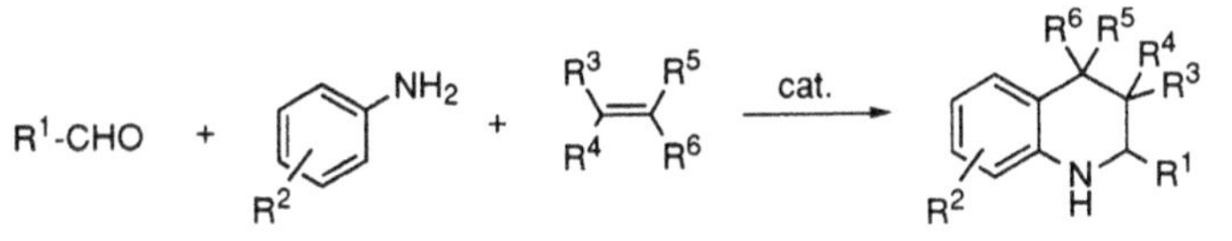

Kobayashi et al.[59] reported the synthesis of immobilised scandium catalyst **94** starting from polyacrylonitrile as shown in *Scheme 3.6.14*.

Scheme 3.6.14

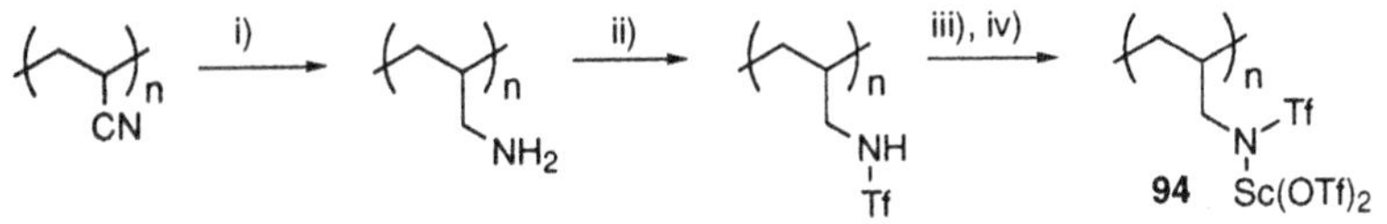

Reagents and conditions: i) BH₃ SMe₂, diglyme, 150°C, 36h; ii) Tf₂O, NEt₃, 1,2-dichloroethane, 60°C, 10h; iii) KH; vi) Sc(OTf)₃

94 is partially soluble in a mixture of dichloromethane and acetonitrile but can conveniently be precipitated by addition of hexane. Using **94** as a catalyst, a quinoline library was generated in high yields (65-100%) and good diastereoselectivities whereby the catalyst could be recovered and re-used without any loss of activity.

245

3.7. Solution chemistry using temporary trapping of intermediates on solid support ("intermediate catch" or "resin capture")

3.7.1. *Ugi* four component condensation using a polymer-bound carboxylic acid

The *Ugi* four component condensation (4CC) using 1-isocyanocyclohexene as a convertible isonitrile has already been mentioned in *Chapter 3.2.3*. In this paragraph we want to discuss the advantages of using one of the components linked to a solid support.

Armstrong et al.[60] reported a modification of the original synthesis with a polymer-bound carboxylic acid as shown in *Scheme 3.7.1*. Succinic anhydride was reacted with either Rink or Wang resin to provide a spacer and a free carboxyl group as one of the 4CC inputs. The *Ugi*-4CC was then performed with an amine, an aldehyde and 1-isocyanocyclohexene at room temperature.

Scheme 3.7.1

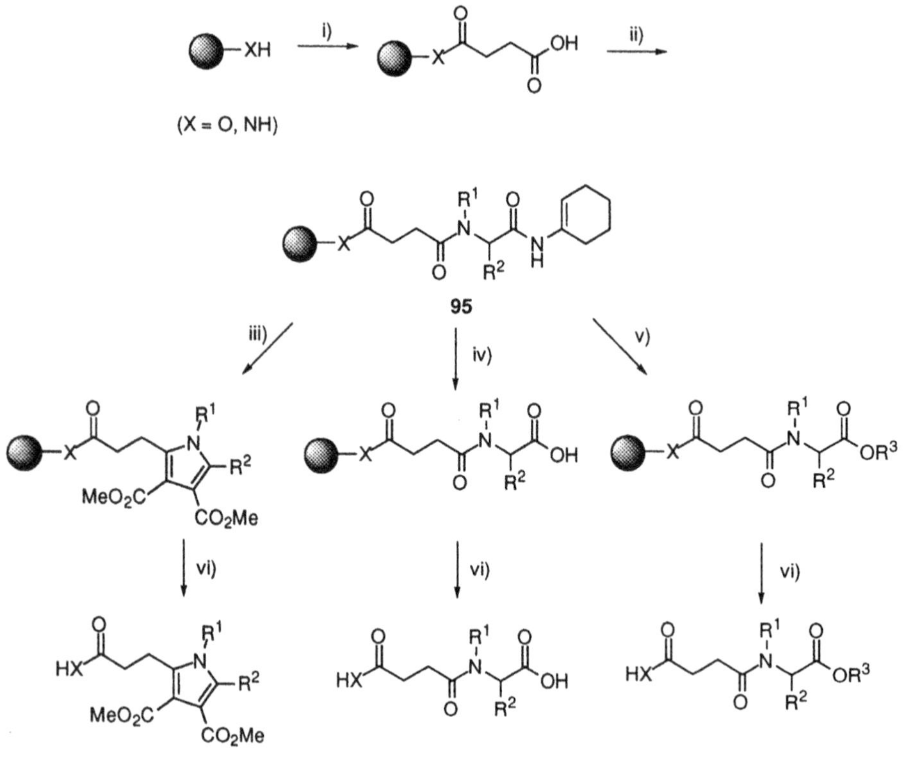

(X = O, NH)

95

Reagents and conditions: i) succinic anhydride, py; ii) R^1-NH_2; R^2-CHO, isonitrile; iii) HCl, DMAD, toluene, 100°C; iv) HCl/THF, r.t.; v) AcCl, R^3-OH, 55°C; vi) TFA/CH$_2$Cl$_2$ (20%)

The advantage of having the product linked to the solid support is obvious. Excess reagent drove the reaction to completion and could easily be removed by a washing procedure. The *Ugi* product **95** could subsequently be modified as discussed previously and then cleaved from the resin.

3.7.2. Sequential *Ugi*-4CC and aza-*Wittig* reaction using a polymer-bound trapping reagent

A. Chucholowski et al.[61] elaborated a novel reaction sequence to produce benzodiazepine and benzodiazocine heterocycles combining an *Ugi*-4CC with an intramolecular aza-*Wittig* reaction. The idea was to have additional functional groups present in the *Ugi* product which would allow subsequent ring closures. Using the polyfunctional building blocks **96**-**99** (*Figure 3.7.1*) as inputs in the *Ugi*-reaction gave the cyclisation precursors of type **100** in variable yields. In order to purify these intermediates from the crude reaction mixture prior to cyclisation, polymer-bound triphenylphosphine was added at -10°C generating the intermediate polymer-bound phosphoryl-imines **101** which were purified by filtration. Unreacted reagents and by-products could subsequently be washed off as the resin-bound phosphorylimine does not further react and is stable at temperatures around -10-0°C. This process represents a nice example of the "intermediate catch" principle.

Figure 3.7.1

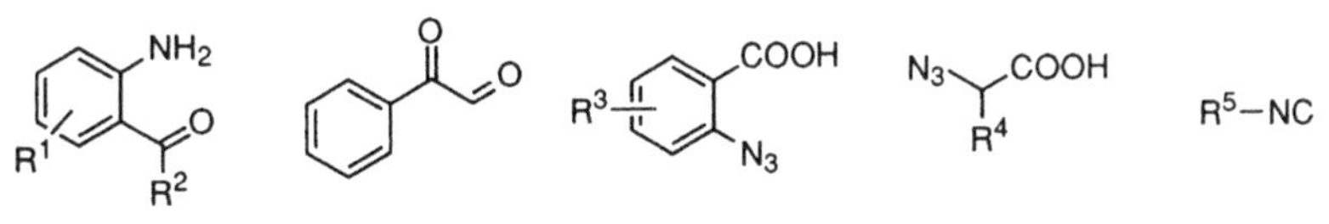

96	**97**	**98**	**99**

Warming up the reaction mixtures to 60-80°C led to intramolecular aza-*Wittig* reaction with simultaneous cleavage from the resin forming benzodiazepine derivatives **102** as outlined in *Scheme 3.7.2*. Thus, the triphenylphosphine functionalised polymer serves several purposes:

1) purification of the *Ugi* product after condensation;

2) conversion of the azido group into an iminophosphorane ready for subsequent aza-*Wittig* reaction; and

3) as a traceless linker.

Scheme 3.7.2

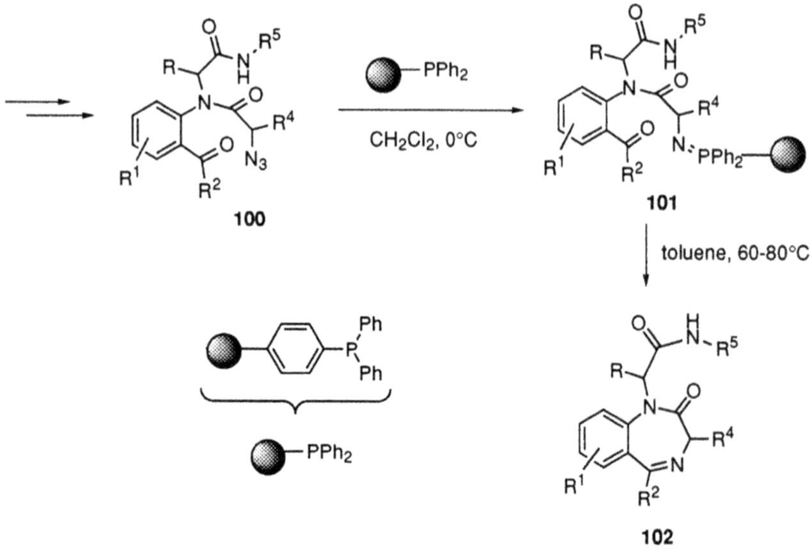

100

101

102

When the *Ugi* reaction was carried out with phenylglyoxal **97** and aromatic azide **98**, an alternative cyclisation to benzodiazepines **105** via intermediates **104** took place as depicted in *Scheme 3.7.3*.

Scheme 3.7.3

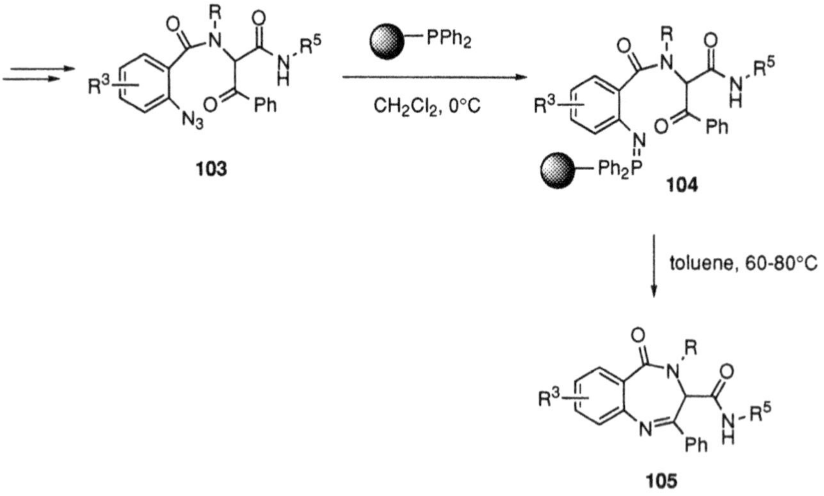

103

104

105

As an extension of this strategy, 8-membered heterocycles **108** were formed after *Ugi* reaction with aniline **96** and aromatic azide **98**, via intermediates **106** and **107** as shown in *Scheme 3.7.4*.

Scheme 3.7.4

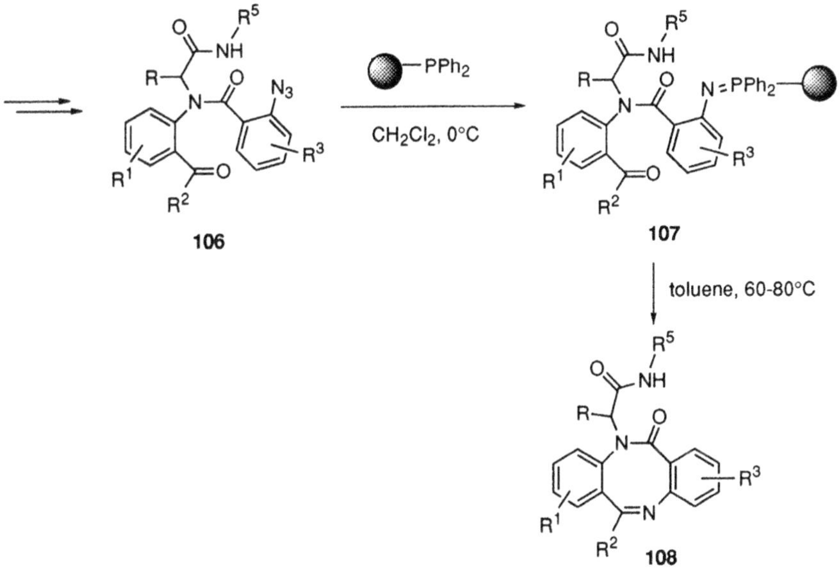

Overall yields of isolated products varied considerably and ranged from 12% to 70%, whereas the purities after removal of the by-products and excesses of reagents from the polymer-bound iminophosphorane by filtration usually exceeded 85% (determined by HPLC).

3.7.3. Multistep *Suzuki*-couplings using resin capture

Recently, *Armstrong et al.*[62,63] elaborated an elegant synthesis for the preparation of tetrasubstituted ethylenes using a resin capture process in the final purification step. Diboration of alkynes, as described previously by *Miyaura* and *Suzuki*,[64,65] gave diboronates **111** in excellent yields (*Scheme 3.7.5*). No isolation was necessary prior to *Suzuki* couplings as the starting alkynes **109** were unreactive under these conditions. Palladium catalysed coupling gave products **112** with high stereocontrol along with some disubstituted derivatives **113**. It is noteworthy that for $R^1 \neq R^2$ regioisomeric mixtures were obtained. Addition of a polymer-bound halide to the reaction mixture induced a second *Suzuki* coupling in which the desired compounds **114** remained linked to the solid support.

Scheme 3.7.5

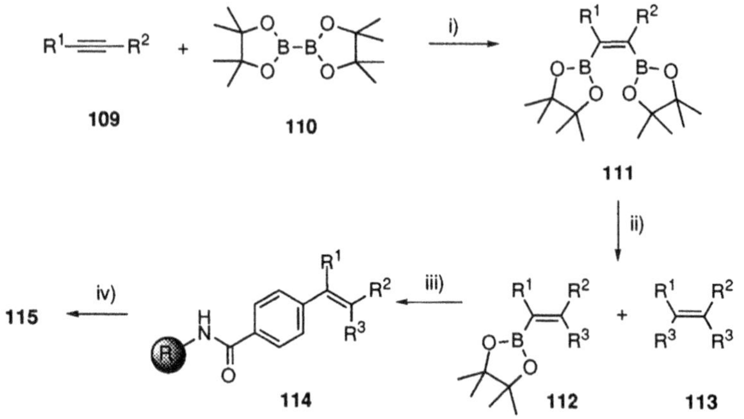

Reagents and conditions: i) Pt(PPh₃)₄, DMF, 0°C; ii) R₃-X (1.5 eq), Pd(PPh₃)₂Cl₂, KOH (3M), DME, 80°C; iii) *p*-iodobenzoate (*Rink* resin bound); iv) TFA

As a result of this resin capture process all by-products and excess reagents stayed in solution and could be removed by simple washing procedures. Furthermore, acidolytic cleavage from the resin with TFA gave products of type **115** which are structurally related to the antitumor agent tamoxifen (**116**) (*Figure 3.7.2*) used in treatments of breast cancer.

Figure 3.7.2

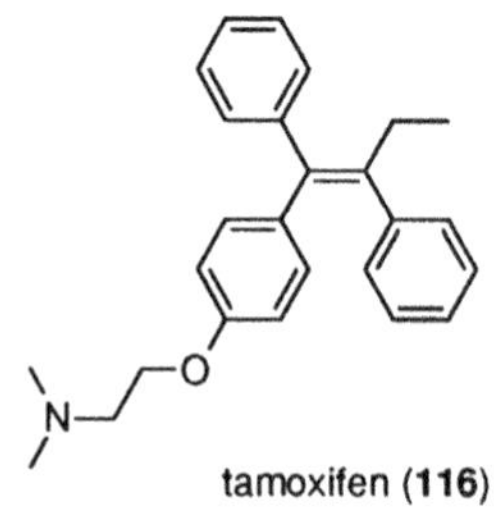

tamoxifen (**116**)

3.7.4. Functionalised polymers as protective groups

A very selective and versatile polymer-grafted protective group for 1,2-diols has been described by *Fréchet et al.*[31,66,67] Polystyryl boronic acid **117** (*Scheme 3.7.6*) reacted selectively with diols to give cyclic boronates. Thus, when **117** was added to a mixture of *cis-* and *trans-*1,2-cyclohexadiol only the *cis-*diol formed a cyclic boronate, whereas the *trans-*diol remained in solution and was removed by washing. Subsequent cleavage was carried out under extremely mild conditions by addition of water or alcohol. The polymer was regenerated in the cleavage step and could be re-used repeatedly.

Scheme 3.7.6

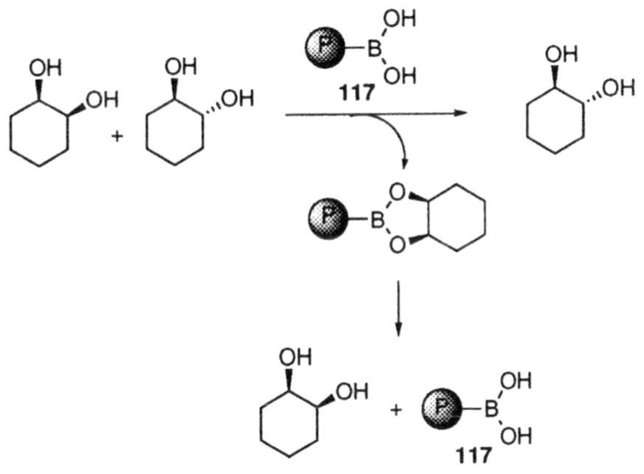

Fréchet et al.[68,69] reported the use of a resin-bound aldehyde as protective group for sugars as outlined in the following example (*Scheme 3.7.7*). In the first step the sugar moiety was anchored to the polymer by formation of an acetal. Subsequent acetylation of the free hydroxyl groups followed by release from the support gave a sugar derivative containing one free primary hydroxyl group at C(6) and a secondary hydroxyl at C(4).

251

Scheme 3.7.7

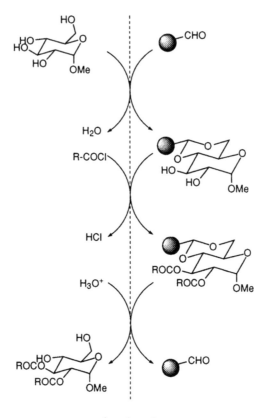

phase boundary

3.8. Organic synthesis on soluble polymeric supports

3.8.1. Introduction

An alternative approach to perform synthesis in solution by still keeping the advantages of macromolecular properties constitutes the use of soluble polymeric supports.[70] Soluble polymers, similarly to insoluble supports, should be commercially available or easily prepared, exhibit good chemical and physical stability to a large variety of reaction conditions and be easily derivatised with a large variety of functional groups in order to attach organic molecules. In contrast to insoluble polymers, soluble supports should show good solubilities in a range of organic solvents irrespective of solubilities of the attached organic entities. Moreover, in some solvents they should

252

show limited solubilities and crystallise in order to be precipitated in the purification process. High loading capacities are equally important for soluble and insoluble supports.

A recent review[70] summarises the most important classes of soluble polymers, comprising supports derived from polyethylenglycol (PEG), polyvinylalcohol, polyethylene imine, polyacrylic acid, polypropylene oxide, cellulose, polyacrylamide to name the commonly used ones.

Purification of the anchored molecules is usually performed by:

* precipitation / crystallisation
* dialysis using semipermeable membranes[71]
* gel permeation chromatography[72]

While precipitation / crystallisation is easily automated, dialysis and gel permeation require more laborious workup procedures. In addition, precipitation / crystallisation seems to be less predictive and reproducible. Notably, it has been shown with soluble polystyrenes in the synthesis of *e.g.* oligosaccharides that the yields of purified polymer-bound products decreased considerably as the oligosaccharide chain grew due to increased solubility of the sugar-polymer conjugate in polar solvents.[73]

An important feature of soluble polymer-bound libraries lies in the fact that biological screening can be performed in the standard format and that reactions can easily be monitored using ^{1}H-NMR- and ^{13}C-NMR spectroscopy. For the synthesis of biopolymers such as peptides CD spectroscopy was employed, *e.g.* for the synthesis of myoglobin 66-73.[74]

3.8.2. Combinatorial synthesis of biopolymers on soluble supports

Similarly to insoluble supports, molecules are attached via a linker unit (see *Chapter 1.7*) to the support. Supports and linkers have to be chemically compatible with the reaction conditions employed for the library synthesis. Conversely, the final products have to be releasable without chemical modifications. Soluble supports have been successfully used for:

Peptide synthesis: non-crosslinked polystyrene (MG 200'000)

                e.g. Val-Tyr-Val-His-Pro-Phe[75]

                Combinatorial library of 1024 pentapeptides[76]
                (composed of amino acids Tyr, Gly, Phe, Leu)

Oligonucleotide synthesis:

Hydrophilic polymeric supports are advantageous[77] *e.g.* polyvinyl alcohol,[78] cellulose,[79,80] polyethyleneglycol.[70]

Oligosaccharide synthesis:

Synthesis performed on non-crosslinked polystyrene,[73] polyacrylamide and copolymers, and polyethylene glycol (examples see review by *Janda*[70]).

The synthesis of a heptaglucoside has been efficiently accomplished on a PEG-support as depicted in *Scheme 3.8.1*. After each coupling step, the product was precipitated and purified by washing of the residue. Subsequent deprotection, cleavage from the resin and purification yielded the final oligosaccharide in 18% yield. The product had the same biological activity as a reference sample prepared previously by standard solution methodology.[81]

Scheme 3.8.1

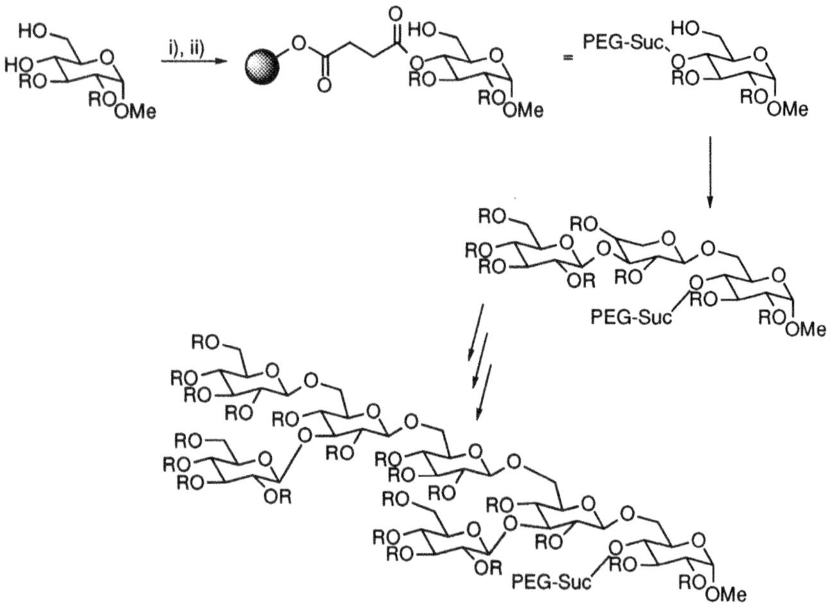

Reagents and conditions: i) DMTCl, py, ii) succinic anhydride, DMAP, py,
iii) PEG-OH, DCC, iv) PhSO$_3$H, MeOH/CH$_2$Cl$_2$

3.8.3. Liquid-phase combinatorial synthesis (LPCS) of small organic molecules

Although in principle any molecule should be amenable to liquid-phase synthesis (synthesis on a soluble polymeric support) as long as the reaction conditions do not interfere with the linker unit and the soluble polymer, LPCS suffers so far from lack of suitable applications. In *Scheme 3.8.2*, a successful recent example of the synthesis of a 6-membered sulfonamide library is shown.[76]

Scheme 3.8.2

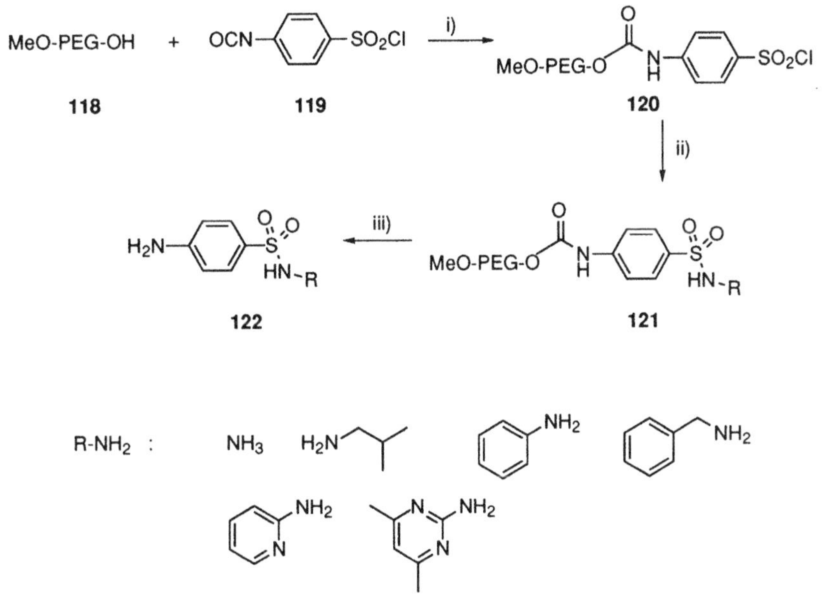

Reagents and conditions: i) dibutyl laureate; ii) R-NH₂; iii) NaOH

In the first step, isocyanate **119** was attached to the support affording urethane derivative **120**. Subsequent treatment with 6 different amines furnished the corresponding polymer-bound sulfonamides **121**. Basic hydrolysis released analytically pure products **122** in high overall yields (95-97%) and purities. Reactions were monitored by NMR spectroscopy.

References

(1) H. Heimgartner, *Angew. Chem. Int. Ed. Engl.* **1991**, *30*, 238.

(2) D. L. Boger, C. M. Tarby, P. L. Myers and L. H. Caporale, *J. Am. Chem. Soc.* **1996**, *118*, 2109.

(3) L. Weber, S. Wallbaum, C. Broger and K. Gubernator, *Angew. Chem. Int. Ed. Engl.* **1995**, *34*, 2280.

(4) A. Rahm, A. Linden, B. R. Vincent, H. Heimgartner, M. Mühlstädt and B. Schulze, *Helv. Chim. Acta* **1991**, *74*, 1002.

(5) T. R. Mihova, A. Linden and H. Heimgartner, *Helv. Chim. Acta* **1996**, *79*, 2067.

(6) M. Schläpfer-Dähler and H. Heimgartner, *Helv. Chim. Acta* **1993**, *76*, 2398.

(7) A. S. Orahovats, S. S. Bratovanov, L. A. and H. Heimgartner, *Helv. Chim. Acta* **1996**, *79*, 1121.

(8) T. A. Keating and R. W. Armstrong, *J. Am. Chem. Soc.* **1996**, *118*, 2574.

(9) H. Maehr and R. Young, *Bioorg. Med. Chem.* **1997**, *5*, 493.

(10) S. Cheng, D. D. Comer, J. P. Williams, P. L. Myers and D. L. Boger, *J. Am. Chem. Soc.* **1996**, *118*, 2567.

(11) D. L. Boger, WO 96/03424, 1996.

(12) A. Studer, S. Hadida, R. Ferritto, S.-Y. Kim, P. Jeger, P. Wipf and D. P. Curran, *Science* **1997**, *275*, 823.

(13) A. Studer, P. Jeger, P. Wipf and D. P. Curran, *J. Org. Chem.* **1997**, *62*, 2917.

(14) D. P. Curran and M. Hoshino, *J. Org. Chem.* **1996**, *61*, 6480.

(15) L. M. Gayo and M. J. Suto, *Tetrahedron Lett.* **1997**, *38*, 513.

(16) T. L. Deegan, O. W. Gooding, S. Baudart and J. A. Porco Jr., *Tetrahedron Lett.* **1997**, *38*, 4973.

(17) S. W. Kaldor, J. E. Fritz, J. Tang and E. R. McKinney, *Bioorg. Med. Chem. Lett.* **1996**, *6*, 3041.

(18) S. W. Kaldor, M. G. Siegel, J. E. Fritz, B. A. Dressman and P. J. Hahn, *Tetrahedron Lett.* **1996**, *37*, 7193.

(19) A. Chucholowski, T. Masquelin, D. Obrecht, J. Stadlwieser and J. M. Villalgordo, *Chimia* **1996**, *50*, 530.

(20) R. M. Dodson, *J. Am. Chem. Soc.* **1948**, *70*, 2753.

(21) R. J. Booth and J. C. Hodges, *J. Am. Chem. Soc.* **1997**, *119*, 4882.

(22) G. M. Atkins and E. M. Burgess, *J. Am. Chem. Soc.* **1968**, *90*, 4744.

(23) P. Wipf and S. Venkatraman, *Tetrahedron Lett.* **1996**, *37*, 4659.

(24) S. Kobayashi, I. Hachiya, S. Suzuki and M. Moriwaki, *Tetrahedron Lett.* **1996**, *37*, 2809.

(25) K. A. Lutomski and A. I. Meyers, in: *Asymmetric Synthesis*, J. D. Morrison (Ed.), Vol. 3, p. 213, Academic: New York (1984).

(26) D. A. Evans, E. P. Ng, J. S. Clark and D. Rieger, *Tetrahedron* **1992**, *48*, 2127.

(27) S. M. Allin and S. J. Shuttleworth, *Tetrahedron Lett.* **1996**, *37*, 8023.

(28) K. Burgess and D. Lim, *J. Chem. Soc., Chem. Commun.* **1997**, 785.

(29) O. Mitsunobu, M. Wada and T. Sano, *J. Am. Chem. Soc.* **1972**, *94*, 679.

(30) H. Staudinger and J. Meyer, *Helv. Chim. Acta* **1919**, *2*, 635.

(31) J. M. J. Fréchet, *Tetrahedron* **1981**, *37*, 663.

(32) G. Wittig and U. Schöllkopf, *Ber.* **1954**, *87*, 1318.

(33) I. Hughes, *Tetrahedron Lett.* **1996**, *37*, 7595.

(34) M. Le Corre, A. Hercourt and H. Le Baron, *J. Chem. Soc., Chem. Commun.* **1981**, 14.

(35) D. Obrecht, M. Altorfer and J. A. Robinson, *Novel Peptide Mimetic Building Blocks and Strategies for Efficient Lead Finding*, in: *Advances in Medicinal Chemistry*, B. E. Maryanoff and A. B. Reitz (Ed.), in press, p. JAI Press Inc.: Greenwich, Connecticut (1998).

(36) T. Holletz and D. Cech, *Synthesis* **1994**, 789.

(37) R. A. Amos and S. M. Fawcett, *J. Org. Chem.* **1984**, *49*, 2637.

(38) C. Zumbrunn, Ph.D. thesis, University of Zürich (1998).

(39) J. J. Parlow, *Tetrahedron Lett.* **1995**, *36*, 1395.

(40) J. Rebek Jr., *Tetrahedron* **1979**, *35*, 723.

(41) B. J. Cohen, M. A. Kraus and A. Patchornik, *J. Am. Chem. Soc.* **1981**, *103*, 7620.

(42) M. Bessodes and K. Antonakis, *Tetrahedron Lett.* **1985**, *26*, 1305.

(43) H. C. Kolb, M. S. VanNieuwenhze and K. B. Sharpless, *Chem. Rev.* **1994**, *94*, 2483.

(44) B. M. Kim and K. B. Sharpless, *Tetrahedron Lett.* **1990**, *31*, 3003.

(45) S. J. Shuttleworth, S. M. Allin and P. K. Sharma, *Synthesis* **1997**, 1217.

(46) C. E. Song, J. W. Yang, H. J. Ha and S. Lee, *Tetrahedron: Asymmetry* **1996**, *7*, 645.

(47) H. Han and K. D. Janda, *Tetrahedron Lett.* **1997**, *38*, 1527.

(48) T. Katsuki and K. B. Sharpless, *J. Am. Chem. Soc.* **1981**, *103*, 464.

(49) L. Canali, J. K. Karjalainen, D. C. Sherrington and O. Hormi, *J. Chem. Soc., Chem. Commun.* **1997**, 123.

(50) S. K. Das, A. Kumar Jr., S. Nadrajog and A. Kumar, *Tetrahedron Lett.* **1995**, *36*, 7909.

(51) E. J. Corey, A. Guzman-Perez and S. E. Lazerwith, *J. Am. Chem. Soc.* **1997**, *119*, 11769.

(52) L. Yu, D. Chen, J. Li and P. G. Wang, *J. Org. Chem.* **1997**, *62*, 3575.

(53) J. M. Fraile, J. A. Mayoral, A. J. Royo, R. V. Salvador, B. Altava, S. V. Luis and M. I. Burguete, *Tetrahedron* **1996**, *52*, 9853.

(54) M. Watanabe and K. Soai, *J. Chem. Soc., Perkin Trans. 1* **1994**, 837.

(55) D. Seebach, M. Overhand, F. N. M. Kühnle, B. Martinoni, L. Oberer, U. Hommel and H. Widmer, *Helv. Chim. Acta* **1996**, *79*, 913.

(56) B. Altava, M. I. Burguete, B. Escuder, S. V. Luis, R. V. Salvador, J. M. Fraile, J. A. Mayoral and A. J. Royo, *J. Org. Chem.* **1997**, *62*, 3126.

(57) P. B. Rheiner, H. Sellner and D. Seebach, *Helv. Chim. Acta* **1997**, *80*, 2027.

(58) Y. Uozumi, H. Danjo and T. Hayashi, *Tetrahedron Lett.* **1997**, *38*, 3557.

(59) S. Kobayashi and S. Nagayama, *J. Am. Chem. Soc.* **1996**, *118*, 8977.

(60) A. M. Strocker, T. A. Keating, P. A. Tempest and R. W. Armstrong, *Tetrahedron Lett.* **1996**, *37*, 1149.

(61) A. Chucholowski, D. Heinrich, B. Mathys and C. Muller, *Generation of benzodiazepine and benzodiazocine libraries through resin-capture of Ugi-4CC*, conference: *214th ACS national meeting*, Las Vegas, 1997.

(62) S. D. Brown and R. W. Armstrong, *J. Am. Chem. Soc.* **1996**, *118*, 6331.

(63) R. W. Armstrong, S. D. Brown, K. T.A. and P. A. Tempest, *Combinatorial Synthesis exploiting Multiplecomponent Condensations, Mircrochip Encoding and Resin Capture*, in: *Combinatorial Chemistry*, S. R. Wilson and A. W. Czarnik (Ed.), p. 153, Wiley & Sons, Inc.: New York (1997).

(64) T. Ishiyama, N. Matsuda, N. Miyaura and A. Suzuki, *J. Am. Chem. Soc.* **1993**, *115*, 11018.

(65) T. Ishiyama, N. Matsuda, F. Ozawa, A. Suzuki and N. Miyaura, *Organometallics* **1996**, *15*, 713.

(66) M. J. Farrall and J. M. J. Fréchet, *J. Org. Chem.* **1976**, *41*, 3877.

(67) E. Seymour and J. M. J. Fréchet, *Tetrahedron Lett.* **1976**, *17*, 3669.

(68) J. M. J. Fréchet and K. E. Haque, *Tetrahedron Lett.* **1975**, *16*, 3055.

(69) J. M. J. Fréchet and G. Pelle, *J. Chem. Soc., Chem. Commun.* **1975**, 225.

(70) D. J. Gravert and K. D. Janda, *Chem. Rev.* **1997**, *97*, 489.

(71) H. Köster, *Tetrahedron Lett.* **1972**, *13*, 1535.

(72) K. E. Geckeler, *Adv. Polym. Sci.* **1995**, *121*, 31.

(73) R. d. Guthrie, A. d. Jenkins and J. Stehlicek, *J. Chem. Soc. (C)* **1971**, 2690.

(74) M. Mutter, H. Mutter, R. Uhmann and E. Bayer, *Biopolymers* **1976**, *15*, 917.

(75) Y. A. Ovchinnikov, A. A. Kiryushkin and I. V. Kozhevnikova, *J. Gen. Chem. USSR (Engl. Transl.)* **1968**, *38*, 2551.

(76) H. Han, M. M. Wolfe, S. Brenner and K. D. Janda, *Proc. Natl. Acad. Sci. USA* **1995**, *92*, 6419.

(77) V. Armanath and A. D. Broom, *Chem. Rev.* **1977**, *77*, 183.

(78) H. Schott, *Angew. Chem. Int. Ed. Engl.* **1973**, *12*, 246.

(79) K. Kamaike, Y. Hasegawa and Y. Ishido, *Tetrahedron Lett.* **1988**, *29*, 647.

(80) K. Kamaike, S. YAmakage, Y. Hasegawa and Y. Ishido, *Nucleic Acids Res.* **1986**, *17*, 89.

(81) R. Verduyn, P. A. M. van der Klein, M. Douwes, G. A. van der Marel and J. H. van Boom, *Recl. Trav. Chim. Pays-Bas* **1993**, *112*, 464.

Chapter 4

Convergent Assembly Strategies on Solid Supports

4.1. Multigeneration assembly strategies with single functional group liberation

4.1.1. Liberation of -OH groups

4.1.1.1. Synthesis of benzimidazoles

Benzimidazoles are an interesting class of heterocycles which are incorporated into several therapeutic agents. There are several standard preparations of benzimidazoles, but many of them require high temperatures and strong acidic reaction conditions. In general, when adapting a solution-phase methodology into a solid-phase approach, very drastic conditions are to be avoided due to potential problems associated with the stability of the solid support. These criteria were taken into account by *Philips* and *Wei* in their approach towards the solid-phase synthesis of a benz-imidazole library.[1]

Scheme 4.1.1

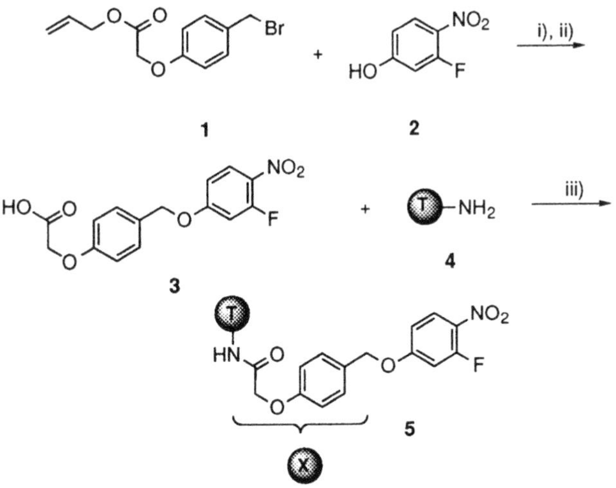

Reagents and conditions: i) KHMDS, DMF; ii) Bu₃SnH, Pd(PPh₃)₄; iii) DIC, HOBt (>90% loading level)

259

As a key step for cyclisation they selected the condensation between bis-donor building-blocks (cf. *Chapter 1.9.1*, *Chart 2*) of the *ortho*-aminoaniline type and bis-acceptors building-blocks of the imidate type.

Thus, allyl (4-bromomethylphenoxy)acetate **1** was reacted with 3-fluoro-4-nitrophenol (**2**) in the presence of a base to give **3**. Subsequent removal of the allylic protective group and tethering to TentaGel® S-NH₂ **4** resin gave the corresponding *ortho*-nitrofluorobenzene **5** linked to the solid support. This linker strategy allows the use of a wide range of reaction conditions and cleavage is usually performed with 95% TFA (*Scheme 4.1.1*). In this particular case, **5** was converted into a masked bis-donor diamine building block by displacement of the activated fluorine atom with different primary amines (**6**). Reduction of the nitro group of **7** with NaBH₄-Cu(acac)₂ gave diamine **8** and treatment with different benzimidates **9** led to the corresponding polymer-bound benzimidazoles **10**. Final cleavage by means of 95% TFA/ H₂O at room temperature afforded hydroxybenzimidazoles **11** in yields ranging between 70-95% and with purities of crude material between 70-98% (*Scheme 4.1.2*).

Scheme 4.1.2

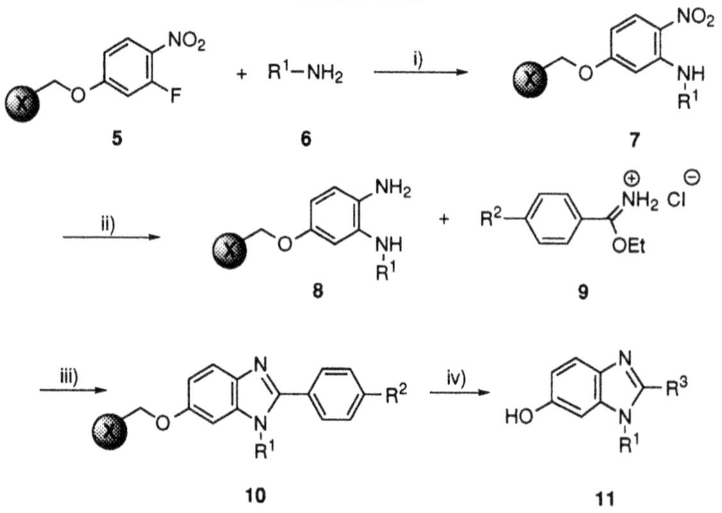

Reagents and conditions: i) DMSO, r.t., 24h; ii) DMF/EtOH, NaBH₄-Cu(acac), r.t., 24h; iii) nBuOH/DMF, 55-90°C, 24-40h; iv) TFA/H₂O (95:5), r.t., 3-12h

4.1.1.2. Synthesis of 1,3-dialkyl quinazolin-2,4-diones

An analogous linker strategy was utilised by *Buckman* and *Mohan* in their reported synthesis of a quinazolinedione library.[2] In this case an acceptor-donor building block, namely an anthranilate derivative **12** (cf. *Chapter 1.9.3, Chart 4*), was attached to TentaGel® S-NH$_2$ resin **4** through the acid cleavable ((4-hydroxymethyl)phenoxy) acetic acid linker, as starting material for the quinazoline library formation. Thus, removal of the Fmoc-group under standard conditions gave free amine **13** which was reacted with different isocyanates to form polymer-bound ureas **15** and cyclised in ethanolic KOH to form 3-alkyl quinazolinediones **16**. The reaction proceeds nearly quantitatively in both the addition and cyclisation steps. Additional molecular diversity could be introduced by alkylation of N-1 by treatment of **16** with lithium oxazolidinone (used as a soluble lithium base), followed by addition of activated alkyl halides, such as alkyl iodides and benzylic or allylic bromides to give **17**.

Scheme 4.1.3

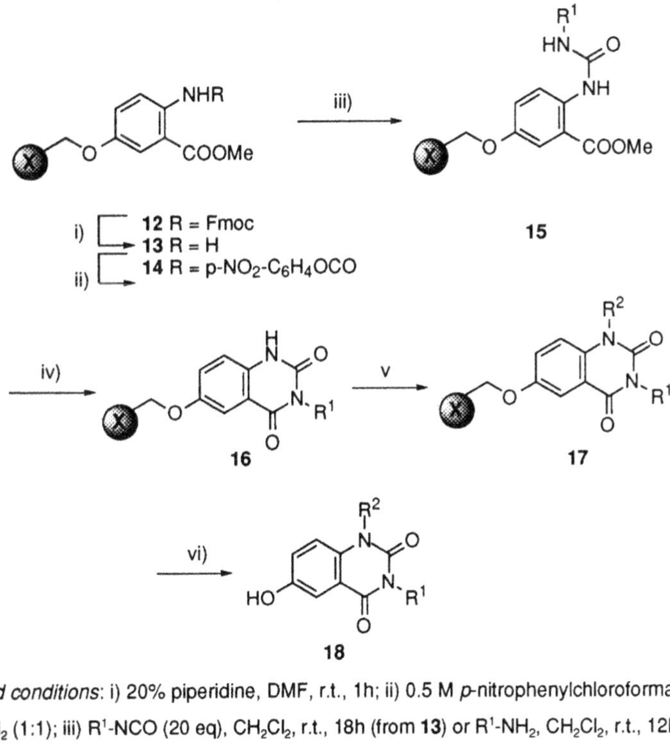

Reagents and conditions: i) 20% piperidine, DMF, r.t., 1h; ii) 0.5 M *p*-nitrophenylchloroformate, 0.5 M NEt$_3$ in THF/CH$_2$Cl$_2$ (1:1); iii) R^1-NCO (20 eq), CH$_2$Cl$_2$, r.t., 18h (from **13**) or R^1-NH$_2$, CH$_2$Cl$_2$, r.t., 12h (from **14**); iv) 1M KOH in EtOH, r.t., 1h; v) Li-benzyloxazolidinone (0.3 M in THF, 15.5 eq), THF, r.t., 1.5h, R^2-X (40 eq), DMF, r.t., 18h, repeat; vi) TFA/H$_2$O (95:5), r.t., 1h

Generally, this alkylation step did not proceed to completion even when using excess base, excess halide, or longer reaction times. However, by re-submitting the mixture to the same alkylating conditions, less than 5% starting material was detected by HPLC. Subsequent acidic cleavage from the resin with 95% TFA afforded 1,3-dialkyl quinazoline-2,4-diones **18** in yields higher than 80%. Alternatively, since the number of different commercially available amines is larger than the corresponding isocyanates, higher molecular diversity could be achieved by generating urea **15** through reactive *p*-nitrophenylcarbamate **14** and primary amines as shown in (*Scheme 4.1.3*).

4.1.1.3. Synthesis of 1,4-benzodiazepine-2-ones

1,4-Benzodiazepines exhibit widespread biological activities and belong to one of the most important classes of bioavailable therapeutic agents. Many derivatives have been identified as highly selective cholecystokinin receptor subtype A and B antagonists,[3] κ-selective opioids,[4] platelet-activation factor antagonists,[5] HIV Tat antagonists,[6] reverse transcriptase inhibitors[7] and as farnesyltransferase inhibitors.[8] Therefore, the development of solid-phase combinatorial approaches toward this class of compounds seems very attractive. *Bunin* and *Ellman* developed[9] a synthetic approach for the preparation of 1,4-benzodiazepin-2-ones based on three key building blocks: acceptor-donor building blocks such as amino benzophenones and amino acids, and alkylating agents as acceptor building blocks. Thus, employing solution chemistry, substituted 2-*N*-Fmoc aminobenzophenones **19** were coupled to the acid-cleavable (4-(hydroxymethyl)phenoxy)acetic acid (HMP) linker through a hydroxyl or carboxylic group located on either aromatic ring of the 2-aminobenzophenone. The linker-derivatised aminobenzophenones were then coupled to the solid support employing standard amide coupling methods as shown in *Scheme 4.1.4*.

Scheme 4.1.4

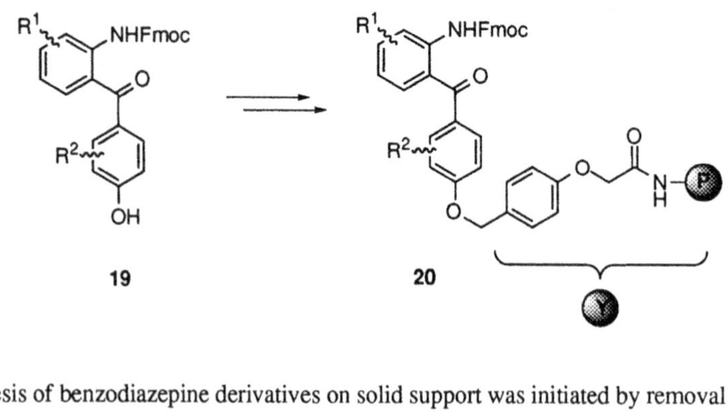

The synthesis of benzodiazepine derivatives on solid support was initiated by removal of the Fmoc protecting group from **20** by treatment with piperidine in DMF followed by coupling of an α-*N*-Fmoc amino acid to the resulting unprotected 2-aminobenzophenone. Standard activation methods for solid-phase peptide synthesis were not successful for this coupling step due to the poor nucleophilicity and basicity of 2-aminobenzophenones. However, activated α-*N*-Fmoc amino acid fluorides[10] proved to be very efficient to yield the corresponding amide products **21** even with electron-deficient 2-aminobenzophenone derivatives. Removal of the Fmoc protecting group with piperidine in DMF and treatment of the resulting free amine with 5% acetic acid in NMP at 60°C provided benzodiazepine derivatives **22** incorporating two of the three elements of molecular diversity. Alkylation of the anilide of **22** by means of lithiated 5-(phenylmethyl)-2-oxazolidinone provided fully derivatised 1,4-benzodiazepines **23**. Final cleavage from the solid support by treatment with TFA/Me$_2$S/H$_2$O (95:5:5) afforded benzodiazepine products **24** in high yields as depicted in *Scheme 4.1.5*.

Scheme 4.1.5

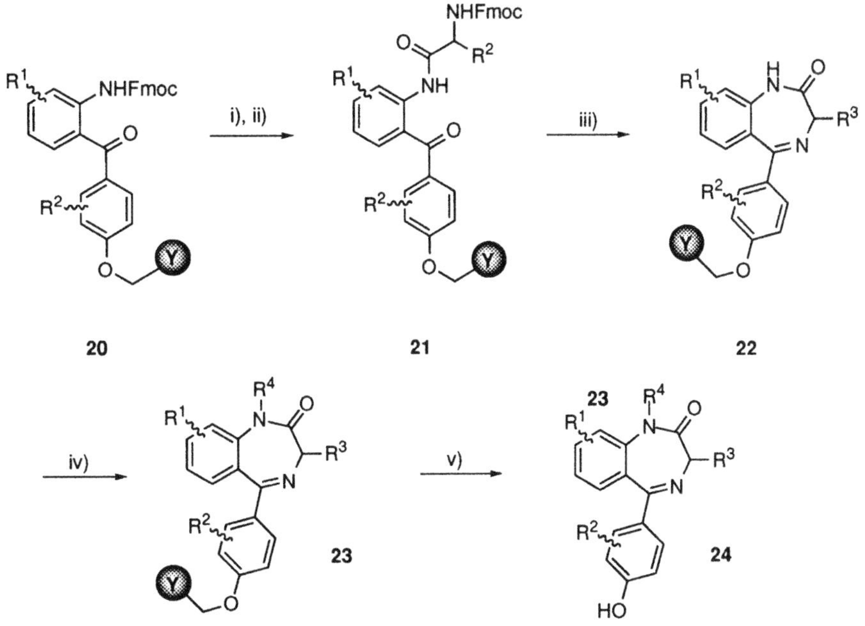

20   **21**   **22**

23   **24**

Reagents and conditions: i) 20% piperidine in DMF; ii) *N*-Fmoc-amino acid fluoride, 4-methyl-2,6-di-*tert*-butylpyridine; iii) 5% AcOH in DMF, 60°C; iv) lithiated 5-(phenylmethyl)-2-oxazolidinone in THF, -78°C, R^4-X/DMF; v) TFA/H$_2$O/Me$_2$S (95:5:5)

While many alkylating agents are commercially available, and N-Fmoc amino acid fluorides can be prepared in a single step without purification from the corresponding N-Fmoc amino acids, few appropriately functionalised 2-aminobenzophenones are readily accessible. Furthermore, since the 2-aminoaryl ketone building block introduces an essential determinant of activity or specificity for many benzodiazepine-based therapeutic agents or candidates, an alternative approach to circumvent this limitation was developed by *Plunkett* and *Ellman*[11] based on the *Stille* coupling reaction.

Thus, the triisopropylsilyl ether of the 4-aminophenol **25** was treated with 2-(4-biphenyl)isopropyl phenyl carbonate in THF followed by addition of excess KH to provide the Bpoc derivative **26**. The trimethyltin group was then introduced by directed *ortho*-metalation followed by reaction with trimethyltin chloride. Treatment of **27** with Bu$_4$NF provided the free phenol, which was then coupled to the cyanomethyl ester of 4-(hydroxymethyl)phenoxyacetic acid under standard *Mitsunobu* reaction conditions to give the active ester **28** (*Scheme 4.1.6*).

Scheme 4.1.6

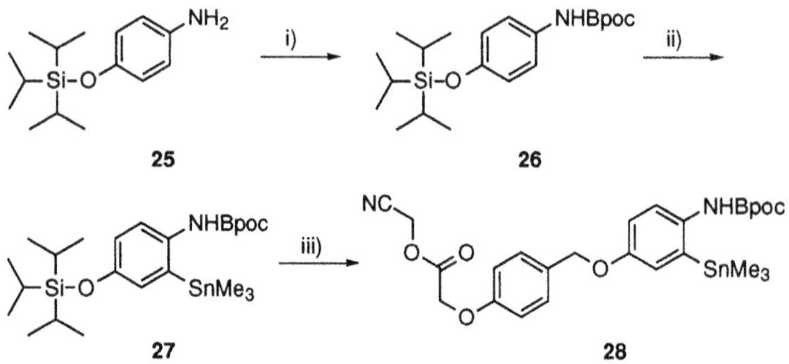

Reagents and conditions: i) 2-(4-biphenyl)isopropyl phenyl carbonate, THF, KH; ii) tBuLi (2.2 eq), Me$_3$SnCl; iii) a) Bu$_4$NF/THF, b) (4-(hydroxymethyl)phenoxy)acetic acid cyanomethyl ester, PPh$_3$, DEAD, THF

Solid-phase synthesis was initiated by coupling **28** to aminomethylated polystyrene resin **29** in NMP with DIEA as base and DMAP as an acylation catalyst. *Stille* reactions of the polymer-bound stannane **30** with different aromatic and aliphatic acid chlorides and subsequent cleavage of the Bpoc protecting group by brief treatment with 3% TFA in CH$_2$Cl$_2$ afforded polymer-bound 2-aminoaryl ketones **31** as shown in *Scheme 4.1.7*.

Scheme 4.1.7

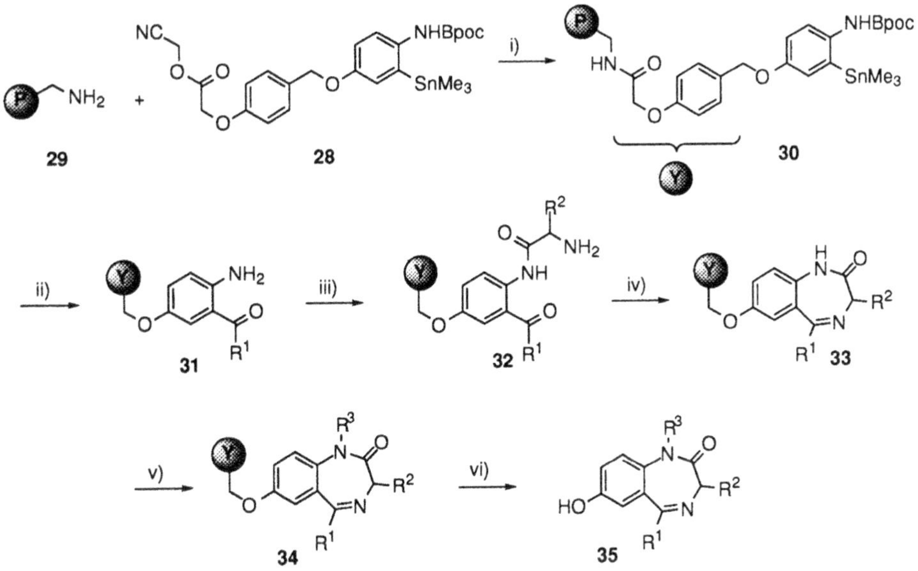

Reagents and conditions: i) NMP, DIEA, DMAP cat.; ii) a) THF, K₂CO₃, DIEA, Pd₂dba₃-CHCl₃, aroyl or acyl chloride, b) 3% TFA in CH₂Cl₂; iii) Fmoc amino acid fluoride, 2,6-di-(*tert*-butyl)-4-methylpyridine; then 20% piperidine in DMF; iv) 5% AcOH in DMF 65°C, 4-8 h; v) lithiated 5-(benzyl)-2-oxazolidinone in THF 0°C, then R³-X in DMF; vi) TFA/Me₂S/H₂O (85:10:5)

Completion of the synthesis of hydroxy-1,4-benzodiazepines was achieved analogously as before: acylation of **31** with Fmoc amino acid fluorides, removal of the Fmoc protecting group, cyclisation and alkylation afforded fully functionalised polymer-supported derivatives **34** that were cleaved from the resin by treatment with TFA/Me₂S/H₂O (85:10:5) at r.t. to give products **35** in purities > 80% and in good overall yields.

4.1.1.4. Synthesis of pyrrolidines

1,3-Dipolar cycloaddition reactions are interesting processes for the preparation of several heterocyclic systems due to the generally mild reaction conditions and the simultaneous formation of several bonds in a single operation. The preparation of pyrrolidines via cycloaddition reactions of azomethine ylides is a well known process and has been extensively studied. A strategy based on the reaction of azomethine ylides with α,β-unsaturated ketones was selected by *Hollinshead* who developed a solid-phase protocol for the preparation of a small library of highly functionalised pyrrolidines.[12] Thus, chlorinated *Wang* resin[13] **36** was coupled to 3-hydroxyacetophenone (**37**) with Cs_2CO_3/NaI in DMF to give polymer-bound acetophenone **38**. *Knoevenagel* condensation with different aromatic aldehydes using a 0.5M solution of MeONa in MeOH, afforded the corresponding enones **39** that were subjected to standard 1,3-dipolar cycloaddition reaction conditions with *N*-metalated azomethine ylides in the presence of DBU and LiBr as a *Lewis* acid to give highly substituted pyrrolidine derivatives **40** with high regio- and diastereoselectivity. These derivatives **40** could also be subsequently reacted with acid chlorides and sulfonyl chlorides to afford **41** (*Scheme 4.1.8*).

Scheme 4.1.8

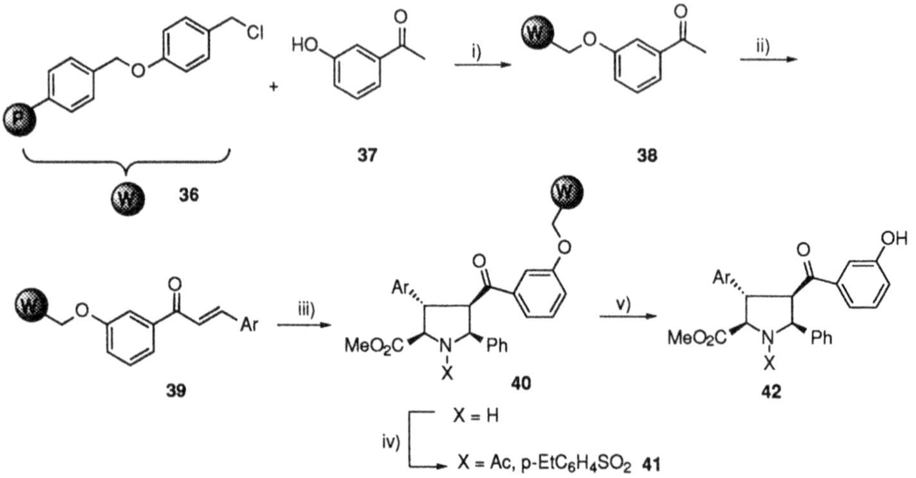

Reagents and conditions: i) Cs_2CO_3, NaI, DMF; ii) Ar-CHO (12 eq), NaOMe (0.5M in MeOH, 12 eq), THF; iii) $PhCH=NCH_2CO_2Me$, LiBr, DBU, THF; iv) acylating agent, py, DMAP, CH_2Cl_2; v) TFA/CH_2Cl_2

Cleavage from the resin (50% TFA/CH_2Cl_2) yielded the highly functionalised crude pyrrolidine products **42** which could be purified by chromatography or crystallisation.

4.1.1.5. Synthesis of aspartic acid protease inhibitors

Attachment of an alcohol group to the solid support constitutes one of the most widely used anchoring strategies. Accordingly, a number of methods have been developed to couple alcohols through ester, silyl ether and trityl ether bond formation and while these methods have proved to be very useful for the synthesis of biopolymers, their utility in the solid-phase synthesis for small organic compound libraries was limited due to their inherent lability towards nucleophilic or basic reagents and/or difficulties in obtaining useful loading levels, particularly for secondary alcohols. To circumvent this problem, *Thompson* and *Ellman*[14] developed a dihydropyran functionalised resin (cf. *Chapter 1.7*) specially designed to attach primary and secondary alcohols. This strategy was used to prepare a library of potential aspartic acid protease inhibitors.[15] Thus, an acceptor-donor building block such as **44**,[16,17] which is known to provide access to some HIV-1 protease inhibitors, was selected as initial building block and coupled to dihydropyran functionalised polystyrene support **43** by employing pyridinium *p*-toluenesulfonate (PPTS) in 1,2-dichloroethane as shown in *Scheme 4.1.9*.

Scheme 4.1.9

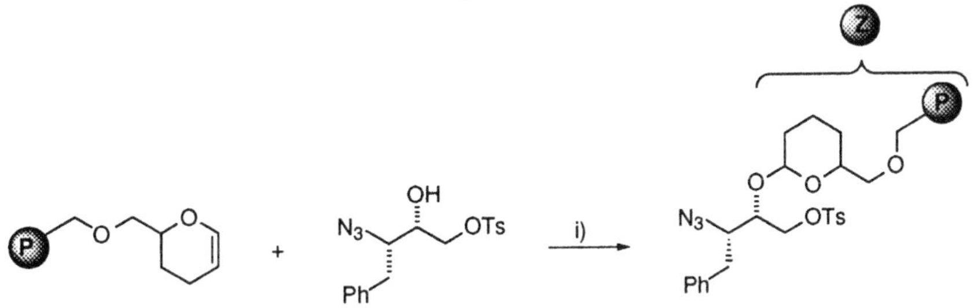

Reagents and conditions: i) PPTS, 1,2-dichloroethane, 80°C

The synthesis was initiated from **45** by displacement of the primary tosyl group with either functionalised or nonfunctionalised primary or secondary amines in NMP at 80°C to give **46**. After coupling of the primary amines, the resulting secondary amines **46** were converted into ureas **47** by reaction with isocyanates or by stepwise treatment with triphosgene followed by amine addition. In addition, acyl chlorides can be employed to provide amides as exemplified for **49** (*Scheme 4.1.10*).

Scheme 4.1.10

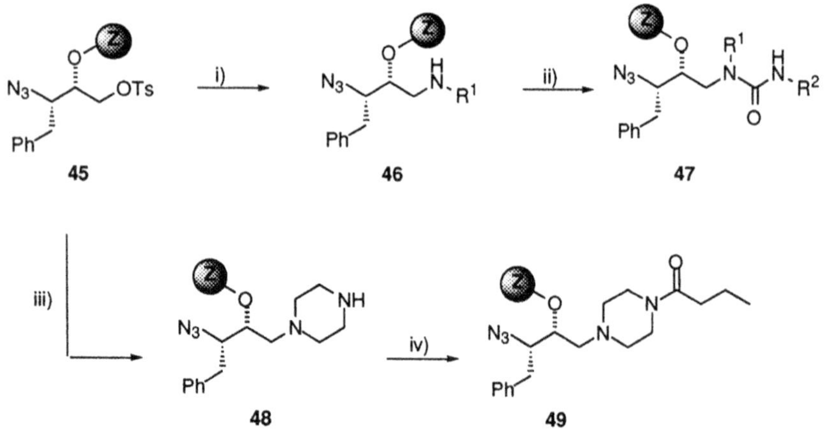

Reagents and conditions: i) R^1-NH$_2$, NMP, 80°C; ii) R^2-NCO, 1,2-dichloroethane or a) triphosgene, Et$_3$N, cat. DMAP, THF, b) R^2-NH$_2$; iii) piperazine, NMP, 80°C; iv) butyryl chloride, DIEA, CH$_2$Cl$_2$

Furthermore, reduction of the azido group of **47** using thiophenol/Et$_3$N/SnCl$_2$ (4:5:1) afforded primary amines **50** that were acylated to provide carbamate or amide products that in turn could be further derivatised. For example, coupling of **50** with protected amino acids under PyBOP/HOBt coupling conditions and subsequent removal of the Fmoc protecting group provided **51** that were coupled with the pentafluorophenyl ester of the quinaldic acid. Cleavage of the material from the solid support with 95% TFA/water provided **52** in 74-85% overall yield for the six-step process. Alternatively, reaction of amines **50** with the activated *N*-succinimidyl carbonate of 3(*S*)-hydroxytetrahydrofuran provided carbamates **53**. Cleavage as before provided analytically pure derivatives **54** isolated after chromatography in 47-86% overall yield as depicted in *Scheme 4.1.11*.

Scheme 4.1.11

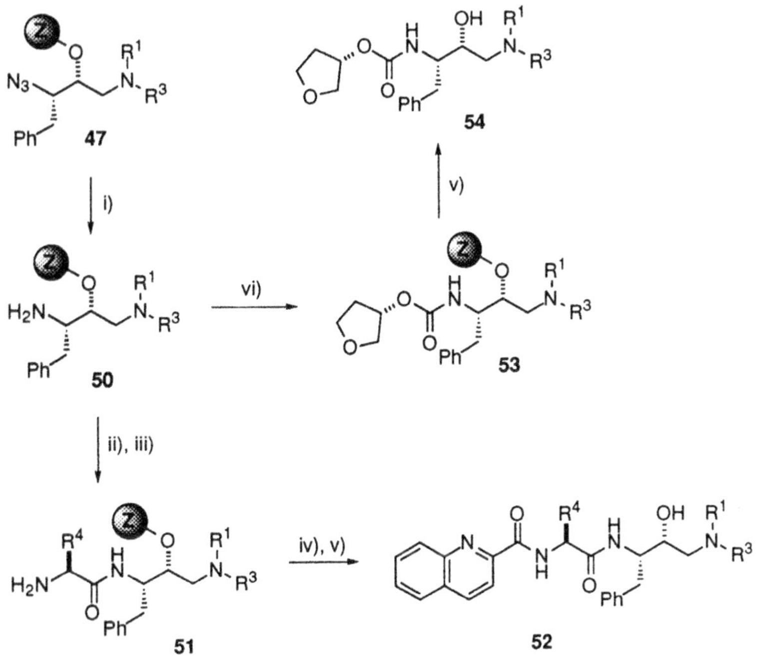

Reagents and conditions: i) SnCl$_2$/HSPh/NEt$_3$ (1:4:5), THF; ii) Fmoc amino acid, PyBOP, HOBt, DIEA

(3 eq), DMF; iii) 20% piperidine in DMF; iv) pentafluorophenylester of quinaldic acid, HOBt, NEt$_3$, DMF;

v) TFA/H$_2$O (95:5); vi) 3(S)- [*N*-succinimidyloxycarbonyl]oxy-tetrahydrofuran, DIEA, CH$_2$Cl$_2$

4.1.2. Liberation of -NH_2 groups

4.1.2.1. Cyclic ureas and thioureas

Many biologically active compounds contain cyclic ureas, including inhibitors of human immunodeficiency virus (HIV) protease (cf. *Chapter 2.7*) and HIV replication.[18] *Houghten et al.* approached the design and solid-phase synthesis of cyclic ureas and thioureas using modified dipeptides as starting materials.[19] p-Methylbenzhydrylamine resin (p-MBHA) **55** was selected as a solid support and a number of different *N*-Fmoc protected amino acids were coupled to the resin under standard conditions (HOBt/DIC). Removal of the Fmoc group to introduce in turn a trityl protective group, allowed the selective *N*-alkylation with lithium *tert*-butoxide in THF, followed by

269

addition of the alkylating agent in DMSO to give **57**. Following removal of the trityl protecting group with 2% TFA in CH$_2$Cl$_2$, the second amino acid was added using standard peptide chemistry. The resulting polymer-bound dipeptide, after N-Fmoc deprotection, was subsequently acylated with a wide range of available carboxylic acids to give **59** (*Scheme 4.1.12*).

Scheme 4.1.12

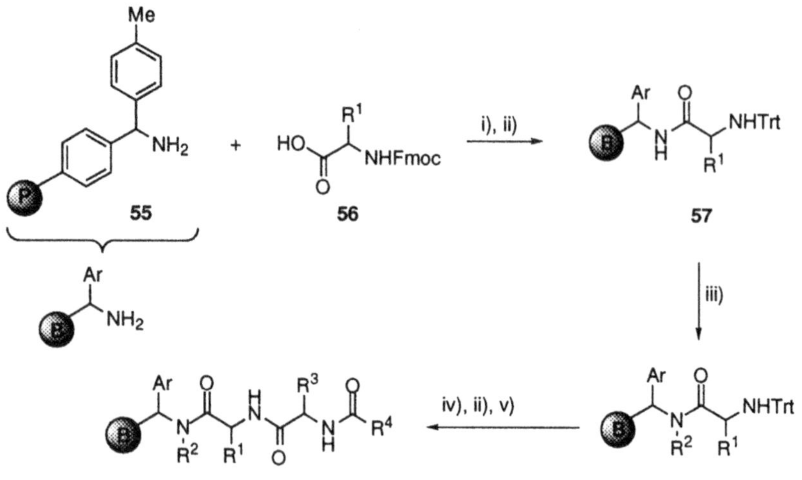

Reagents and conditions: i) HOBt/DIC; ii) a) 25% piperidine in DMF, b) trityl chloride, DIEA in CH$_2$Cl$_2$/DMF (9:1) overnight, r.t.; iii) a) 1M LiOtBu in THF, b) R^2-X in DMSO; iv) 2% TFA in CH$_2$Cl$_2$, second Fmoc amino acid, HOBt/DIC; v) R^4-COOH, HOBt/DIC

Reduction of the amide groups using diborane in THF at 65°C generated polymer-bound triamines **60** that were cyclised to five-membered cyclic ureas **61** and thioureas **62** using carbonyldiimidazole and thiocarbonyldiimidazole in anhydrous CH$_2$Cl$_2$, respectively. The desired products **63** and **64** were obtained in good yields and high purity (>90%) after cleavage from the resin with anhydrous HF in the presence of anisole (*Scheme 4.1.13*). The cyclisation step could also be carried out using triphosgene or thiophosgene.

Scheme 4.1.13

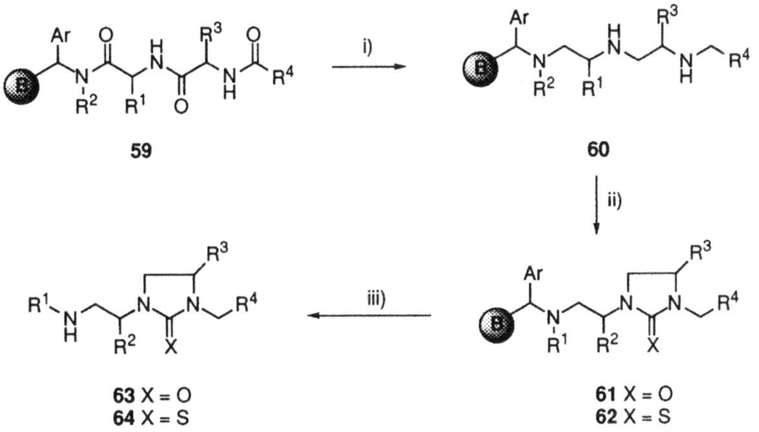

59 → 60

63 X = O
64 X = S

61 X = O
62 X = S

Reagents and conditions: i) B$_2$H$_6$, THF, reflux; ii) carbonyldiimidazole (X = O)
or thiocarbonyldiimidazole (X = S); iii) HF, anisole

4.1.2.2. Synthesis of 1,4-dihydropyridines

The dihydropyridine core structure is found in many bioactive compounds which include various vasodilator, antihypertensive, bronchodilator, antiarteriosclerotic, hepatoprotective, antitumor, antimutagenic, geroprotective and antidiabetic agents.[20-25] Examples of existing dihydropyridine therapeutics include potent calcium channel blocker drugs such as Nifedipine, Nitrendipine, and Nimodipine. Second-generation calcium antagonists include dihydropyridine derivatives with improved bioavailability, tissue selectivity and/or stability such as for instance the antihypertensive/antianginal drugs Elgodipine[26] Furnidipine,[27,28] Darodipine,[29] Pranidipine,[30] Lemildipine[31] and Dexniguldipine.[32] Following discovery of the compound Bay K 8644,[33] a number of dihydropyridine calcium agonists have been introduced as potential drug candidates for treatment of congestive heart failure.[34,35] Among dihydropyridines with other types of bioactivity, Cerebrocrast[36] has been recently introduced as a neuroprotectant and cognition enhancer lacking neuronal-specific calcium antagonist properties. In addition, a number of dihydropyridines with platelet antiaggregatory activity have also been discovered.[37] These recent examples highlight the level of ongoing interest toward new dihydropyridine derivatives and aimed *Gordeev* and coworkers to explore the solid-phase synthesis of this important pharmacophore as a fertile source of new bioactive molecules.[38,39] The selected strategy toward the preparation of 1,4-dihydropyridines on solid support is based on the classical *Hantzsch* two- or three-component cyclocondensation reaction of enamino esters with 2-arylidene β-ketoesters or β-ketoesters and aldehydes. Thus, solid-phase synthesis started by reacting polystyrene-based acid cleavable PAL[40] or *Rink* amine resin **65** with β-ketoesters **66** to afford polymer bound aminocrotonate **67** followed by reaction with either 2-arylidene β-ketoesters **68** (previously prepared by standard *Knoevenagel* condensation) or directly with β-ketoesters or β-diketones **69** and aldehydes **70** in dry pyridine to generate resin-bound intermediate **72**, which after cleavage with TFA cyclised to provide the corresponding 1,4-dihydropyrimidine derivatives **74** as shown in *Scheme 4.1.14*.

Scheme 4.1.14

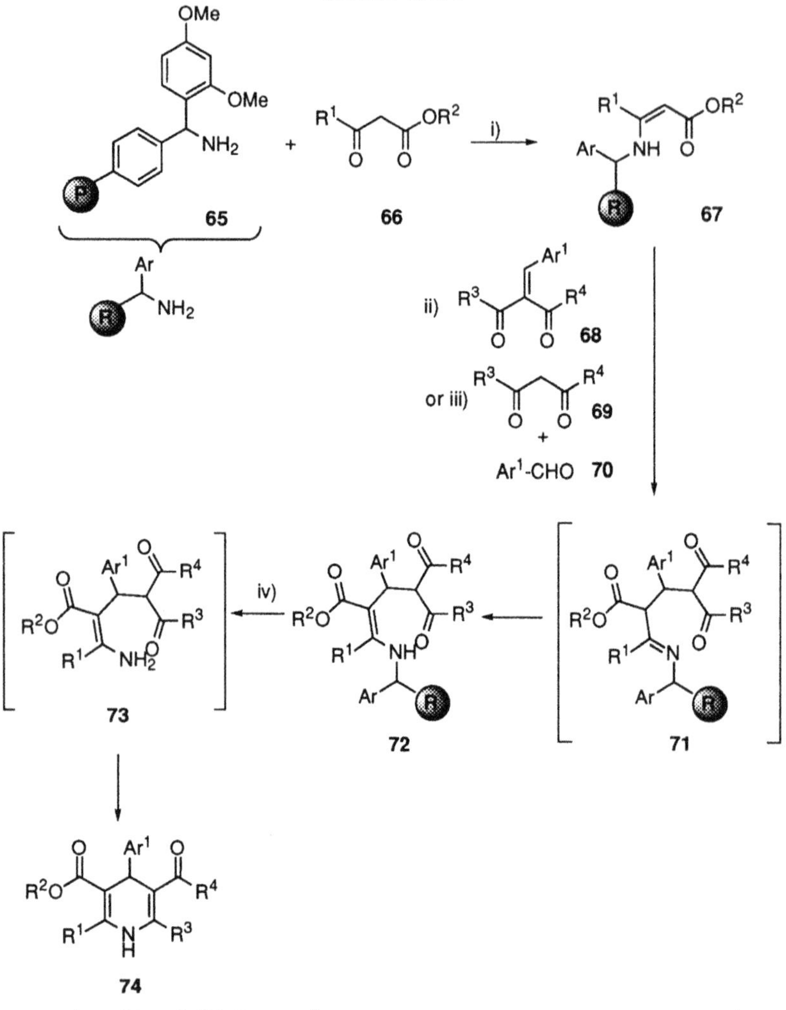

Reagents and conditions: i) CH$_2$Cl$_2$, r.t., 4Å molecular sieves; ii) aryliden β-ketoesters **68**, 4Å molecular sieves, py, 45°C; iii) β-ketoester or β-diketone **69**, arylaldehyde **70**, 4Å molecular sieves, py, 45°C; iv) 3% TFA/CH$_2$Cl$_2$

Useful insight into the mechanism of the formation of **74** was obtained during optimisation of the solid-phase synthesis. The presence of pyridine in the reaction media turned out to be essential for successful transformation of immobilised enamino esters **67** into **74**, probably by favouring isomerisation of **71** into thermodynamically more stable conjugated enamines of type **72**. Studies with ^{13}C-labeled immobilised **67** showed that cleavage to **73** had to take place prior to cyclisation to **74**.

4.1.2.3. 2,9-Disubstituted purines

The purine ring is a common structural element of the substrates and ligands of many biosynthetic, regulatory, and signal transduction proteins including cellular kinases, G proteins and polymerases playing key roles in many celullar processes. Therefore, it is reasonable to expect that combinatorial libraries of purine derivatives may provide inhibitors of these processes that are useful biological probes or lead molecules for drug development efforts. Taking as a reference the relatively selective inhibitor Olomoucine **75**, *Schultz et al.*[41] developed a method for the combinatorial synthesis of purine analogues. Since it has been established that the 6-benzylamino group contributes significantly to the specificity and binding affinity of Olomoucine, in the selected synthetic strategy, the benzylamino group was incorporated at an early stage into the purine core both as a key structural element and as an attachment point to the solid support. Thus, 2-fluoro-6-(4-aminobenzylamino) purine **78** was chosen as the central template. Synthesis of **78** was accomplished by converting commercially available 2-amino-6-chloropurine **76** into **77** by diazotisation in aqueous fluoroboric acid with sodium nitrite followed by monoamination at the 6-position with 4-nitrobenzylamine and subsequent hydrogenation (*Scheme 4.1.15*).

Scheme 4.1.15

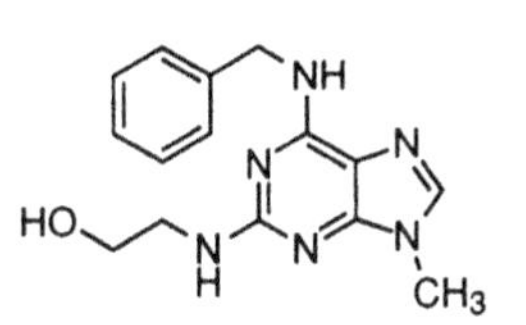

Olomoucine, **75**

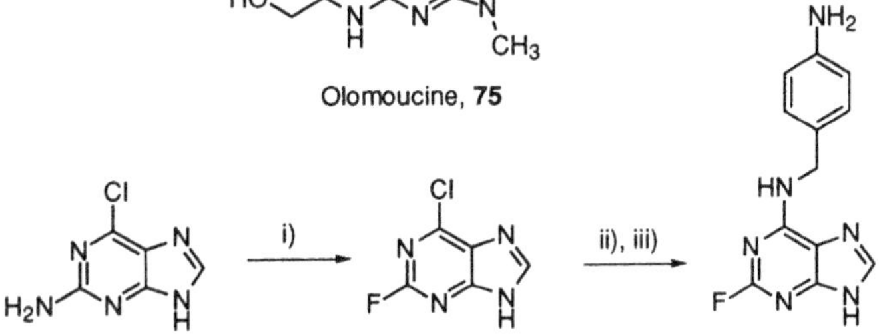

76   **77**   **78**

Reagents and conditions: i) 0.3 M NaNO$_2$, HBF$_4$ (48% in H$_2$O), -15°C;
ii) 4-nitrobenzylamine hydrochloride, DIEA, nBuOH, 50°C; iii) H$_2$, Pd/C

Solid-phase synthesis was initiated by coupling 5-(4-formyl-3,5-dimethoxyphenyloxy)valeric acid (PAL linker,[40] cf. *Chapter 1.7*) to amine derivatised crowns using diisopropylcarbodiimide hydroxybenzotriazole in DMF to give **79**. Subsequent reductive alkylation of **78** by means of triacetoxyborohydride in DMF containing 1% of AcOH led to the desired central template linked to the solid support **80** as depicted in *Scheme 4.1.16*.

Scheme 4.1.16

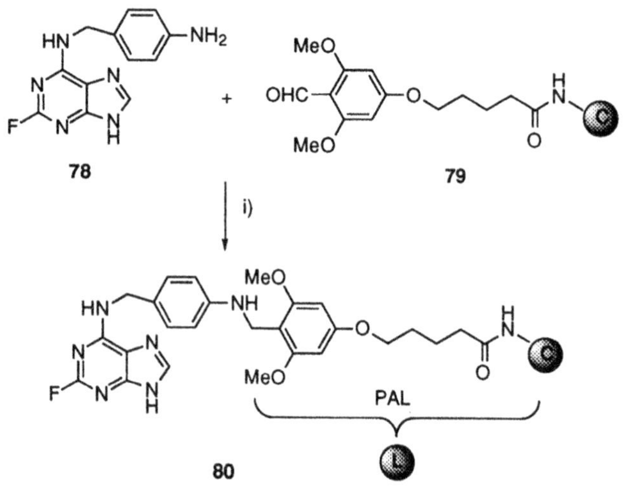

Reagents and conditions: i) NaBH(OAc)$_3$, 1% AcOH in DMF.

The first combinatorial step was performed by alkylation of the N-9 position with a variety of alcohols under *Mitsunobu* conditions to yield **81**. The alkylation reaction was monitored by cleavage of the product from the support followed by analytical reverse-phase HPLC and characterisation by FAB-MS. Good conversion was observed with a variety of alcohols including hindered secondary alcohols. While primary and secondary aliphatic alcohols alkylated exclusively at the N-9 position, reaction with benzylic alcohols resulted in partial alkylation of the N-6 position. Because mild reaction conditions were employed, alcohols containing base sensitive functional groups could be incorporated. In addition, the large number of commercially available alcohols allowed the introduction of considerable diversity in the library at the N-9 position. A second element of diversity was introduced by substitution of the fluorine at the C-2 position of the purine ring with a wide range of different primary and secondary amines in nBuOH/DMSO (4:1) at 90-100°C to yield **82**. Final products **83** were released from the solid support under standard conditions for PAL linkers (95% TFA/H$_2$O) and analysed by HPLC and FAB-MS giving good HPLC yields of the desired compounds (*Scheme 4.1.17*).

Scheme 4.1.17

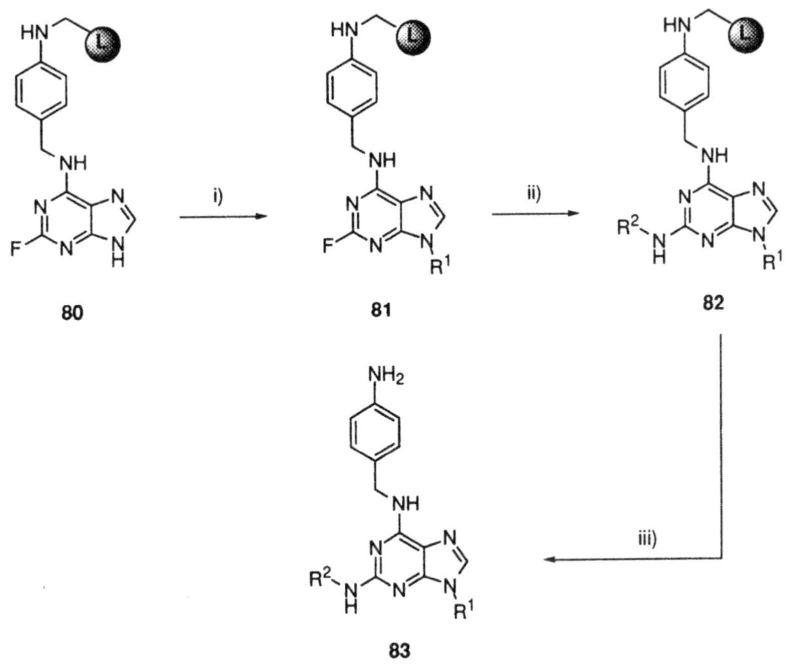

80   81   82

83

Reagents and conditions: i) R^1-OH, 1.0 M PPh_3, 0.5M DEAD, 1:1 (v/v) THF/CH_2Cl_2; ii) R^2-NH_2, nBuOH/DMSO (4:1), 90-100°C, 48h; iii) $TFA/H_2O/Me_2S$ (95:5:5)

4.1.3. Liberation of -*COOH* groups

4.1.3.1. Synthesis of quinolones

Quinolones are known antibacterial agents which represent a class of highly potent, broad-spectrum antibiotics. Their mode of action is believed to involve inhibition of DNA gyrase,[42] and more recently they have been shown to have *in vitro* activity against *M. Tuberculosis*.[43,44]

A solid phase approach for the synthesis of a small library of quinolones (including the known drug Ciprofloxacin) using the diversomer® technology has been reported by *MacDonald* and coworkers.[45] Thus, starting from 2,4,5-trifluorobenzoylacetic acid ethyl ester **85**,[46] transesterification with *Wang* resin **84** by heating the mixture in toluene with a catalytic amount of DMAP led to the corresponding *O*-tethered β-ketoester **86**. Following *Knoevenagel* condensation with dimethylformamide dimethyl acetal resulted in the formation of the bis-acceptor intermediates

87 which were treated with a series of primary amines to produce a number of different enamines **88** that underwent cyclisation by intramolecular nucleophilic aromatic substitution to the corresponding polymer-bound quinolones **89**. Further nucleophilic aromatic substitution allowed the replacement of a second fluorine atom by cyclic secondary amines to yield **90**. Final treatment with 40% TFA produced the final quinolone carboxylic acids **91**. Combining three benzoylacetic acids, three primary amines and four secondary cyclic amines, a focused library of 36 different quinolones of type **91** was prepared[47] (*Scheme 4.1.18*). It is interesting to mention, that the biological activity of Ciprofloxazin, synthesised *via* this route could only be confirmed after HPLC purification. This fact underscores the importance of purity of the synthesised compounds.

Scheme 4.1.18

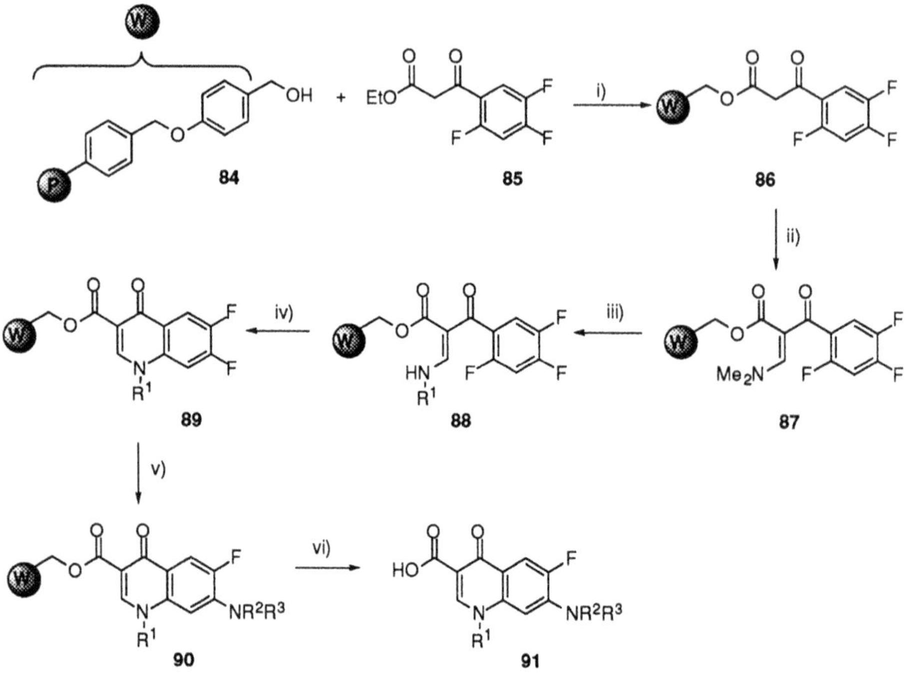

Reagents and conditions: i) DMAP, toluene, 110°C, 18h; ii) $(CH_3O)_2CHN(CH_3)_2$, THF, 25°C, 18h; iii) R^1-NH_2, THF, 25°C, 72h; iv) tetramethylguanidine, CH_2Cl_2, 55°C, 18h; v) R^2R^3NH, NMP, 110°C, 4h; vi) TFA/CH_2Cl_2 (4:6), 25°C, 1h

4.1.3.2. Synthesis of pyridines and pyrido[2,3-b]pyrimidines

The pyridine core structure is a key element of various drugs, including numerous antihistamines, as well as antiseptic, antiarrhytmic, antirheumatic, and other pharmaceuticals, and in fact, pyridines are amongst the most frequent heterocyclic compounds present in commercial pharmaceutical agents.[48] Based on *Knoevenagel* and *Hantzsch* condensation chemistry, *Gordeev et al.*[39] developed a protocol for the solid-phase synthesis of highly molecularly diverse pyridines and pyrido[2,3-b]pyrimidines. Thus, starting from hydroxy functionalised polymers such as *Wang* or *Sasrin* resin **92**, reaction with diketene **93** in the presence of a catalytic amount of DMAP resulted in the formation of *O*-tethered β-ketoester **94**. *Knoevenagel* condensation of **94** with different aldehydes gave rise to the bis-acceptor polymer-bound benzylidene derivatives **96**. Next, a *Hantzsch*-type heterocyclisation with different α-oxo enamine building blocks of type **97** was carried out yielding the corresponding 1,4-dihydropyrimidines **98**. Oxidation with ceric ammonium nitrate (CAN) in dimethylacetamide afforded the expected immobilised pyridines **99** that were cleanly cleaved from the polymeric support with 95% TFA or 3% TFA (for *Wang* or *Sasrin*, respectively) to afford pyridines **100**.

Scheme 1.4.19

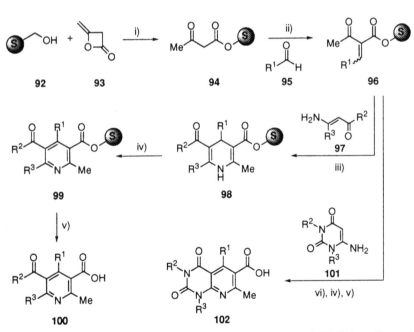

Reagents and conditions: i) DMAP, CH₂Cl₂; ii) aldehyde **95**, cat. piperidine, iPrOH/C₆H₆, Δ; iii) enamine **97**, DMF, Δ; iv) CAN, Me₂NCOMe; v) TFA, CH₂Cl₂; vi) heterocyclic enamine **101**, DMF, Δ

Analogously, employing 6-aminouracils **101** as the enamino component for cyclocondensation provided a facile entry into another biologically important class of nitrogen heterocycles, namely the pyrido[2,3-*b*]pyrimidines **102** as shown in *Scheme 1.4.19* (Synthetic sequence, shown for *Sasrin* resin).

4.1.3.3. Synthesis of mercaptoacyl prolines

Based on a 1,3-dipolar cycloaddition reaction between resin-bound azomethine ylides and electron-poor olefins (*e.g.* acrylates, cinnamates, conjugated enones, maleinimides, etc.) as reactive dipolarophiles, *Gallop et al.*[49] developed a protocol for the solid-phase synthesis of highly substituted pyrrolidines and applied this strategy to prepare a library of ~500 mercaptoacyl prolines, structurally related to the clinically important antihypertensive agent captopril. By screening this library against angiotensin converting enzyme (ACE), an unusually potent inhibitor of this therapeutically important metalloprotease was identified. Since in this case, a carboxyl group was a constant element of the target structures, acid-labile *Sasrin* or TentaGel® resins **103**, preloaded with Fmoc-protected aminoacids, were used as supports for the solid-phase pyrrolidine synthesis. Deprotection of the Fmoc group with piperidine provided the corresponding α-aminoesters **104**. Condensation reaction with different aromatic and heteroaromatic aldehydes **105** at room temperature in neat trimethylorthoformate as dehydrating agent, smoothly afforded the corresponding resin-bound aryl imines **105**. *Lewis* acid-promoted formation of *N*-metalloazomethine ylides, and subsequent cycloaddition reaction with different electron-deficient olefins **106** under basic conditions, afforded pyrrolidines **107**. Characterisation of representative proline analogues **108** was achieved by acidic cleavage from the resins by means of 10% TFA in yields ranging between 50-80% and with diastereoselectivities ranging from 2.5:1 to greater than 10:1 (*Scheme 4.1.20*).

Scheme 4.1.20

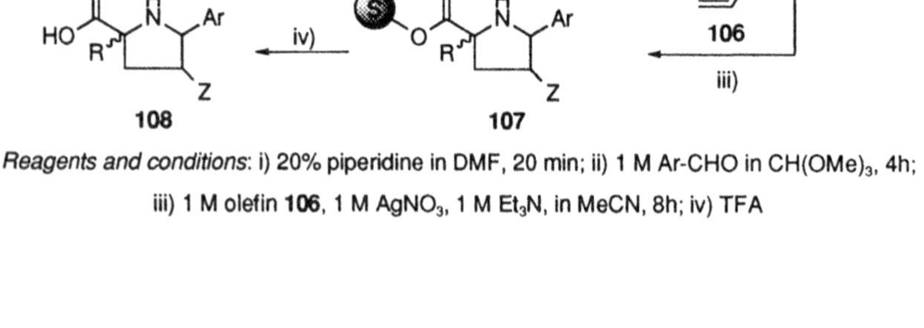

Reagents and conditions: i) 20% piperidine in DMF, 20 min; ii) 1 M Ar-CHO in CH(OMe)$_3$, 4h; iii) 1 M olefin **106**, 1 M AgNO$_3$, 1 M Et$_3$N, in MeCN, 8h; iv) TFA

Since functionalised prolines and proline analogues are frequently found as C-terminal residues in numerous ACE inhibitors, *Gallop et al.* used this solid-phase chemistry to generate a library of mercaptoacyl prolines. Thus the library was prepared by the split synthesis[50] using four amino acids, four aldehydes, five olefins, and three mercaptoacyl chlorides as shown in *Scheme 4.1.21*.

Scheme 4.1.21

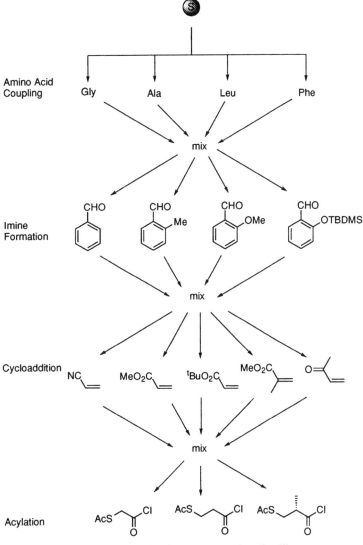

Split-mixed synthesis of a mercaptoacyl proline library

279

4.1.3.4. Synthesis of β-lactams

In a similar fashion, resin-bound imines **109** were employed to prepare a library of structurally diverse β-lactams by [2+2] cycloaddition reactions with different ketenes. Thus, as shown in *Scheme 4.1.22*, amino acids tethered to the acid labile *Sasrin* resin (**103**) were condensed quantitatively to imines **109** by using a large excess of alkyl, aryl, or α,β-unsaturated aldehydes in a mixture of trimethylorthoformate and dichloromethane. Optimisation studies of the [2+2] cycloaddition step, showed that conversion to β-lactams **110** could only take place by slow addition of acid chlorides to a suspension of the imine resin at 0°C in the presence of triethylamine. By using a large excess of ketene at high concentration, the cycloaddition of imines derived from even sterically hindered amino acids (*e.g.* valine) could be carried out with full conversion. After mild TFA cleavage from the resin and preparative HPLC purification, the β-lactams **111**, **112** were isolated in yields of 55-97%.

Scheme 4.1.22

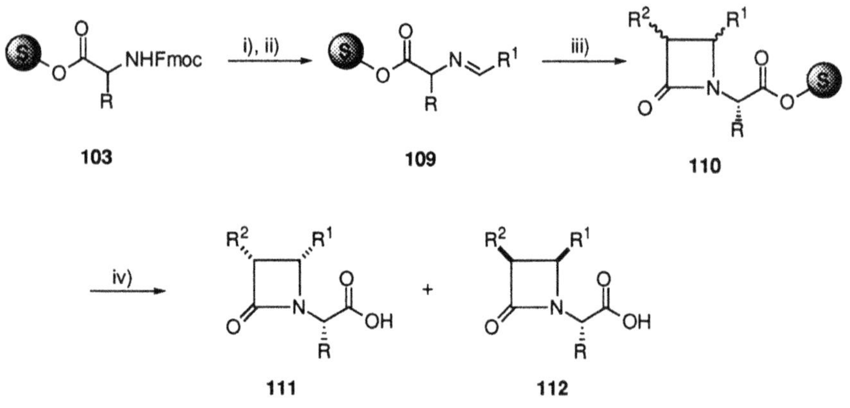

Reagents and conditions: i) 30% piperidine in NMP, 45 min (→**104**); ii) 0.8 M R¹-CHO in (MeO)₃CH/CH₂Cl₂ (1:1), 3 h; iii) 0.8 M R²CH₂COCl, 1.1M Et₃N in CH₂Cl₂, 0°C to r.t., 16 h; iv) 3% TFA in CH₂Cl₂

Although the solid-phase formation of this clinically valuable pharmacophore occurred with high *cis* selectivity, the diastereoselectivity induced by the asymmetric center of the amino acid was only very moderate. Induction could be achieved, however, with an optically active ketene. By using this approach, a model library of 25 β-lactams of type **111/112** was initially formed in higher than 80% purity, and this chemistry was subsequently used to generate a combinatorial library of thousands of 3,4-bis-substituted 2-azetidinones of sufficient purity to be directly submitted for bioassays.

4.1.3.5. Synthesis of quinazolin-2,4-diones

Fused pyrimidine-2,4-diones are important heterocyclic pharmacophores: examples include natural purine bases, alkaloids, and pteridines. Benzo derivatives in this series (quinazoline-2,4-diones) bear ample precedence as potent ligands and inhibitors of receptors and enzymes of pharmaceutical interest. Based on these precedents, *Gordeev et al.*[51] reported a solid-phase protocol for the synthesis of fused pyrimidine-2,4-diones.

The developed solid-phase approach started with the preparation of immobilised ureas **114** obtained *via* acylation of amino acids linked to acid-labile *Sasrin* resin **104** with *ortho*-methoxycarbonyl aryl isocyanates or activated *para*-nitrophenyl carbamates **113**. Alternatively, ureas of type **114** could also be generated by reaction of anthranilic esters **115** with immobilised amino acid-derived isocyanates (easily generated from tethered amino acids with triphosgene or phosgene in toluene in excess of a base such as 2,6-lutidine) or activated carbamates **116** (*Scheme 4.1.23*).

Scheme 4.1.23

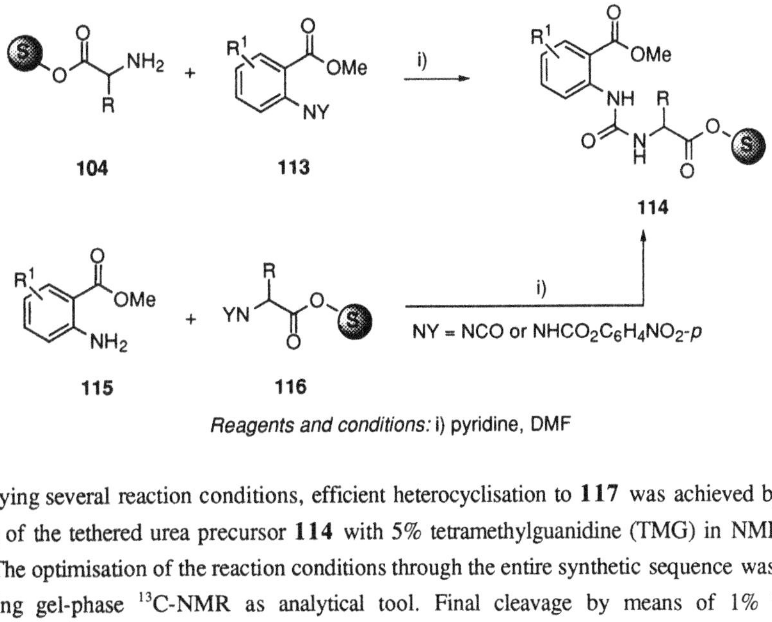

Reagents and conditions: i) pyridine, DMF

After trying several reaction conditions, efficient heterocyclisation to **117** was achieved by gentle heating of the tethered urea precursor **114** with 5% tetramethylguanidine (TMG) in NMP at 55-65°C. The optimisation of the reaction conditions through the entire synthetic sequence was carried out using gel-phase ^{13}C-NMR as analytical tool. Final cleavage by means of 1% TFA in dichloromethane afforded quinazoline-2,6-diones **118** in high yields and high levels of purity (typically 90-98% as determined by HPLC analysis of the crude reaction mixture). Moreover, additional molecular diversity was introduced by alkylation of immobilised N-1-unsubstituted quinazolindiones **117**. These transformations could be accomplished under mild reaction conditions, either by using alkyl halides in the presence of TMG in NMP at room temperature, or by using a *Mitsunobu* reaction type with alcohols in the presence of diisopropyl azodicarboxylate and Ph$_3$P in THF. In both cases, the corresponding N-1-substituted products **119** were obtained (after TFA cleavage from resin) in high yields and purities (*Scheme 4.1.24*).

Scheme 4.1.24

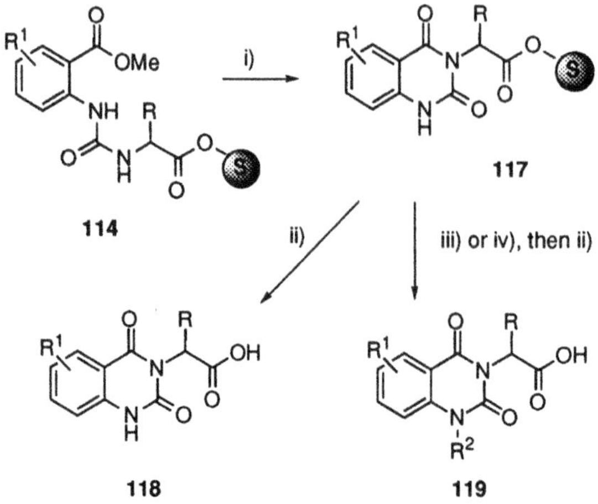

Reagents and conditions: i) tetramethylguanidine, NMP, 55-65°C; ii) 1% TFA in CH$_2$Cl$_2$; iii) tetramethylguanidine, NMP, R^2-X; iv) R^2-OH, DIAD, PPh$_3$, THF

This sequence of transformations was repeated using different amino acid and isocyanate or activated carbamate building blocks to generate a collection of molecularly diverse quinazolindiones **118** and **119**. Since this solid-phase synthesis employed a strong base (TMG) under thermal conditions in the cyclisation step and in order to prove that the process took place with no racemisation, immobilised dipeptide H-L-Phe-L-Ala-*Sasrin* was converted to the corresponding quinazolindione **120**, showing by HPLC and ^{1}H-NMR analysis that only one diastereoisomer was obtained. In addition, this methodology was successfully extended to the preparation of fused heterocyclic pyrimidine-2,4-diones by employing heterocyclic amino esters in place of the anthranilic acid derivatives, as demonstrated by the synthesis of thieno[2,3-*d*]pyrimidine-2,4-dione **121** from *Sasrin*-linked L-alanine and *p*-nitrophenyl carbamate derivative of methyl 3-aminothiophene-2-carboxylate (*Figure 4.1.1*).

Finally, due to the mild reaction conditions employed for the alkylation step, incorporation of hydrophobic as well as hydrophilic functionalities was readily achieved by employing alkylating agents with appropriately protected functional groups (acids, bases, amides, esters, alcohols, etc.).

Figure 4.1.1

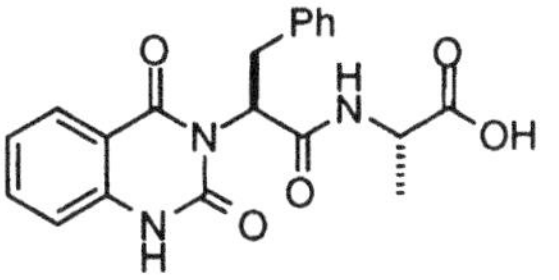

120 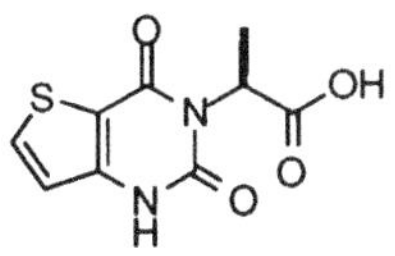   **121**

4.1.3.6. Synthesis of dihydro- and tetrahydroisoquinolines

Based on a *Bischler-Napieralski* reaction[52] *Meutermans* and *Alewood* developed a solid-phase synthesis of tetrahydroisoquinolines.[53] The synthetic route is illustrated in *Scheme 4.1.25*. Thus, polystyrene-bound, *N*-Boc protected 3,4-dimethoxyphenylalanine **122** was treated with TFA to give **123**. Acylation with acetic acid derivatives **124** using HBTU as coupling reagent, afforded **125** which were subsequently reacted with POCl$_3$ under optimised reaction conditions to furnish dihydroisoquinolines **126**.

Scheme 4.1.25

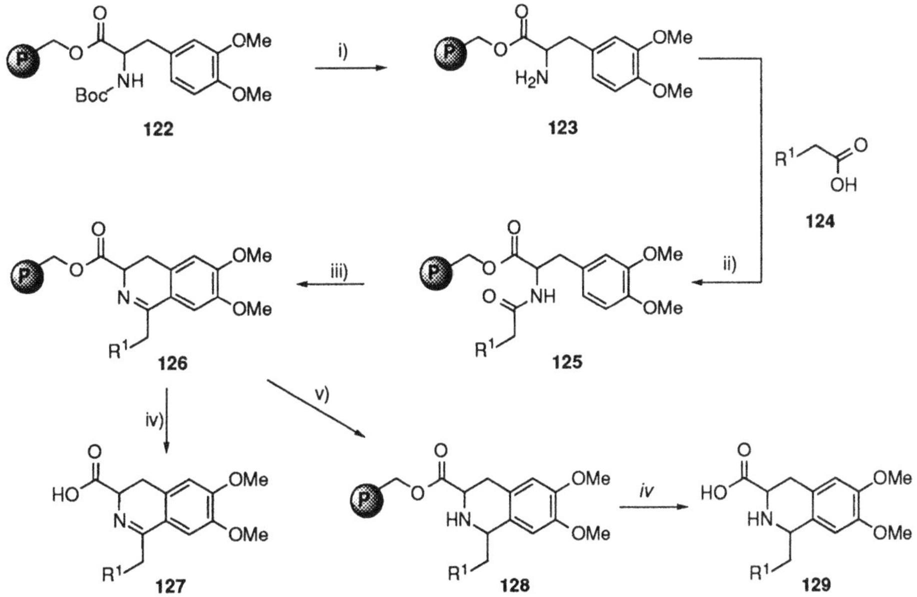

Reagents and conditions: i) TFA, CH$_2$Cl$_2$; ii) **124**, HBTU, DIEA; iii) POCl$_3$, toluene, 80°C; iv) HF, *p*-cresol; v) NaCNBH$_3$, MeOH/HCl

283

Best results were obtained using a large excess (30 eq) of freshly distilled $POCl_3$ at 80°C in toluene for 8 hours. The desired dihydroisoquinolines **127** were isolated following cleavage with HF in *p*-cresol and HPLC purification. To further extend this strategy to the synthesis of tetrahydroisoquinolines, the dihydroisoquinolines **126** were treated with $NaBH_3CN$ and yielded, after HF cleavage, tetrahydroisoquinolines **129**.

4.1.4. Liberation of $-CONH_2$ groups

4.1.4.1. Synthesis of 3,4-dihydro-2(1H)-quinolinones

Based on the tea-bag technology,[54] *Pei et al.*[55] developed a solid phase synthesis of 3,4-dihydro-2(1*H*)-quinolinones via β-lactam intermediates. Monocyclic β-lactams, readily available from simple starting materials through well established chemistry, have been shown to be versatile and reactive intermediates in the synthesis of heterocycles, amino acids and their derivatives.[56-60] Hence, initially *N*-Boc protected amino acids **130** were attached onto polystyrene MBHA resin **55** using standard peptide coupling conditions (DIC/HOBT/DIEA) either in CH_2Cl_2 or DMF depending on the solubility of the amino acids. Complete coupling with each amino acid was followed with the ninhydrin test. Removal of the Boc protecting group by treatment with 55% TFA in CH_2Cl_2 and subsequent condensation with *ortho*-nitrobenzaldehyde **132** in CH_2Cl_2 in the presence of anhydrous sodium sulfate as dehydrating agent furnished imines **133**. After washing with CH_2Cl_2 and drying under high vacuum over P_2O_5, [2+2] cycloaddition of **133** with ketenes was carried out in CH_2Cl_2 at -78°C. The ketenes were generated *in situ* from the corresponding acetyl chloride in the presence of triethylamine. To monitor the reaction sequence up to this point, β-lactam intermediates **134** from each amino acid were cleaved from the resin using HF/anisole (95/5) and analysed by ^{1}H-NMR. In all cases, *cis* β-lactams were obtained as single products in almost quantitative yields. The nitro group of the β-lactams intermediates **134** was reduced to an amino group using $SnCl_2$ (2.0 M) in DMF at room temperature. Under these reaction conditions, the β-lactam ring underwent rearrangement to give the corresponding 3,4-dihydro-2(1*H*)-quinolinones **135**, through intramolecular nucleophilic attack of the β-lactam amide moiety by the newly generated amino group. Dihydroquinolinones **136** were obtained in excellent yields after cleavage using HF/anisole (95/5) and with high levels of purity (>85%) as determined by HPLC-MS (*Scheme 4.1.26*). After validation of this methodology, a library containing 4140 dihydroquinolinones (2070 pairs of enantiomers) of type **136** using the split-mixed synthesis[61] was prepared by combining 69 amino acids, 6 *ortho*-nitrobenzaldehydes and 5 acid chlorides.

Scheme 4.1.26

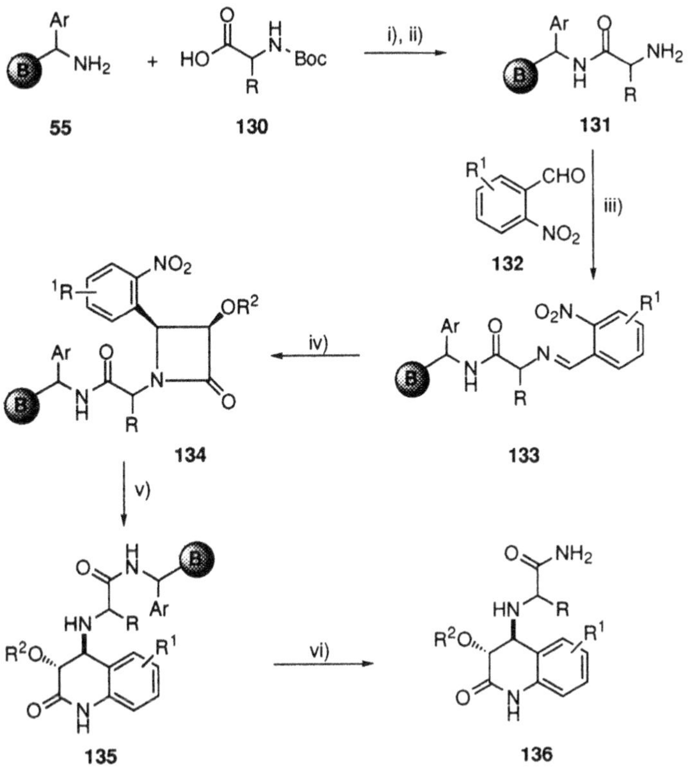

Reagents and conditions: i) DIC, HOBt, DIPEA in CH_2Cl_2 or DMF; ii) 55% TFA in CH_2Cl_2;
iii) o-nitrobenzaldehyde **132**, Na_2SO_4, CH_2Cl_2; iv) R^2OCH_2COCl, Et_3N, CH_2Cl_2, -78°C
v) $SnCl_2$, DMF, r.t.; vi) HF/anisole

4.1.4.2. Synthesis of 1,2,3,4-tetrahydroquinoxalin-2-ones

2-Quinoxalinones were recently found to be inhibitors of aldose reductase[62] and partial agonists of the γ-aminobutyric acid (GABA)/benzodiazepine receptor complex.[63,64] In addition, the corresponding N-oxides were also shown to be potent angiotensin II receptor antagonists.[65] *Lee et al.*[66] developed a solid-phase methodology towards the synthesis of this class of interesting N-heterocycles.

285

Scheme 4.1.27

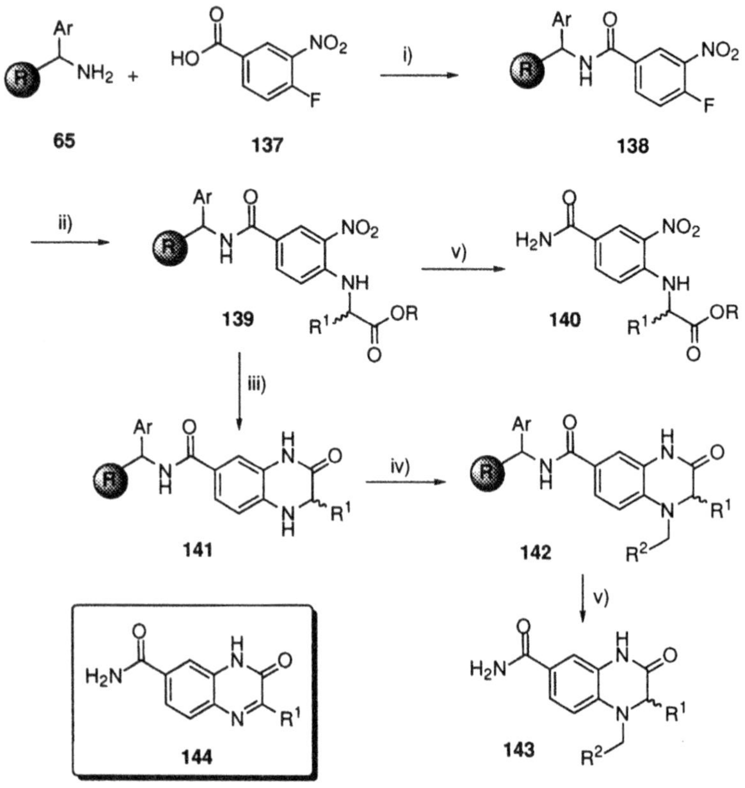

Reagents and conditions: i) HATU, DIEA, DMF, r.t., 24h; ii) H$_2$NCH(R^1)COOR, (R = Me or Et), DIEA, DMF, r.t., 3d; iii) SnCl$_2$H$_2$O, DMF, r.t, 24h; iv) R^2-CH$_2$-X, K$_2$CO$_3$, acetone, 55°C, 24h; v) TFA/H$_2$O (95/5)

Starting from *Rink* resin **65**, solid-phase synthesis began with attachment of 4-fluoro-3-nitrobenzoic acid **137** using HATU and DIEA as coupling reagents to give **138**. The resin beads **138** were partitioned into five equal portions, and each portion was then allowed to react with five different α-amino esters to give **139**. The aromatic substitution reaction of the activated aryl fluoride was accomplished in the presence of DIEA in DMF for 3 days. To verify the extent of aromatic substitution, small portions of the resins **139** were treated with 95% TFA over a period of 50 min. The cleaved products **140** were analysed by LC, MS and ^{1}H-NMR showing purities greater than 95% and complete absence of starting material. Reduction of the nitro group with monohydrated SnCl$_2$ in DMF[67] directly afforded to the resin-bound quinoxalinones **141**. TFA treatment of **141** and analysis of the cleaved products showed no traces of noncyclised products.

The only by-products so far observed were oxidised compounds **144** in amounts up to 25%. Since acidic treatment of quinoxalinones lacking substitution at the N-4 position are known to produce oxidation of the 3,4-C,N bond of the quinoxalinone,[68] and assuming that the oxidation was due to the TFA cleavage conditions, the resins **141** were again divided into five equal portions and subjected to alkylation with different alkyl halides in refluxing acetone in the presence of K_2CO_3 affording **142**. Cleavage under standard conditions released the desired quinoxalinones **143** with no traces of oxidation products, in yields ranging between 32-93% (*Scheme 4.1.27*).

4.1.4.3. Synthesis of 1,4-benzodiazepine-2,5-diones

1,4-Benzodiazepine-2,5-diones represent a versatile pharmacophore with a wide range of pharmaceutical utility which can function as opiate receptor antagonist,[69] anticonvulsant agents[70] and glycoprotein mimics.[71] *Goff et al.*[72] developed a solid-phase synthesis of 1,4-benzodiazepine-2,5-diones to prepare a combinatorial library of hybrid molecules combining a benzodiazepinedione nucleus with an appended *N*-substituted glycine (peptoid) side chain. Thus, solid-phase synthesis started with acylation of *Rink* resin **65** with bromoacetic acid followed by amination with isobutylamine. A second bromoacetylation and nucleophilic displacement with the amino acid methyl or ethyl esters in DMSO gave intermediates **147** (*Scheme 4.1.28*).

Scheme 4.1.28

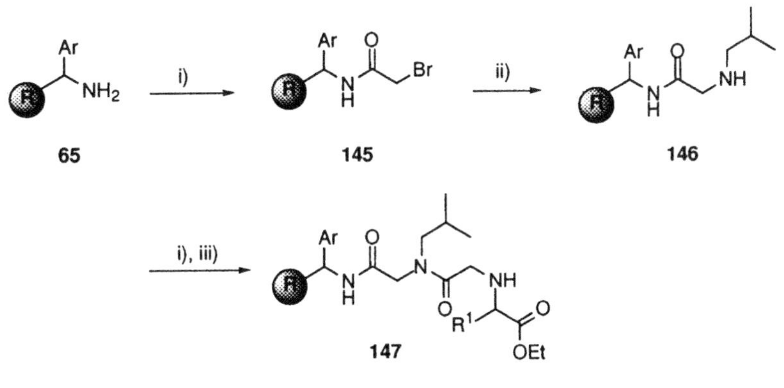

Reagents and conditions: i) 0.6 M bromoacetic acid, 0.6 M DIC in DMF, 1h, r.t.;
ii) 2.0 M isobutylamine in DMSO, 2h, r.t.; iii) amino acid ester

Scheme 4.1.29

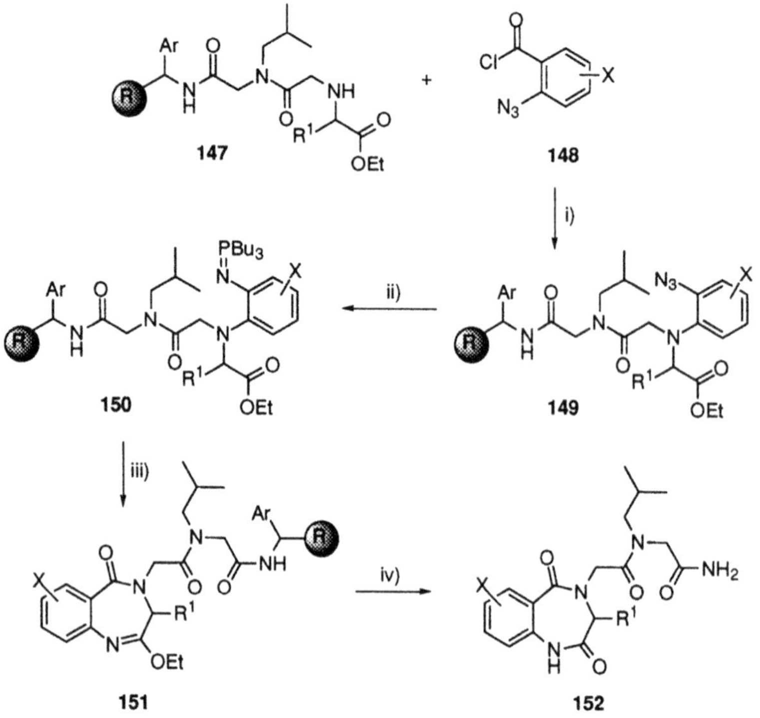

Reagents and conditions: i) 0.5 M **148** in 1,2-dichloroethane, Et$_3$N (1eq), r.t., 2x30 min; ii) 0.6 M tributylphosphine in toluene, 2x30 min, r.t.: iii) 130°C, *p*-xylene, 5-7h; iv) 95/5 TFA/H$_2$O, r.t., 20 min.

Compounds of type **147** were directly acylated with the appropriate freshly prepared *o*-azidobenzoyl chloride **148**. Following treatment of **149** with tributylphosphine in toluene at room temperature afforded immobilised iminophosphoranes **150** that cyclised at 130°C presumably to the corresponding polymer-bound benzodiazepinediones **151**. Acidolytic cleavage with 95% TFA in H$_2$O and two lyophilisations from glacial acetic acid afforded crude **152** (*Scheme 4.1.29*).

4.1.5. Multigeneration assembly strategies with mono-functional liberation of miscellaneous groups

4.1.5.1. Synthesis of substituted 1,2,3-triazoles

Based on an aminoalkylurethane linker attached to the *Wang* resin **155**, *Zaragoza et al.*[73] developed a solid-phase synthesis of 1,2,3-triazoles. Thus, as shown in *Scheme 4.1.30*, *Wang* resin **84** was primarily treated with 4-nitrophenyl chloroformate **153** in the presence of pyridine to give **154** and then reacted with piperazine in DMF to produce **155**. Subsequent reaction with a freshly prepared solution of 3-oxobutyric acid phenyl ester[74] afforded resin-bound 3-oxobutyryl piperazine **156**. In the presence of triethylorthoformate, the condensation of **156** with primary aliphatic amines readily produced the corresponding 3-amino-2-butenoic acid amides attached to the solid support (**157**).

Scheme 4.1.30

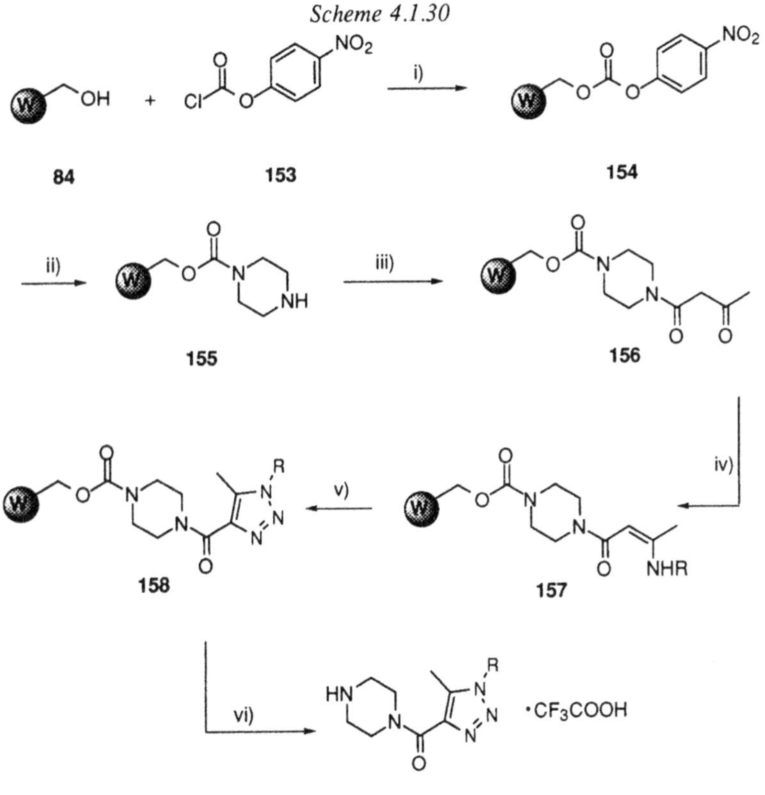

Reagents and conditions: i) CH$_2$Cl$_2$, pyridine, then **153**; ii) piperazine, DMF, r.t.,13h;
iii) 3-oxobutyric acid phenyl ester (5 eq), toluene, DIEA, 2h; iv) R-NH$_2$, DMF/HC(OEt)$_3$ (1:1), 24h;
v) tosylazide, DIEA, DMF; vi) 60% TFA in CH$_2$Cl$_2$

The cyclisation to 1,2,3-triazoles **158** ocurred after treatment with tosyl azide in the presence of a tertiary amine. Following removal of solvents, the treatment of the resin-bound 1,2,3-triazoles **158** with 60% TFA in CH$_2$Cl$_2$ afforded triazoles **159** as trifluoroacetates salts in good purities. The mild reaction conditions used in this protocol enabled its application to building blocks with a large number of different functionalities. Only when using electron rich benzylic amines in the enamine forming step, mixtures of products were obtained.

4.1.5.2. Synthesis of thiophenes and dihydrothiazoles

A similar linker strategy was used by *Zaragoza*[75] in a solid-phase synthesis of substituted 3-aminothiophenes and 2-methylene-2,3-dihydrothiazoles by adapting the *Laliberté* thiophene synthesis.[76,77]

Scheme 4.1.31

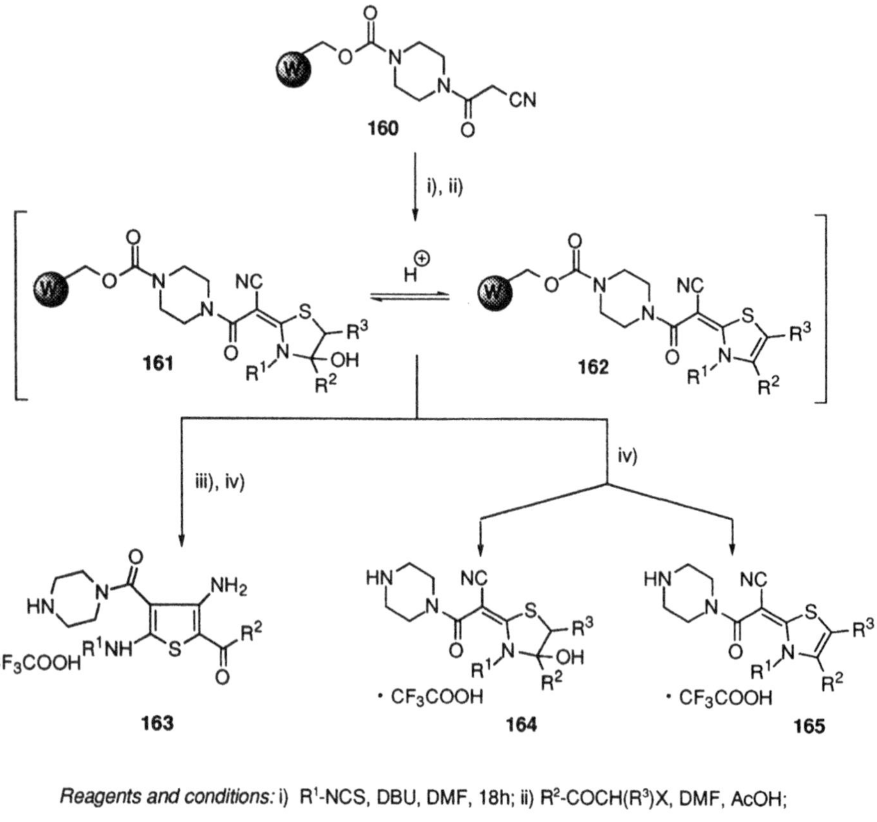

Reagents and conditions: i) R^1-NCS, DBU, DMF, 18h; ii) R^2-COCH(R^3)X, DMF, AcOH;
iii) DBU, DMF, 20h; iv) 50% TFA in CH$_2$Cl$_2$, r.t., 3h

Thus, treatment of the resin bound (cyanoacetyl)piperazine **160** with aliphatic or aromatic isothiocyanates in the presence of DBU, followed by *S*-alkylation with α-haloketones under slightly acidic or neutral conditions resulted in the formation of the intermediates **161** and **162** (which one being the predominant form was determined by the electronic properties of the substituents R^1-R^3). The treatment of these intermediates **161/162** with DBU in DMF following acidolytic cleavage of the resin with TFA yielded 3-aminothiophene derivatives **163** as trifluoroacetates. This synthetic sequence towards **163** encountered however some limitations. For instance, complex mixtures of products were obtained in those cases, where strongly electron-donating isothiocyanates or α-haloketones were used, and in general, no thiophenes resulted from aliphatic haloketones, except for 3-bromo-1,1,1-trifluoro-2-propanone.

On the other hand, direct treatment of the intermediates **161, 162** with TFA yielded the 2-methylene-2,3-dihydrothiazoles **164** and **165** of unknown configuration. The reaction sequence leading to **164** and **165** showed higher tolerance towards variation of the substitution pattern. Generally pure products were obtained for both, electron-donating and electron-withdrawing isothiocyanates or α-haloketones. For most of the studied cases, the dehydrated 2-methylenethiazoles **165** were obtained as single products. 4-Hydroxythiazolidines of type **164** resulted only in those cases where R^2 was a strongly electron-withdrawing group (*Scheme 4.1.31*).

4.2. Multigeneration assembly strategies using traceless linkers

4.2.1. Synthesis of 1,4-benzodiazepines

In the previous described synthetic strategies of 1,4-benzodiazepines, the benzodiazepine nucleus was attached to the solid support through a phenolic functionality. After cleavage from the support, the residual hydroxyl group may have a negligible, positive or negative effect on the biological or chemical activity. For certain applications however, it would be desirable to have a linker strategy that after cleavage from the support would leave behind no trace of the solid-phase synthesis sequence. For this purpose *Ellman et al.*[78] developed a silicon-based linker that could serve for the traceless solid-phase linkage of aromatic compounds. As the first demonstration for this strategy, the synthesis of 1,4-benzodiazepines[79] was undertaken. The synthesis of an appropriately silicon-containing (aminoaryl) stannane was initiated with protection of 4-bromoaniline **166** using the 2-(4-biphenylyl)isopropyloxycarbonyl (Bpoc) group.[80] Carbamate deprotonation with KH, lithium-halogen exchange with *tert*-butyl lithium and quenching with 3-butenylchlorodimethylsilane gave the protected arylsilane **167** in high yield.

Scheme 4.2.1

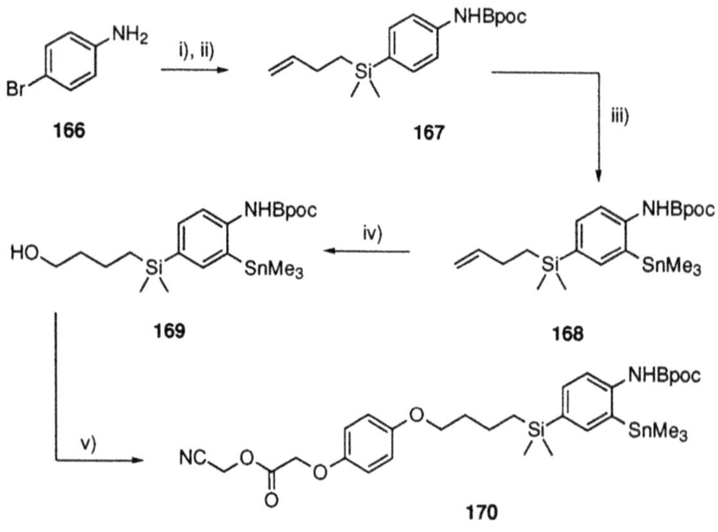

Reagents and conditions: i) Bpoc-OPh, KH; ii) a) KH, b) tBuLi, 3-butenyl-chlorodimethylsilane; iii) a) nBuLi, b) tBuLi, c) Me₃SnCl; iv) 9-BBN, then H₂O₂; v) cyanomethyl 4-hydroxyphenoxyacetate, PPh₃, DEAD

The stannane was introduced using the directed *ortho*-metalation reaction followed by addition of trimethyltin chloride to give **168**. Hydroboration of arylstannane **168** with 9-BBN and oxidative work-up with basic peroxide provided the primary alcohol **169**. *Mitsunobu* reaction of this alcohol with the cyanomethyl ester of 4-hydroxyphenoxyacetic acid afforded preactivated ester derivative **170** in good yield (*Scheme 4.2.1*).

Solid-phase synthesis was initiated with acylation of **170** onto (aminomethyl)polystyrene resin using DMAP and DIEA to give **171**. A synthetic sequence analogous to the one described in *Scheme 4.1.7* afforded 1,4-benzodiazepines **172** linked to the solid support. The aryl-silicon bond was then cleaved in the final step with anhydrous HF, delivering benzodiazepines **173** in good yields (*Scheme 4.2.2*).

Scheme 4.2.2

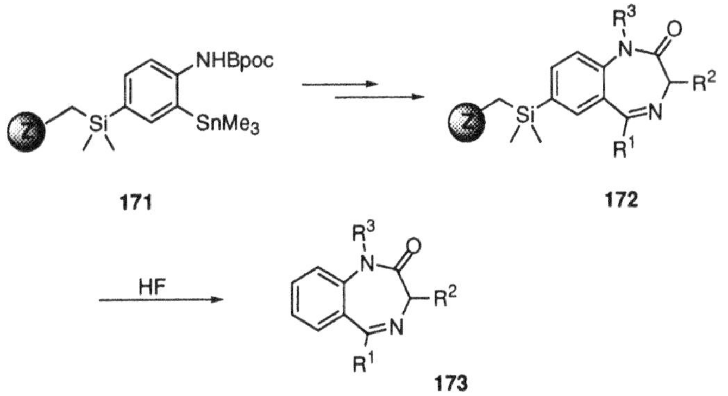

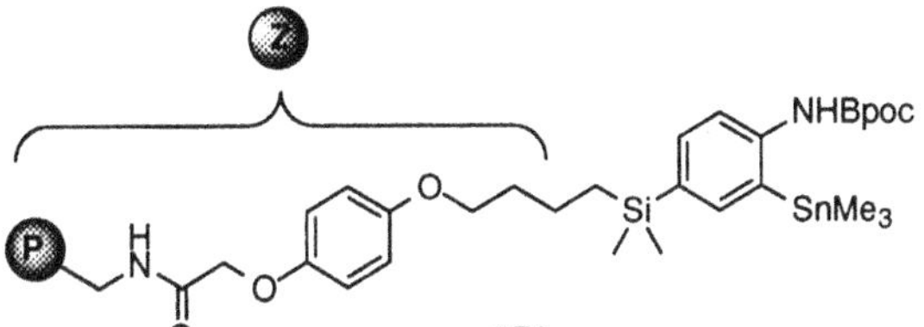

Furthermore, to obtain a linking element that could be easily cleaved with trifluoroacetic acid, thus avoiding the need of using highly toxic HF, the same research group investigated a similar linker strategy based on germanium.[81] The synthesis of the appropriately functionalised germanium-linked resin is shown in *Scheme 4.2.3*.

Protection of 4-bromoaniline **166** as the 2-(4-methylphenyl)isopropyl carbamate (Mpc group) and sequential deprotonation, lithium-halogen exchange and trapping of the aryl anion with chlorodimethyl{6-[(triethylsilyl)oxy]hexyl}germane gave silyl ether **174**. The directed orthometalation reaction was used as before to introduce the trimethyltin group to give **175**. Removal of the silyl group with TBAF afforded alcohol **176** that was allowed to react with cyanomethyl (4-hydroxyphenoxy)acetate under *Mitsunobu* conditions to afford preactivated ester **177**. (Aminomethyl)polystyrene resin **29** was then directly acylated with **177** in NMP to give resin **178**.

Scheme 4.2.3

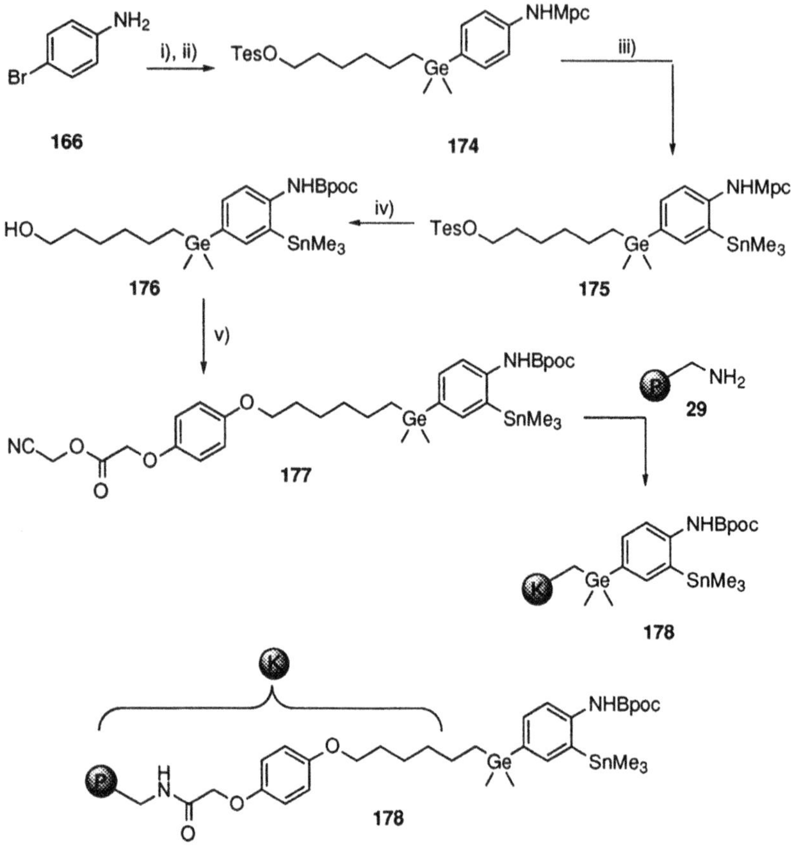

Reagents and conditions: i) Mpc-OPh, KH; ii) KH, then tBuLi, 6-(TesO-hexyl)-Me$_2$GeCl; iii) a) nBuLi, b) tBuLi, c) Me$_3$SnCl; iv) TBAF, THF, 5 min; v) cyanomethyl 4-hydroxyphenoxyacetate, PPh$_3$, DEAD, THF

The solid-phase route to germanium-linked benzodiazepines has been described before (*Scheme 4.1.7*), except for the carbamate deprotection of the Mpc group. While treatment of silyl-linked **180** with 3% TFA caused no detectable 2-aminobenzophenone cleavage from the resin, **181** showed less stability under acidic conditions. Cleavage of the Bpoc carbamate (as in the silicon linker system), produced more than 10% cleavage of 2-aminobenzophenone under a variety of acids and reaction times. The use of a Mpc carbamate group as protecting group (which cleaves four times faster) brought the amount of undesired cleavage to less than 5%. In addition, treatment of **181** with the *Lewis* acid *B*-chlorocatecholborane and 0.5 equivalents of DIEA resulted in a fast and clean deprotection without any undesired 2-aminobenzophenone cleavage. Next, acylation of the aniline, Fmoc deprotection, cyclisation and alkylation afforded resin-bound germanium-linked benzodiazepines **184** which were cleaved from the solid support by treatment with neat TFA at 60°C to afford fully functionalised benzodiazepine derivatives **185**. In addition, electrophilic cleavage of the germanium-linked benzodiazepines **184** was investigated. Thus, it was found that treatment of **184** with elemental bromine for 5 min afforded bromobenzodiazepines **186** through cleavage of the germanium-aryl bond without detectable overbromination (*Scheme 4.2.4*). By using the germanium route, numerous functional groups, *e.g.* amides, esters, acids, phenols, acetals, acids and halides were tolerated. In addition, the ease of electrophilic demetalation, makes this route a significant improvement over the silyl method for traceless 1,4-benzodiazepine synthesis.

Scheme 4.2.4

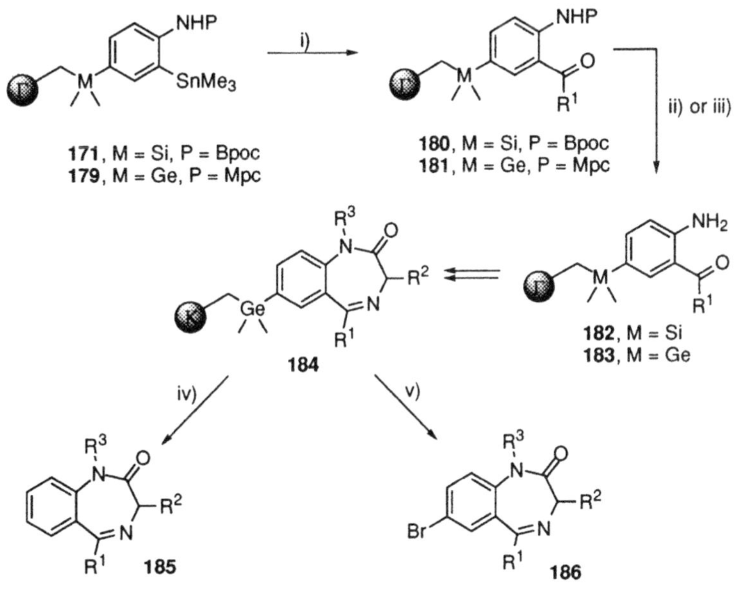
NHP
SnMe3
M
i)
NHP
O
M
R1
ii) or iii)
171, M = Si, P = Bpoc
179, M = Ge, P = Mpc
180, M = Si, P = Bpoc
181, M = Ge, P = Mpc
R3
N
O
R2
N
Ge
R1
184
NH2
O
M
R1
182, M = Si
183, M = Ge
iv)
v)
R3
N
O
R2
N
R1
185
R3
N
O
R2
Br
N
R1
186

Reagents and conditions: i) R^1-COCl, Pd_2dba_3-$CHCl_3$, K_2CO_3, DIEA,THF;
ii) 3% TFA in CH_2Cl_2 (for **180**); iii) B-chlorocatecholborane/DIEA (2:1), CH_2Cl_2 (for **181**);
iv) TFA, 60°C, 24h; v) Br_2 (4 eq), CH_2Cl_2

4.2.2. Synthesis of furans

A traceless linker strategy for the synthesis of functionalised furans based on mesoionic isomünchnones generated on solid support, was developed by *Gallop et al.*[82] The key step of this protocol takes advantage of an efficient [3+2] cycloaddition reaction with electron deficient acetylenes, followed by a thermally promoted cycloreversion reaction.[83-87] As shown in *Scheme 4.2.5*, TentaGel®-NH₂ resin was primarily acylated with different carboxylic acids using diisopropylcarbodiimide (DIC) in the presence of catalytic amounts of DMAP. Conversion of amide **187** to imide **189** was accomplished by treating the resin twice with a 1:1 (v/v) mixture of malonyl chloride **188** in benzene at 60°C. Quantitative diazo-transfer reaction to diazoimide **190** was effected at room temperature using tosyl azide in CH₂Cl₂/Et₃N. Optimisation of this reaction sequence was facilitated by using gel-phase ¹³C-NMR, monitoring these transformations with initially acetylated resin with 2-¹³C-labeled Ac₂O.[88] The analysis indicated that both the imide and diazoimide formation proceeded with >95% conversion.

Scheme 4.2.5

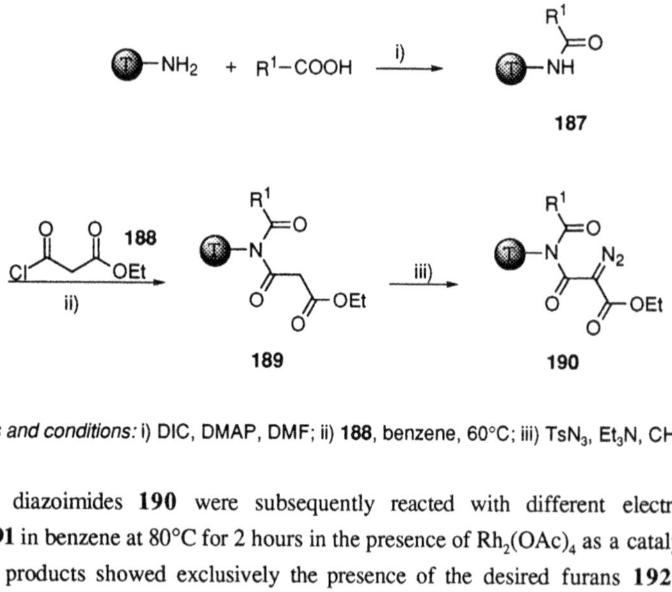

Reagents and conditions: i) DIC, DMAP, DMF; ii) **188**, benzene, 60°C; iii) TsN₃, Et₃N, CH₂Cl₂, 18h

Resin-bound diazoimides **190** were subsequently reacted with different electron deficient acetylenes **191** in benzene at 80°C for 2 hours in the presence of Rh₂(OAc)₄ as a catalyst. Analysis of the crude products showed exclusively the presence of the desired furans **192** and excess unreacted acetylene, which, when sufficiently volatile (*e.g* propiolate esters), were eliminated *in vacuo* to provide furans of high purity. To avoid contamination of the desired furan product with residual, non-volatile acetylene, a two step sequence was implemented for the cycloaddition reaction. Thus, ¹³C-labelled diazoimide **193** was allowed to react with a large excess (10 eq) of dimethyl acetylenedicarboxylate (DMAD) **194** in the presence of Rh₂(OAc)₄ at room temperature in anticipation of trapping the bicyclic intermediate **196** on the polymeric support. After washing the beads in order to remove excess acetylene, resin **196** was suspended in fresh solvent and heated to 80°C to promote cycloreversion. HPLC analysis of the crude product from this reaction showed the presence of pure furan **197** with neither starting acetylene nor other impurities being observed. Resin washings did not contain any ^{13}C label, indicating that no cycloreversion occurred at room temperature, while gel-phase ^{13}C NMR of the resin after thermolysis showed no resonances from enriched carbons and suggested that cycloaddition to the polymer-bound isomünchnone **195** proceeded efficiently at room temperature (*Scheme 4.2.6*).

Scheme 4.2.6

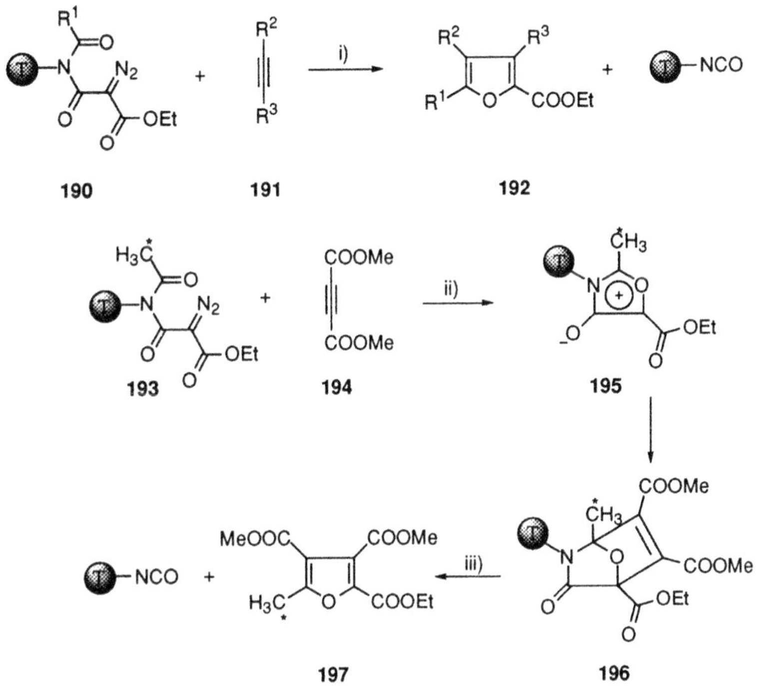

Reagents and conditions: i) Rh$_2$(OAc)$_4$, benzene, 80°C; ii) Rh$_2$(OAc)$_4$, benzene, r.t.; iii) benzene, 80°C

The stepwise (room temperature/thermal) cycloreversion sequence failed to provide furans from acetylene derivatives activated by just a single electron withdrawing moiety (*e.g.* propiolate esters). These derivatives, less reactive than DMAD, did not undergo cycloadditions to the immobilised dipoles at an appreciable rate at room temperature.

To further illustrate the utility of this solid-phase synthesis, a small (32 member) combinatorial library of furans was generated *via* split synthesis by using the stepwise cycloaddition sequence. From eight carboxylic acids, two malonyl chlorides and two acetylenes, a library of four pools comprising eight compounds each was synthesised.

4.2.3. Synthesis of prostaglandins

Based on a modification of the *Noyori* prostaglandin synthesis, *Ellman et al.*[89-91] developed a solid-phase synthesis of this interesting class of carbocycles (*Scheme 4.2.7*). Thus, after addition of alkylzinc reagents **202** to resin-bound 4-hydroxy-2-cyclopentanones **201**, alkylation was carried out with alkyl triflates **203**. This synthesis was originally performed with the DHP linker, but the cleavage conditions were not compatible with the target molecule. Due to the severe reaction conditions employed for coupling, the hydroxycyclopentane **199** had to be protected with the dimethoxytrityl group when using silyl linker **198**.

Scheme 4.2.7

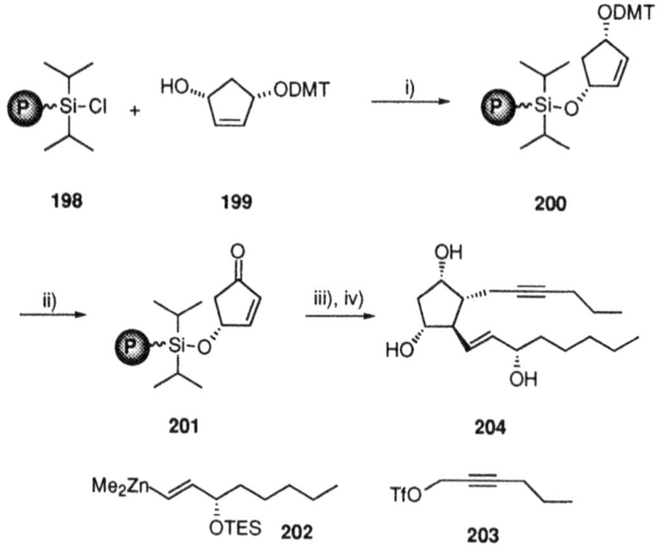

Reagents and conditions: i) NaH, DMF, r.t., 4 h; ii) a) 3% Cl₃CCOOH in CH₂Cl₂,
b) PDC, DMF, 40°C, 3h; iii) a) **202**, THF, -78°C to r.t., b) **203**, THF, -78°C, 1h;
iv) a) L-Selectride, -78°C, b) 1M TBAF, THF, r.t., 3h

After two reaction steps, the prostaglandin **204** was isolated in a yield greater than 50% as a single diastereoisomer. Furthermore, the alkylation reaction on the *Merrifield* resin gave only poor results. A macroporous resin proved to be more appropriate for this reaction sequence.

On the other hand, it is well known in prostaglandin chemistry that the lack of reactivity of the initially formed enolate as well as the danger of proton exchange and subsequent elimination of the alkoxy group can lead to the corresponding cyclopentenone. To overcome this problem, *Ellman et al.* introduced an alternative solid-phase prostaglandin synthesis. As shown in *Scheme 4.2.8*, a resin-bound iodocyclopentenone **205**, formed in three steps from **200**, was treated with an organoborane under palladium catalysis. Subsequent *Michael* addition of a vinyl cuprate led to the introduction of the ω side chain. Analogous chemical manipulations as described before afforded prostaglandins **206**.

Scheme 4.2.8

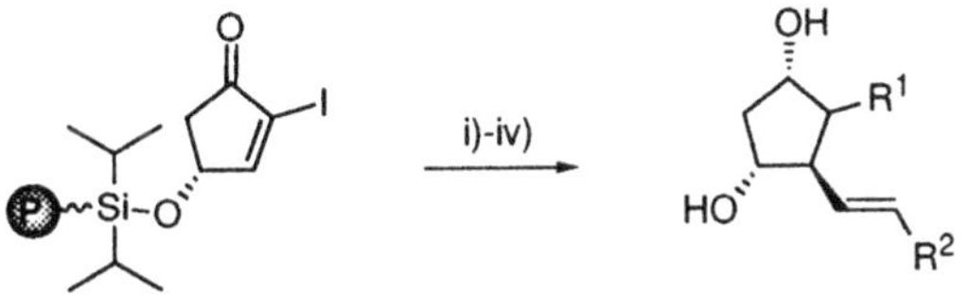

205   **206**

Reagents and conditions: i) R^1-BBN, Pd(0), K_2CO_3; ii) vinyl cuprate; iii) L-selectride; iv) TBAF

299

4.3. Multicomponent assembly strategies with single functional group liberation

4.3.1. Synthesis of imidazoles

Many biologically active therapeutic agents contain five-membered heterocycles.[92] The imidazole ring system is of particular interest since it is a component of histidine and its decarboxylation metabolite histamine. The wide applicability of the imidazole pharmacophore can be attributed to its hydrogen bond donor-acceptor capability as well as its high affinity for metals which are present in many protein active sites (*e.g.* Zn, Fe, Mg). Furthermore, improved pharmacokinetics and bioavailability of peptide-based protease inhibitors have been observed by replacing an amide bond with an imidazole.

Based on a four component condensation reaction (the *Ugi* reaction[93]), *Mjalli et al.*[94] developed a solid-phase protocol for the synthesis of tetrasubstituted imidazoles. Because of the limited number of commercially available isocyanides, the selected strategy started with the generation of an isocyanide component tethered to *Wang* resin.

Scheme 4.3.1

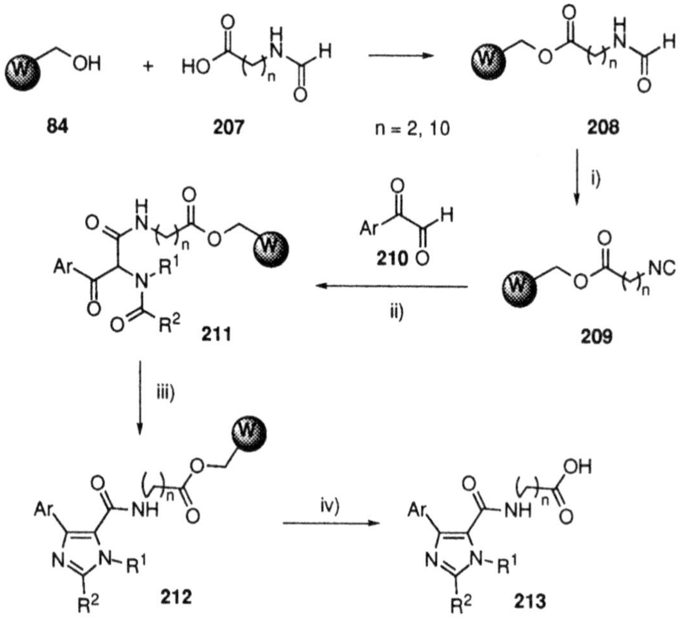

Reagents and conditions: i) Ph$_3$P, CCl$_4$, Et$_3$N, CH$_2$Cl$_2$; ii) arylglyoxals **210**, R^1-NH$_2$, R^2-COOH; iii) NH$_4$OAc, AcOH, 100°C; iv) 10% TFA in CH$_2$Cl$_2$

A series of *N*-formylated aliphatic amino acids[95] **207** were subsequently attached to the *Wang* resin as shown in *Scheme 4.3.1*, affording **208**. Dehydration of **208**[96] provided resin-bound isocyanides **209** quantitatively as shown by ¹H-NMR of the resin. Multicomponent condensation of resin **209** with arylglyoxals **210**, primary amines and carboxylic acids afforded resin-bound α-(*N*-acyl-*N*-alkylamino)-β-ketoamides **211** that were further reacted with a large excess of NH₄OAc in AcOH to give immobilised imidazoles **212**. Final cleavage with 10% TFA provided imidazoles **213**.

The length of the isocyanide linker, the electronic nature of the aryl glyoxals, the use of different primary amines (with exception of aniline), as well as the use of both aliphatic and aromatic carboxylic acids, showed no effect on the yield of the imidazoles **213**.

Extension of this synthetic protocol was further investigated by *Sarshar et al.*[97] by attaching the aldehyde or amine component to the *Wang* resin. Hence, functionalised polymers **215** and **217** were prepared by standard coupling methods of carboxybenzaldehyde **214** and N-Fmoc-6-aminohexanoic **216**, respectively. Resin **219** was prepared via a modified *Mitsunobu* coupling of 4-hydroxybenzaldehyde **218** to *Wang* resin **84** (*Scheme 4.3.2*).

Scheme 4.3.2

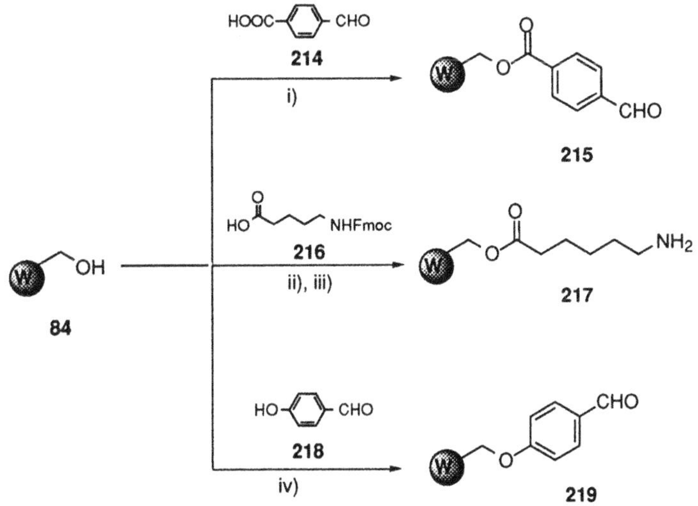

Reagents and conditions: i) DIC, DMAP, THF, 23°C, 48h; ii) DIC, DMAP, CH₂Cl₂, 23°C, 24h; iii) 20% piperidine, DMF; iv) N-ethylmorpholine, DIAD, Ph₃P, sonicate for 1h, then stir at 23°C, 16h.

The protocol for the preparation of solid supported imidazoles **220** and **221** involved the condensation of resins **215** or **217** with a large excess of 1,2-dicarbonyl compounds, NH$_4$OAc and primary amines or aldehydes. Imidazoles of type **222** were prepared in a similar manner by replacing primary amines with NH$_4$OAc. Following resin cleavage with 20% TFA afforded imidazoles **223**, **224** and **225** in high yields and purities (*Scheme 4.3.3*).

Scheme 4.3.3

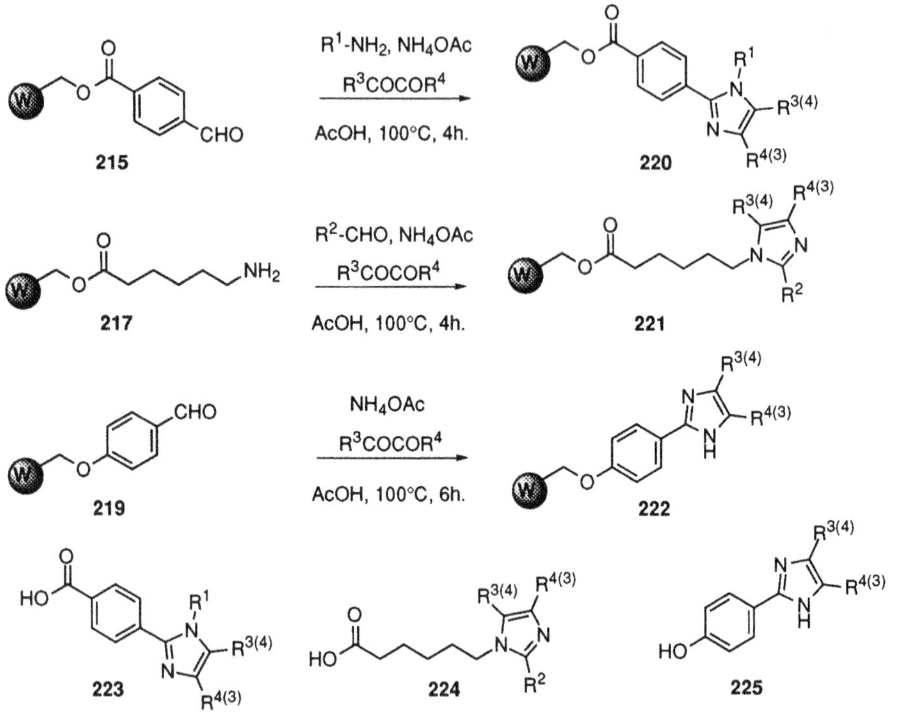

4.3.2. Synthesis of proline derivatives

By using a three component 1,3-dipolar cycloaddition reaction,[98] *Hamper et al.*[99] developed a solid-phase synthesis of highly substituted pyrrolidines *via* resin-bound azomethine ylides.

As shown in *Scheme 4.3.4*, several substituted hydroxybenzaldehydes **226** were attached to *Wang* resin **84** *via* an alkylaryl ether linkage through a *Mitsunobu* reaction to give **227**. These polymer-supported aldehydes were allowed to react with an α-amino ester **228** and maleimide **229** in DMF to give immobilised proline analogues **230** that were liberated from the resin by treatment with 50% TFA to afford **231**. The dipolar cycloaddition provided mixtures of diastereoisomers, which could be separated by HPLC.

Scheme 4.3.4

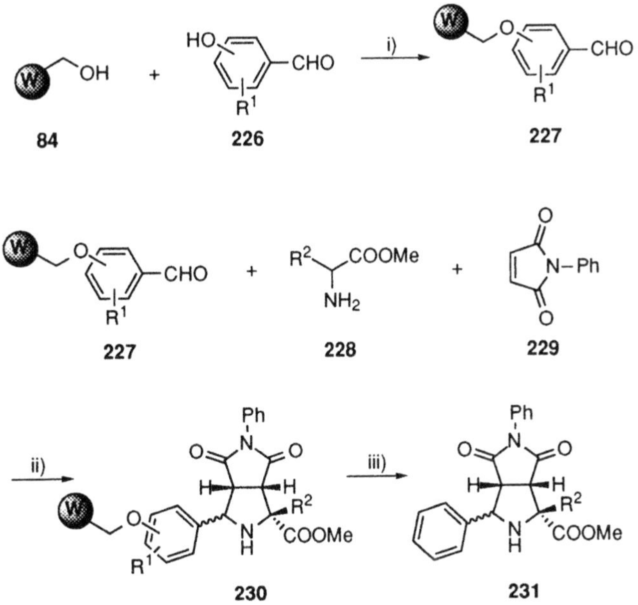

Reagents and conditions: i) Ph$_3$P, DEAD, THF; ii) DMF, 80-100°C; iii) 50% TFA in CH$_2$Cl$_2$

4.3.3. Synthesis of 2-arylquinoline-4-carboxylic acid derivatives

2-Arylquinoline-4-carboxylic acids and derivatives were shown to possess a wide variety of biological effects as antimalarial, antimicrobial, antitumor, antioxidants and cardiovascular agents.[100] In addition, other derivatives have recently been shown to be potent tachykinin NK3 receptor antagonist[101] or to exhibit analgesic activity.[102]

Based on a multi-component condensation approach related to the *Doebner* reaction,[103] *Gopalsamy et al.*[104] developed a solid-phase synthesis for this clinically useful pharmacophore. Thus, starting from *Rink* resin **65**, acylation with the required *N*-Fmoc-amino acid **232** and deprotection with piperidine gave the polymer-supported amine **233** (90% yield) which was acylated with pyruvyl chloride (**234**). Immobilised pyruvic amide **235** was refluxed with excess of preformed benzylidene aniline **239** or alternatively condensed with an excess of an equimolecular mixture of aldehyde **236** and anilines **237**. Cleavage of **238** with 45% TFA afforded compounds **240** in good yields and with high purities (>90%) (*Scheme 4.3.5*).

Scheme 4.3.5

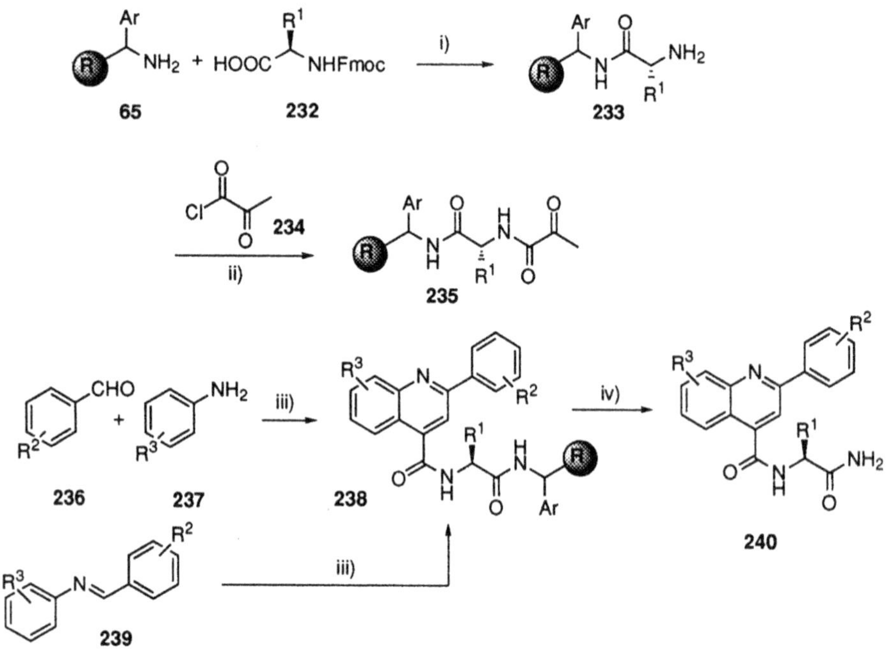

Reagents and conditions: i) HOBt, HBTU, DIEA, DMF, then 20% piperidine in DMF; ii) CH$_2$Cl$_2$, py, 0°C; iii) **235**, benzene, 80°C, 8h; iv) 50% TFA in CH$_2$Cl$_2$

Although electron-withdrawing groups in the aldehyde building-block improved the overall yield, the reaction proceeded well for benzaldehyde. Furthermore there was no significant effect on the cyclisation by varying the amino acid employed.

4.3.4. Synthesis of dihydropyrimidines

The *Biginelli* dihydropyrimidine synthesis[105,106] is another relevant multicomponent condensation reaction that has been recently adapted to the solid phase by *Wipf et al.*[107] to prepare a small library of this interesting pharmacophore. In fact, many examples of this class of *N*-heterocycles have been shown to display a wide range of biological effects as for instance antiviral,[108] calcium channel blocking, [109] or antihypertensive activity.[110]

In this sequence, γ-aminobutyric acid derived urea **241**, prepared from GABA and KOCN in H$_2$O/THF,[111] was attached to *Wang* resin **84** to provide **242** in high yield. Subsequent multicomponent *Biginelli* condensation with excess of β-ketoesters **243** and aldehydes **244** in the presence of a catalytic amount of HCl afforded immobilised dihydropyrimidines **245** which, after cleavage of the benzyl ester group with 50% TFA, provided dihydropyrimidines **246** (*Scheme 4.3.6*).

Scheme 4.3.6

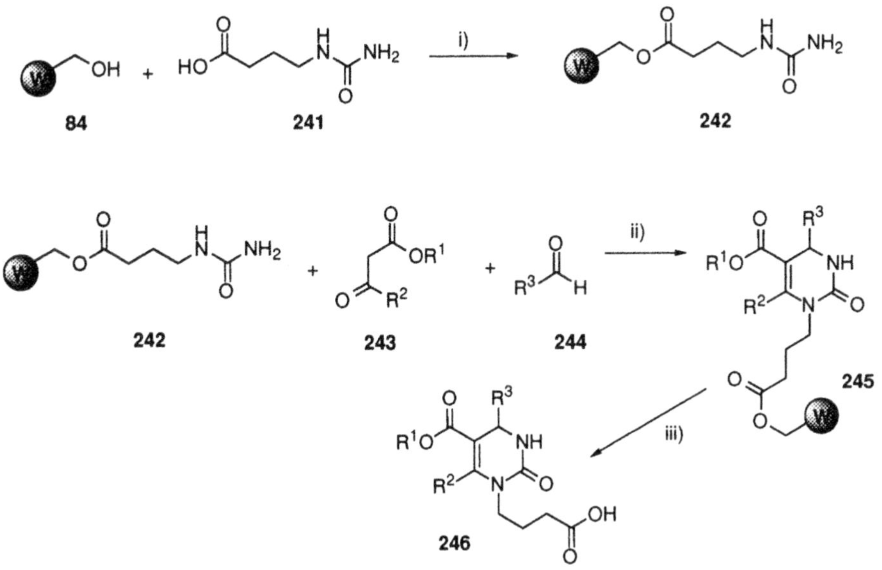

Reagents and conditions: i) EDCI, DMAP, DMF, r.t., 24h; ii) THF/conc. HCl (4:1), 55°C, 36h; iii) TFA, CH$_2$Cl$_2$

The individual products **246** were obtained in high yields and excellent purity (>95%). Since β-oxoesters **243** are readily obtained in one step and many aldehydes **244** are commercially available, this reaction is applicable to the parallel synthesis of large compound libraries. The GABA-derived acid linker increases the water solubility of the products and offers a convenient handle for additional functionalisation.

4.4. Multigeneration assembly strategies with cyclisation-assisted cleavage

4.4.1. Synthesis of 1,4-benzodiazepine-2,5-diones

Featuring a cyclisation-assisted cleavage strategy (cf. *Chapter 1.7*), first described by *Camps et al.*,[112] *Mayer et al.*[113] developed a solid-phase synthesis of 1,4-benzodiazepinediones.

Scheme 4.4.1

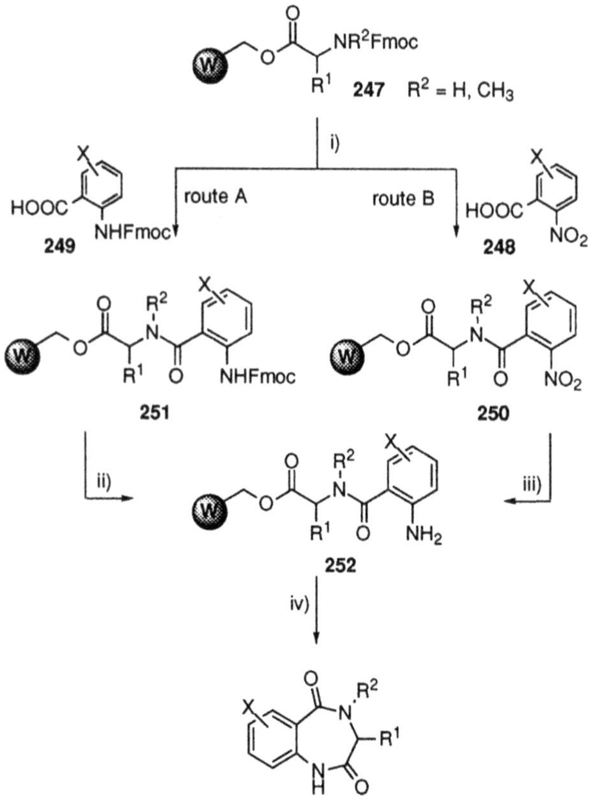

Reagents and conditions: i) DIC, DMF, then **249** (route A), **248** (route B); ii) 20% piperidine in DMF; iii) 2M SnCl₂ in DMF, 5h; iv) NaOtBu, THF, 60°C, 24h

As outlined in *Scheme 4.4.1*, starting from commercially available Fmoc-amino acid derivatised *Wang* resins **247**, or those prepared by known loading procedures, deprotection with 20%

piperidine in DMF and coupling with *o*-nitrobenzoic acids **248** or protected *o*-anthranilic acids **249** afforded immobilised derivatives **250** (route A) and **251** (route B). Fmoc deprotection (route A) or reduction of the nitro group (route B) gave quantitatively polymer-bound amido-anthranilate **252** as determined by TFA cleavage of a small sample of intermediates **251** and **252**. The option of using either synthon within the reaction sequence permits a wider range of aromatic substituents to be incorporated than would otherwise be possible through a single route. Cyclisation of the common aminoamide intermediate **252** with concomitant release from the support furnished the desired 1,4-benzodiazepin-2,5-diones **253** in high yields and excellent purities. Optimal base-promoted cyclisation results were obtained by heating the resin precursor in THF with sodium *tert*-butoxide (*Scheme 4.4.1*).

4.4.2. Synthesis of benzoisothiazolones

The *Parke-Davis* research group developed a solid-phase synthesis of benzoisothiazolones **258**,[47] which are known for their antibacterial, antimycotic and antifungal properties.[114] As outlined in *Scheme 4.4.2*, aromatic thiocarboxylic acids **254** were tethered to the *Merrifield* resin in the presence of Et$_3$N to form polymer-bound thioethers **255**. Conversion into active esters by the BOP group followed by treatment with different amines and hydrazides yielded the corresponding immobilised amide derivatives **256**. Oxidation of the sulfide link to the corresponding sulfoxide **257** and activation with trichloroacetic anhydride led to the expected ring closure with simultaneous release of the product benzoisothiazolones **258** in up to 60% yield in a *Pummerer*-type rearrangement.

The oxidative key step to the sulfoxide **257** provided an interesting problem. In solution, this activation can be accomplished using a stoichiometric amount of oxidant to avoid over oxidation to the sulfone. In practice, it is difficult to control the reagent-to-starting material ratio in solid-phase organic synthesis. Generally, the possibility to use a large excess of reagent and to remove it easily, is considered as an advantage of the solid-phase synthesis method. After a survey of several potential oxidants for selective oxidation of sulfides to sulfoxides, *N*-(phenylsulfonyl)-3-phenyloxaziridine (*Davies* reagents) was selected as the reagent of choice. Selective monooxidation was achieved by limiting the reaction time and removal of excess reagent.

Scheme 4.4.2

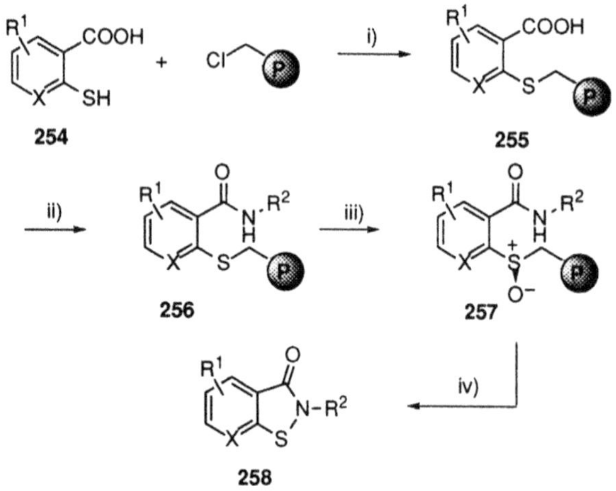

Reagents and conditions: i) 0.4 M Et₃N in dioxane, 96h; ii) *N,N*-carbonyldiimidazole, dioxane, DIEA, then R²-NH₂; iii) 0.13 M NaBrO₂ in dioxane/H₂O (8:1), or *N*-(phenylsulfonyl)-3-phenyloxaziridine), CHCl₃; iv) trichloroacetic anhydride, CH₂Cl₂, 0°C

4.4.3. Synthesis of pyrazolones

An efficient and general method to prepare diverse β-ketoesters linked to a solid support was developed by *Tietze et al.* using readily available acid chlorides **259** and haloalkanes as building blocks. The obtained polymer-bound 1,3-dicarbonyl compounds served as substrates toward the preparation of a library of substituted pyrazolones, which are well known for their widespread biological activity as analgesics, antipyretics, antiphlogistics, antirheumatics, antiarthritics and uricosurics.[115]

Using solution chemistry, reaction of acid chlorides **259** with *Meldrum's* acid **260** in the presence of pyridine gave the corresponding acyl *Meldrum's* acid **261** almost quantitatively. Heating an excess of **261** with the spacer-modified polystyrene-resin **262**[52] afforded polymer-bound β-ketoesters **263** with concomitant release of carbon dioxide and acetone. The reaction could easily be monitored by FT-IR spectroscopy of the resin (KBr-pellet), showing in every case a strong appearence of the characteristic carbonyl stretching of β-ketoesters. Selective alkylation of the α-position of **263** to afford resin-bound **264**, was achieved at room temperature using a large excess of haloalkanes in the presence of TBAF (*Scheme 4.4.3*).

Scheme 4.4.3

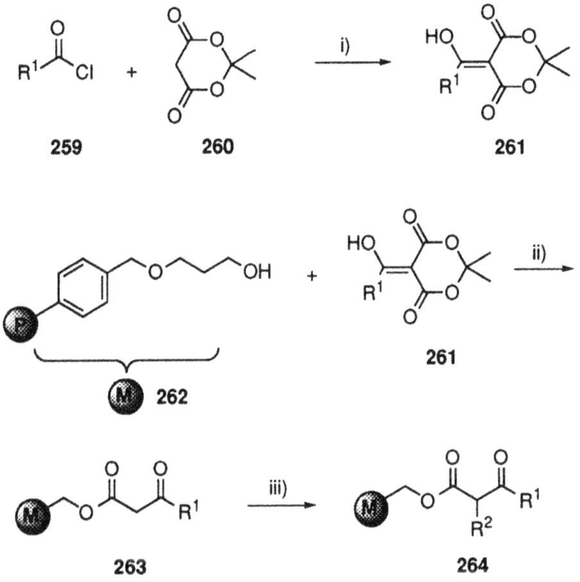

259   **260**   **261**

262

261

263   **264**

Reagents and conditions: i) py, CH$_2$Cl$_2$, 0°C - r.t.; ii) THF, reflux; iii) R^2-X, TBAF, THF, 25°C - 66°C

By following the above described reaction sequence, a set of immobilised β-ketoesters of type **264** were synthesised and directly used to prepare a collection of 1-phenylpyrazolones **266**. Thus, addition of a large excess of phenylhydrazine in the presence of trimethylorthoformate as dehydrating agent afforded polymer-bound phenylhydrazones **265**. Cyclisation of **265** with concomitant release of the 3,4-disubstituted-*N*-phenylpyrazolones **266** was accomplished by mild acidic treatment with 2% TFA in good overall yields and excellent levels of purity (*Scheme 4.4.4*).

Scheme 4.4.4

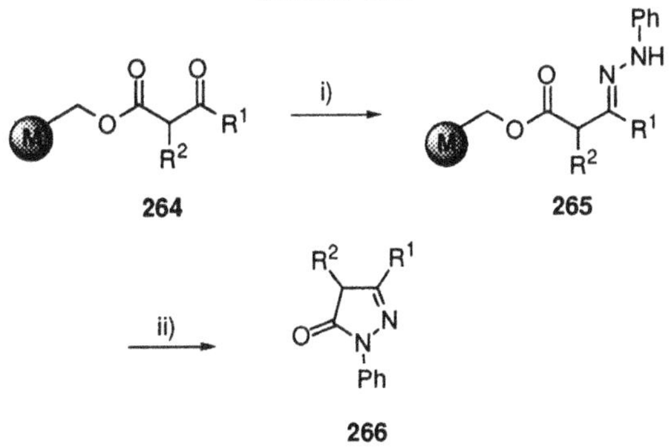

264   **265**

266

Reagents and conditions: i) phenylhydrazine, THF, trimethylorthoformate, r.t.; ii) 2% TFA in MeCN, r.t.

Alternatively, γ-alkylation of polymer-bound acetoacetate **267**[116] provided the starting materials **268** for the synthesis of 3-substituted-*N*-phenylpyrazolones **269** *via* cyclisation-assisted cleavage (*Scheme 4.4.5*).

Scheme 4.4.5

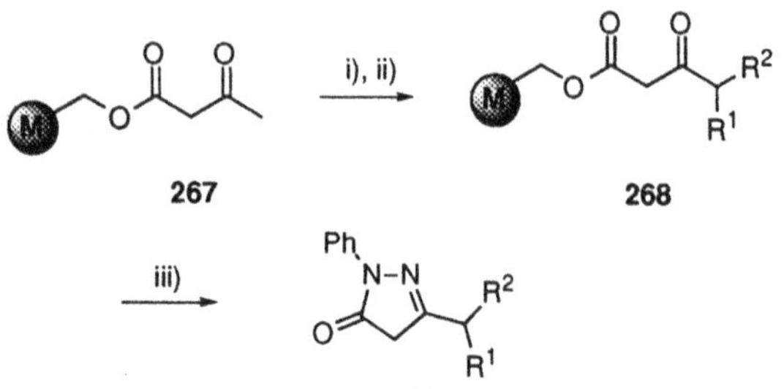

Reagents and conditions: i) LDA, R^1-X; ii) nBuLi, R^2-X;
iii) a) phenylhydrazine, THF, r.t., 3h, b) toluene, 100°C

4.4.4. Synthesis of hydantoins

Hydantoins have been reported to possess a wide range of biological activities as anticonvulsants, antiarrythmics and antidiabetics. Herbicidal and fungicidal activities have also been noted.[117] In a solid-phase synthesis aproach of hydantoins developed by the *Parke-Davis* group[47] different C-terminally coupled amino acids **270** were first deprotected and then treated with different isocyanates to give resin-bound ureas **271**. Heating of **271** in 6N hydrochloric acid allowed cyclisation with simultaneous cleavage from resin, affording hydantoins **274**. To achieve a greater diversity in the products, the amino acids may be reductively alkylated with aldehydes prior to reaction with the isocyanates (*Scheme 4.4.6*).

Dressman et al.[118] described an alternative solid-phase synthesis in which the amino acids were *N*-terminally linked to a hydroxymethyl resin through a carbamate linker. Amide formation gave products **273** which, after treatment with base cyclised to yield hydantoins of type **274** (*Scheme 4.4.6*).

Scheme 4.4.6

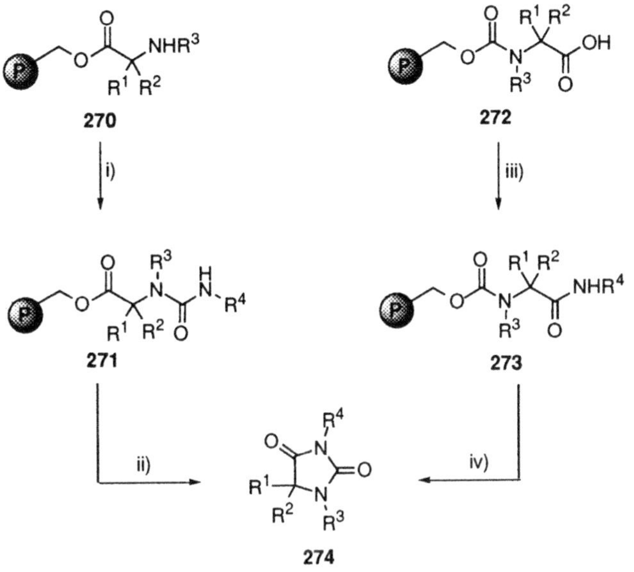

Reagents and conditions: i) R⁴-NCO, DMF, 6h; ii) 6 N HCl, 100°C, 2h;
iii) R⁴-NH₂, DCC, HOBt, DMF, r.t., 24h; iv) Et₃N, MeOH, 48h

In a similar approach, *Kim et al.*[119] coupled *N*-Fmoc protected amino acids **275** to the *Wang* resin **84** under standard conditions and, after deprotection, reductively alkylated the free amino group using different aldehydes and NaBH₃CN to give polymer-bound amines **277**. Reaction with different isocyanates provided immobilised ureas **278**. Upon treatment with neat diisopropylamine, hydantoin derivatives **274** were rapidly formed with simultaneous cleavage from the resin in high yields and purities (*Scheme 4.4.7*).

Scheme 4.4.7

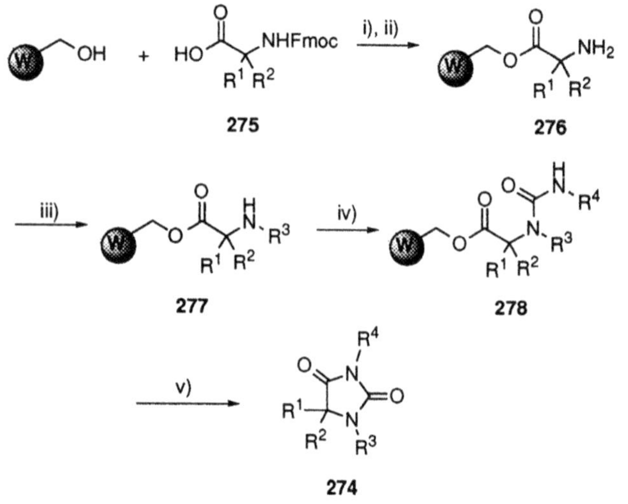

Reagents and conditions: i) DIC, DMF, DMAP, 12h; ii) 20% piperidine, DMF, 1h;
iii) R³-CHO, DMA, 1% AcOH, 5h, then NaBH₃CN; iv) R⁴-NCO, DMF/toluene (1:1);
v) neat diisopropylamine, r.t., 1h

4.4.5. Synthesis of 5-alkoxyhydantoins

A protocol for the solid-phase synthesis of 5-alkoxyhydantoins **283** incorporating three sites of functional diversity on the hydantoin nucleus was developed by *Hanessian et al.*[120] After optimisation of the synthetic sequence in solution, the protocol was extended to the solid support. Thus, the cesium salts of α-hydroxy acids (**279**) were linked to the *Merrifield* resin to give polymer-bound α-hydroxyesters **280** which were efficiently transformed into the corresponding N-benzyloxyamino esters **281** by treatment with trifluoromethanesulfonic acid anhydride in the presence of lutidine, followed by *in situ* addition of O-benzylhydroxylamine. Condensation of **281** with different individual aryl isocyanates gave immobilised ureas **282**. Treatment of **282** with potassium *tert*-butoxide in an alcoholic solution led to sequential cyclisation and detachment from the polymer to give in high purities the desired 5-alkoxyhydantoins **283** via an elimination-addition mechanism (*Scheme 4.4.8*).

Scheme 4.4.8

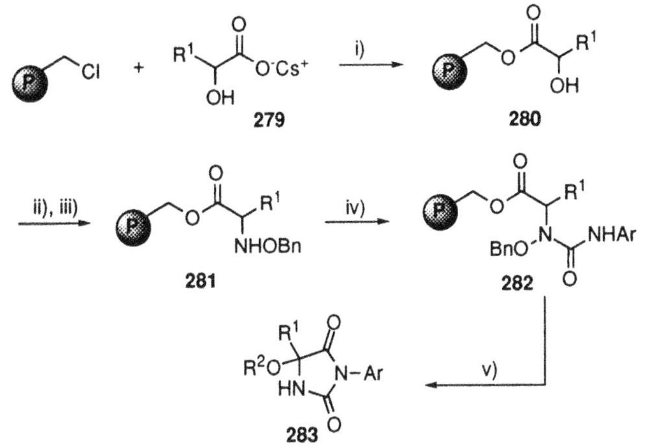

Reagents and conditions: i) DMF, 80°C, 8h; ii) (CF$_3$SO$_2$)$_2$O, lutidine,CH$_2$Cl$_2$, -78°C - 0°C; iii) BnONH$_2$, 0°C; iv) Ar-NCO, 1,2-dichloroethane, reflux, 24h; v) KOtBu, R^2-OH, r.t.

4.4.6. Synthesis of 1,3-disubstituted quinazolin-2,4-diones

When seeking compounds with activity in the central nervous system (CNS), one has to consider that highly polar functional groups generally prevent molecules from easily crossing the blood-brain barrier. On the other hand, it is known that the quinazolinedione core structure occurs in a large number of bioactive compounds including serotonergic, dopaminergic and adrenergic receptor ligands and inhibitors of aldose reductase, lipoxygenase, cyclooxygenase, collagenase and carbonic anhydrase. A library based upon this structure would therefore be expected to provide lead compounds in a wide range of bioassays.

Taking into account these arguments, *Smith et al.*[121] aimed to develop a solid-phase approach to 1,3-disubstituted quinazolinediones which would not leave an extraneous polar resin-tethering substituent on the resulting molecules which might compromise CNS penetration.

Thus, as outlined in *Scheme 4.4.9,* a chloroformate-functionalised polystyrene resin **284** was treated individually with a wide range of anthranilic acid derivatives **285** in the presence of *Hünig's* base to give the urethane-linked system **286**. Structurally diverse primary amines were readily coupled to the free carboxylic acid of **286** using standard PyBOP conditions to give anthranilamides **287**. Thermal treatment of **287** caused the amide nitrogen to cyclise onto the

313

urethane to generate, with concomitant resin release, heterocycles of type **288** lacking any additional tethering substituents and with very high levels of purity (usually higher than 95%).

Scheme 4.4.9

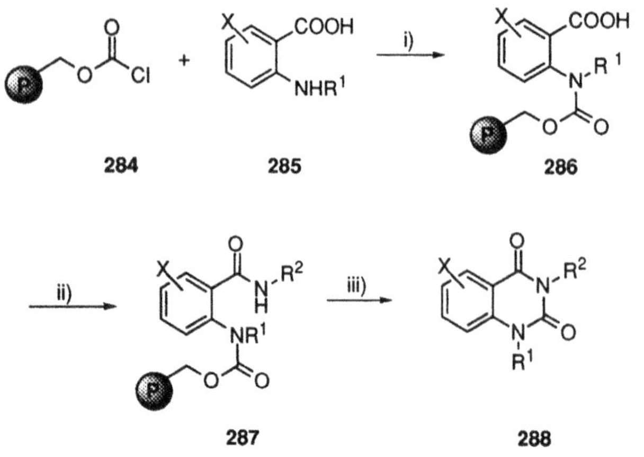

Reagents and conditions: i) DIEA, CH$_2$Cl$_2$, r.t., 1h; ii) R^2-NH$_2$ (3 eq), PyBOP; iii) DMF, 125°C, 16h

Whilst there are several commercially available anthranilic acids of type **285**, much greater structural diversity can be achieved if the substiuents at N-1 in **286** could be incorporated from primary amines **290**. Using a modified literature procedure[122] a number of different anthranilic acid derivatives of type **285** were synthesised (*Scheme 4.4.10*) and incorporated in the above described protocol.

Scheme 4.4.10

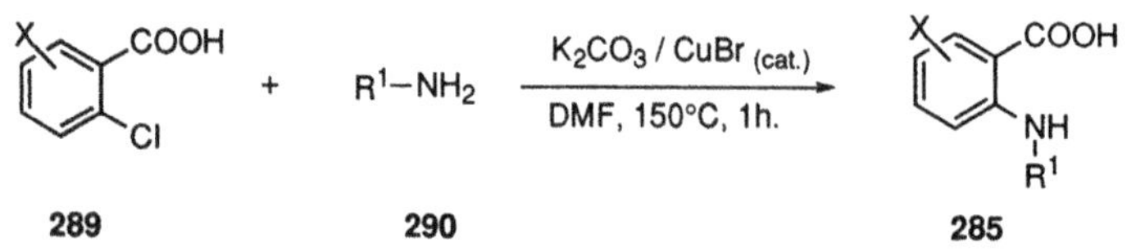

4.4.7. Synthesis of pyrazolo[1,5-*a*]pyrrolo[3,4-*d*]pyrimidinediones

In a further example where cleavage is induced during final cyclisation step, *DeWitt et al.*[47] developed a solid-phase protocol for the synthesis of tricyclic systems of type **295** which in addition are known to be inhibitors of cholesterol-*O*-acetyltransferase.

Thus, as depicted in *Scheme 4.4.11*, starting from acrylic acid substituted resin **291**, polymer-bound pyrrolidinones **292** were easily prepared by sequential *Michael* addition of amines followed by cyclisation with diethyloxalate. After substitution reaction with different 3-aminopyrazoles at 100°C, subsequent treatment with 5N HCl at 80°C afforded with concomitant cleavage from the resin, the corresponding pyrazolopyrrolopyrimidinediones **294**. Additional molecular diversity could be introduced by alkylation of the liberated compounds **294** in solution to give **295**.

Scheme 4.4.11

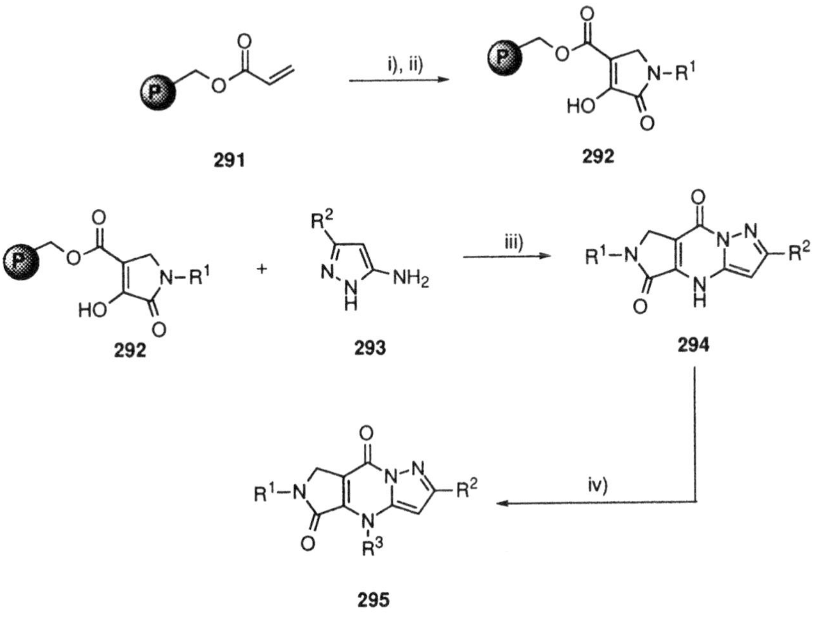

Reagents and conditions: i) R^1-NH$_2$, DMF, r.t., 24h; ii) (EtOOC)$_2$, Na$_2$CO$_3$, DMF, 60°C, 2h; iii) AcOH, 100°C, 2h; iv) R^3-X, K$_2$CO$_3$, DMF, 60°C, 2h

4.4.8. Synthesis of quinazolinones

3H-Quinazolin-4-ones of type **300/301** represent interesting pharmacophores displaying a wide range of biological properties.[123-125] Therefore, the development of efficient strategies for the combinatorial and parallel synthesis on solid support of this interesting target allowing the introduction of a high degree of molecular diversity has attracted considerable attention.[126]

One example of a successful application of solid-phase chemistry constitutes the highly versatile synthesis of quinazolinones **300/301** which efficiently combines an aza-*Wittig* reaction with a multidirectional cyclisation cleavage. Iminophosphoranes were shown to be useful intermediates in organic synthesis, particularly for the preparation of different heterocyclic systems containing an endocyclic C,N double bond. In these cases, an aza-*Wittig*-mediated anellation reaction was involved as the key step.[127-129]

Scheme 4.4.12

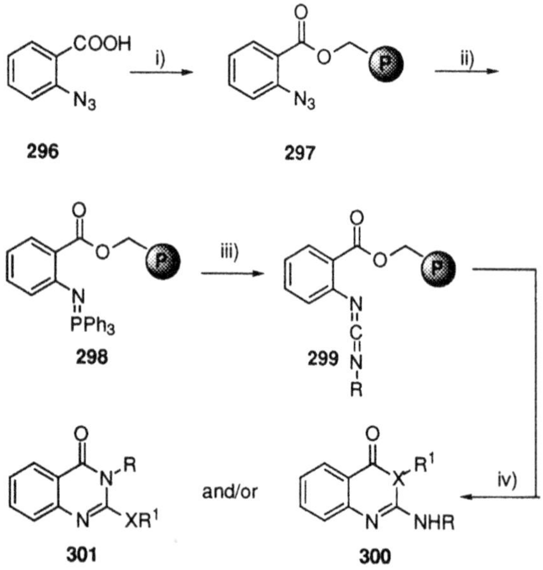

Reagents and conditions: i) *Merrifield* resin (3.4 mmol/g), Cs_2CO_3, DMF, KI, 80°C, 8h; ii) Ph_3P (1M in THF), r.t., 6h; iii) R-N=C=O, CH_2Cl_2, r.t., 8h; iv) R^1-XH, THF, 50°C, 4h

Alkylative esterification of **296** *via* the corresponding cesium salt[130] with highly loaded *Merrifield* resin (3.4 mmol/g) resulted in polymer-bound o-azido ester **297**.[131,132] Treatment of **297** with a fivefold excess of PPh_3 in THF at room temperature produced the corresponding iminophosphorane **298** attached to the resin (evolution of N_2 was clearly detected). Partitioning of the beads and subsequent aza-*Wittig* reaction with different isocyanates at room temperature smoothly formed the corresponding reactive carbodiimides **299**. Additional partitioning and treatment with different nucleophiles (*e.g.* amines, thiols) led, *via* intramolecular cyclisation and simultaneous cleavage from the resin, to quinazolines **300** (and/or **301** when primary, non-sterically hindered amines were used) in good yields and excellent purities (>97%) (*Scheme 4.4.12*).

The formation of the polymer-bound compounds was routinely monitored by FT-IR of the resin beads. The product composition **300:301** was strongly influenced by differences in nucleophilicity and/or steric hindrance. When those differences were minimal, both possible compounds **300** and **301** were generally formed in 1:1 ratio.[131,132]

4.5. Multicomponent assembly strategy with cyclisation-assisted cleavage

4.5.1. Synthesis of diketopiperazines

The diketopiperazine scaffold proved to be a versatile template in combinatorial chemistry due to four ring atoms which can be centers for the generation of molecular diversity.[133] Based on the *Ugi* four component condensation reaction of a polymer-bound amino acid, followed by cyclisation-assisted cleavage, *Chucholowski et al.*[132] developed a novel solid-phase synthesis that allows the generation of diketopiperazine libraries with four centers of diversity. As outlined in *Scheme 4.5.1*, *Rink* amine resin was charged with a Fmoc protected amino acid to afford after deprotection amines **303**. Next, the resin-bound amino acid was divided into separate reaction vessels and the *Ugi* reaction was performed treating each vessel individually with an aldehyde, an isocyanide, and an Fmoc-protected amino acid[134] to afford **304**. Deprotection of the Fmoc group under standard conditions to give dipeptides **305** and subsequent warming up in dioxane, resulted in cyclisation-assisted cleavage and concomitant release of very pure diketopiperazines **306**.[135] Although diketopiperazines **306** were obtained with very high levels of purity, the yields of isolated material were only moderate, probably due to the alternative cyclisation mode leading to resin-bound **307**.

Scheme 4.5.1

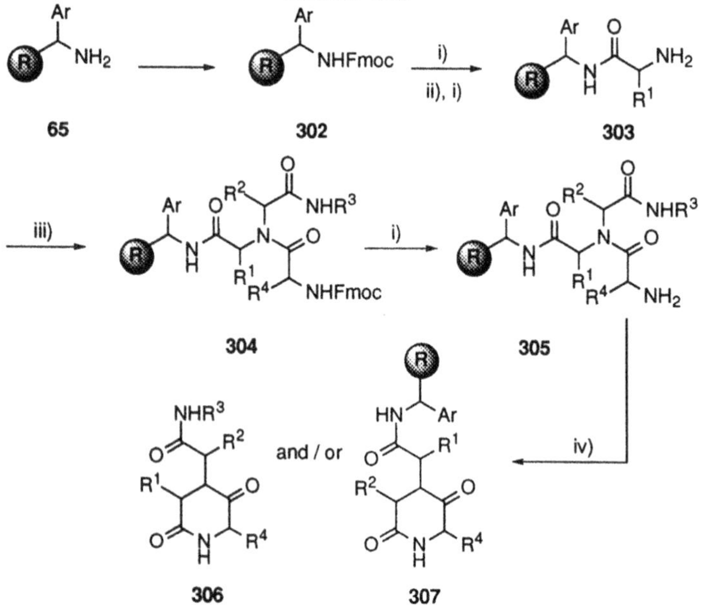

Reagents and conditions: i) 20% piperidine/DMF, r.t.; ii) TPTU, Fmoc-NHR^{1}CHCO$_2$H, DIEA, DMF, r.t.; iii) R^{2}CHO, R$_3$NC, Fmoc-NR^{4}CHCO$_2$H, dioxane/CH$_2$Cl$_2$/MeOH (4:1:1); iv) dioxane, 102°C

318

4.6. Multigeneration assembly strategies with multidirectional cleavage

4.6.1. Without activation of the linker moiety

4.6.1.1. Synthesis of ketoureas

The concept of introducing additional molecular diversity while liberating final compounds from the resin was used by *DeWitt et al.*[47] by employing resin-bound *Weinreb* amides.[136] Hence, simple bifunctional templates were derivatised on solid support to produce a library of ketoureas of type **312**. Thus, as outlined in *Scheme 4.6.1*, *Kornblum* oxidation with sodium hydrogen carbonate in DMSO converted the *Merrifield* resin into a support-bound benzaldehyde **308**, which gave an *O*-methylhydroxylamine resin **309** by heating with hydroxylamine and subsequent reduction with cyanoborohydride. Coupling with different *N*-Boc protected aminobenzoic acids **310**, deprotection and reaction with different isocyanates afforded the corresponding ureas **311**. Treatment of **311** with different alkyl- and arylmagnesium bromides led, via nucleophilic attack, to cleavage of ketoureas **312** from the resin.

Scheme 4.6.1

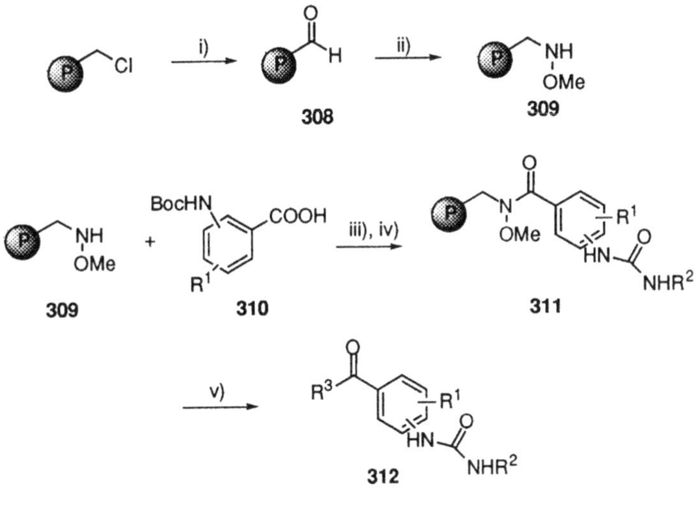

Reagents and conditions: i) NaHCO$_3$, DMSO, 155°C, 20h; ii) a) MeONH$_2$, py, EtOH, 80°C, b) NaBH$_3$CN, EtOH, HCl, r.t., 3h; iii) DIEA, BOP, then TFA; iv) R^{2}NCO; v) R^{3}MgX, THF, CH$_2$Cl$_2$, -78°C to -10°C, 2h

4.6.1.2. Synthesis of bicyclo[2.2.2]octanes

Employing enolate chemistry, *Ley et al.*[137] developed a solid-phase synthesis of bicyclo[2.2.2] octanes of type **316** based on a double *Michael* addition strategy. As depicted in *Scheme 4.6.2*, polymer-bound acrylate **313** was treated with cyclohexenone enolates derived from **314** to give **315**. After reductive amination, resin-bound bicyclo[2.2.2]octanes **316** were cleaved under a set of different conditions. Aminolysis with different amines and acidolysis using TFA led to the formation of **317** and **318**, whereas reduction with DIBAH gave alcohols of type **319**.

Since more than 250 different acrylates and enones are currently commercially available and the number of amines available for reductive amination is enormous, this synthetic sequence offers great potential for the synthesis of a highly diverse library.

Scheme 4.6.2

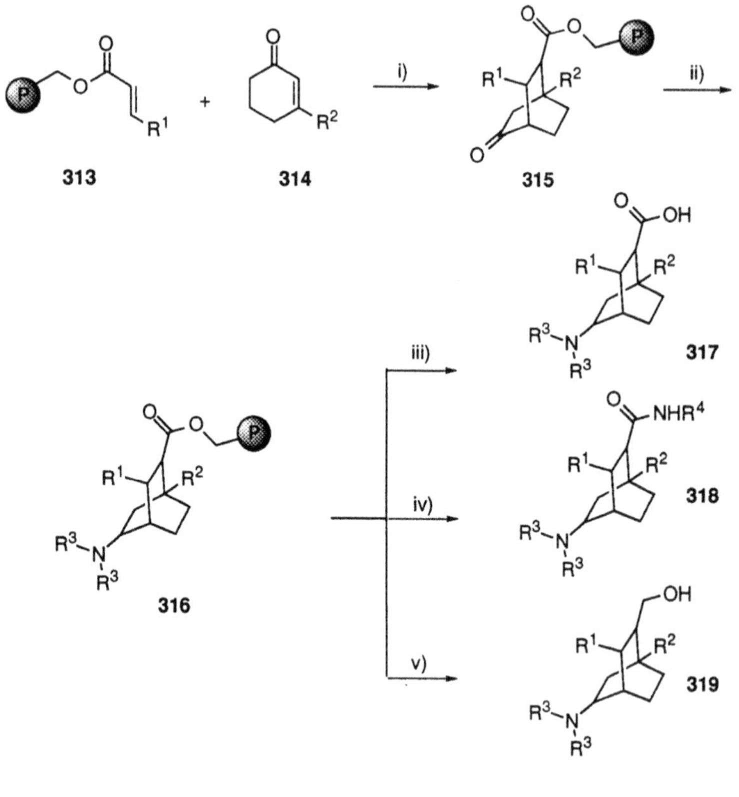

Reagents and conditions: i) LDA, THF, -78°C - r.t.; ii) (R³)₂NH, Na₂SO₄, NaBH(OAc)₃, CH₂Cl₂, AcOH; iii) TFA; iv) R⁴-NH₂; v) DIBAH

4.6.1.3. Synthesis of cyclopentane and cyclohexane derivatives

Based on a domino *Knoevenagel/ene* reaction, *Tietze* and *Steinmetz*[52] developed a stereoselective solid-phase synthesis of cyclopentane and cyclohexane derivatives of type **326** and **327** using a *Merrifield* resin modified with a propandiol linker **320** as shown in *Scheme 4.6.3*. Subsequent reaction with monomethyl malonoyl chloride **321** afforded the polymer-bound malonate **322**, which, in a two-component domino reaction was treated with unsaturated aldehydes **323** in the presence of a catalytic amount of piperidinium acetate and zinc chloride. Except for α-substituted aldehydes, the initial *Knoevenagel* condensation occurred without addition of dehydrating agents and the subsequent intramolecular ene reaction gave cyclopentane and cyclohexane derivatives **325** after cleavage from the resin either by reduction or transesterification.

Scheme 4.6.3

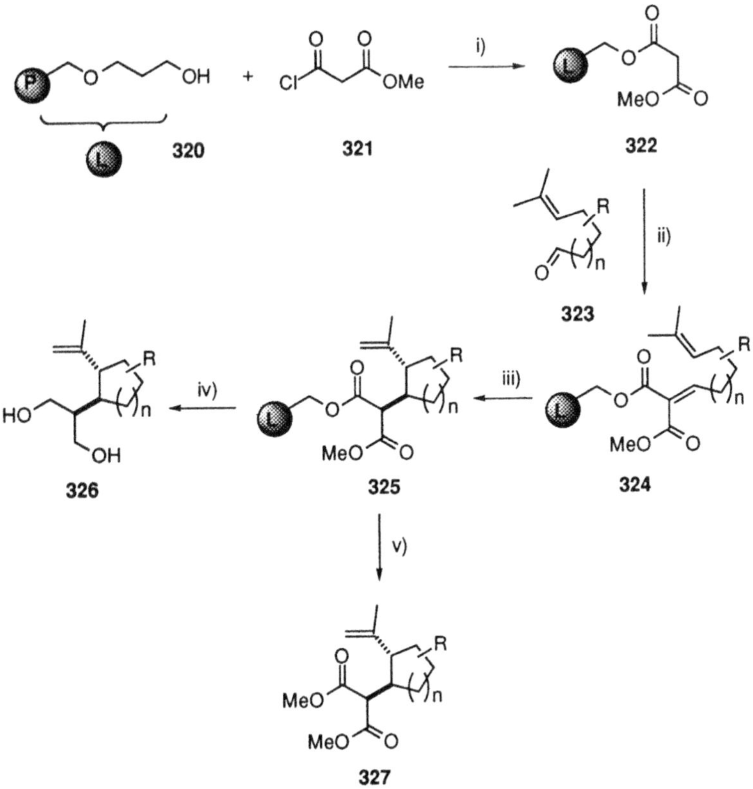

Reagents and conditions: i) DIEA, CH$_2$Cl$_2$, 0°C, 2h; ii) PipOAc, CH$_2$Cl$_2$, r.t.; iii) ZnBr$_2$, CH$_2$Cl$_2$, 3d; iv) DIBAH, toluene, 0°C, 12h; v) Ti(OEt)$_4$, AcOEt, 80°C, 3d

In either case the products were obtained in good yields, high purity and excellent levels of diastereoselectivity. Only modest diastereoselectivities were observed when geminally disubstituted aldehydes were used.

4.6.2. With activation of the linker moiety ("safety-catch" linkers)

4.6.2.1. Synthesis of arylacetic acids

The development of general strategies for the formation of carbon-carbon bonds is of particular importance for the construction of small organic compound libraries. As an example, *Ellman et al.*[138] developed a solid-phase synthesis of substituted arylacetic acid derivatives **335, 336** which represent an important class of cyclooxygenase inhibitors. As key steps for carbon-carbon bond-forming reactions in this synthetic sequence, enolate alkylation[139] and palladium-mediated *Suzuki* cross-coupling[140] were selected. The acylsulfonamide linker developed by *Kenner* was selected as linkage that would be compatible with the basic enolate formation and the *Suzuki* reaction conditions, yet at the end of the synthesis labile enough for nucleophilic cleavage of the final products from the solid support.[141] Under basic conditions, the acylsulfonamide (pK$_a$~2.5) is deprotonated, preventing nucleophilic cleavage. Once solid-support synthesis is completed, however, alkylation *e.g.* by treatment with diazomethane results in the formation of the *N*-substituted derivatives that are activated for nucleophilic displacement.

Thus, as shown in *Scheme 4.6.4*, starting from sulfonamide-derivatised support **329**, readily prepared by treating aminomethylated resin with 4-carboxybenzenesulfonamide **328** under standard coupling conditions, treatment with the pentafluorophenyl ester of 4-bromophenylacetic acid **330** provided acylsulfonamide **331**. Addition of excess LDA to **331** at 0°C gave the corresponding trianions which were treated with activated or unactivated alkyl halides to yield monoalkylated derivatives **332** (overalkylation to the bisalkylated products was minimal, and partial alkylation of the carboxamide group is irrelevant since the linker remains attached to the solid support after cleavage). In addition, the alkylated product is protected from cleavage under the basic reaction conditions since the acylsulfonamide remains deprotonated. A subsequent palladium-catalysed conversion with alkyl-9-BBN derivatives or alkylboronic acids under *Suzuki* conditions yielded the coupling products **333**. Again, under the basic conditions employed deprotonation of the acylsulfonamide moiety prevented hydrolysis. Following rinsing of the resin with 5% TFA to ensure complete protonation and *N*-methylation of the sulfonamide linker with diazomethane gave **334**, thus activating the linker moiety towards nucleophilic displacement. Cleavage from the resin with a series of reactive nucleophiles afforded products **336** and **337**. Cleavage reaction with weaker nucleophiles such as aniline succeeded only when the acylsulfonamide was activated by cyanomethyl substitution **335**.[142]

Scheme 4.6.4

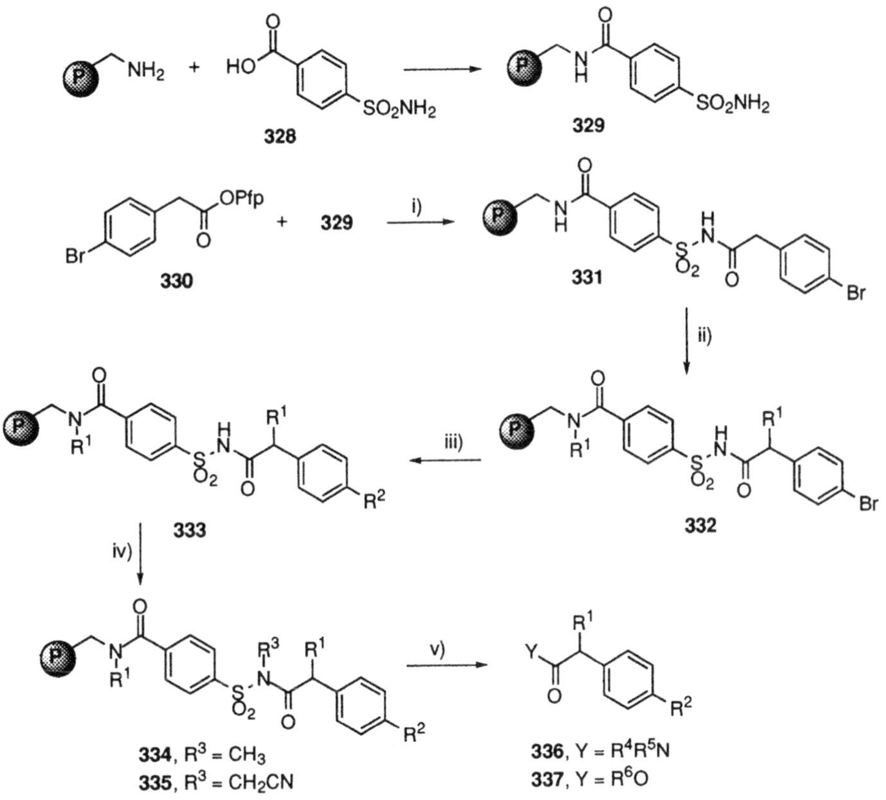

Reagents and conditions: i) DMAP; ii) LDA, THF, 0°C, then R^1-X; iii) R^2-B(OH)$_2$, Pd(PPh$_3$)$_4$);
iv) CH$_2$N$_2$ or BrCH$_2$CN; v) R^6-OH or R^{4}R^{5}NH

4.6.2.2. Synthesis of pyrimidines

Pyrimidines are an interesting class of nitrogen-containing heterocycles which display a broad range of useful properties.[143,144] Based on a novel cyclocondensation reaction between acetylenic ketones[145] of type **339** and polymer-bound isothiourea **338**, *Obrecht et al.*[132] developed a versatile and efficient solid-phase synthesis of polymer-bound 2-(alkylthio)-4,6-disubstituted pyrimidines **340**, and combined this protocol with the nucleophilic displacement reaction of the 2-sulfonyl group of pyrimidines[146,147] as the key cleavage reaction. In addition, simple transformations of the pyrimidine nucleus allowed the preparation of structurally highly diverse pyrimidines. Thus, as outlined in *Scheme 4.6.5*, reaction of resin-bound thiouronium salt[148] **338** with acetylenic ketones **339** in DMF in the presence of DIEA, followed by cleavage of the *tert*-butyl ester group with TFA, afforded the polymer-bound pyrimidine carboxylic acids **340** in high yields (as determined by cleavage of the 2-alkylsulfonyl moiety from the resin with pyrrolidine to give **341**). Conversion of the carboxylic acid **340** into the corresponding pentafluorophenyl esters **342** or hydroxysuccinimide derivatives **343** under standard conditions proceeded smoothly. Partition of the resin beads and parallel treatment with different primary and secondary amines gave the polymer-bound pyrimidine amides **344**. This reaction sequence could easily be followed by ATR-IR (attenuated total reflexion method). As a key step in this strategy, the 2-thioalkyl moiety was activated by oxidation with *m*-CPBA yielding the corresponding sulfones **345**. Partitioning of the beads and multidirectional cleavage with different O-, N-, S-, and C-nucleophiles gave the final products **346** in high yields and excellent purities. This oxidation and cleavage protocol constitutes a novel type of safety-catch linker strategy[141] allowing the introduction of additional structural diversity onto the pyrimidine ring under simultaneous release of the final compounds **346**.

Scheme 4.6.5

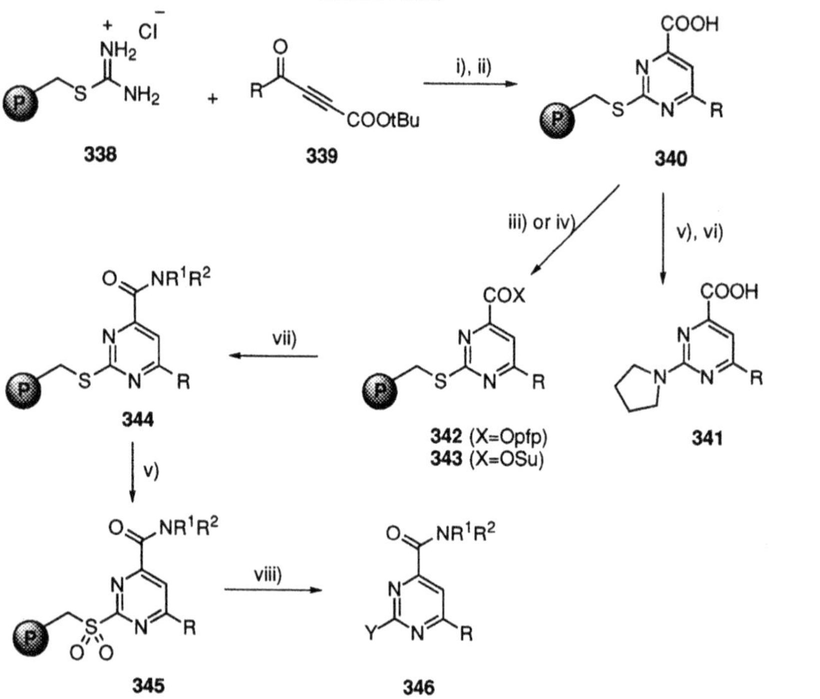

Reagents and conditions: i) DIEA, DMF, r.t.; ii) 50% TFA, CH$_2$Cl$_2$, r.t.; iii) EDCI, DMF, C$_6$F$_5$OH; iv) EDCI, HOSu, CH$_2$Cl$_2$; v) *m*-CPBA, CH$_2$Cl$_2$; vi) pyrrolidine, dioxane, r.t.; vii) R^{1}R^{2}NH, CH$_2$Cl$_2$; viii) YH, dioxane, 50°C

In view of achieving the highest possible degree of molecular diversity, the developed solid-phase pyrimidine-assembly strategy was combined with the attractive features of multicomponent reactions.[134] Thus, starting from the polymer-bound pyrimidine carboxylic acid **340** as the acid component, reaction with an excess of aldehydes, amines, and isocyanides afforded in an *Ugi*-type reaction[149] the corresponding resin-bound α-(acylamino)amides **347**. Oxidation to the corresponding sulfones **348** and cleavage with pyrrolidine led to the corresponding *Ugi* products **349** in good yields and excellent levels of purity (*Scheme 4.6.6*).

Scheme 4.6.6

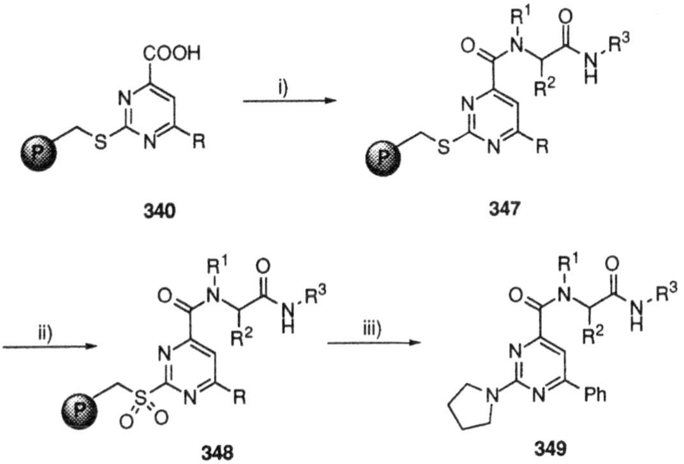

Reagents and conditions: i) R^1-NH_2, R^2-CHO, R^3-NC, (5 eq), dioxane/MeOH (4:1), 55°C, 48h; ii) *m*-CPBA (3 eq), CH_2Cl_2, r.t., 15h; iii) pyrrolidine (1 eq), dioxane, r.t., 6h

Scheme 4.6.7

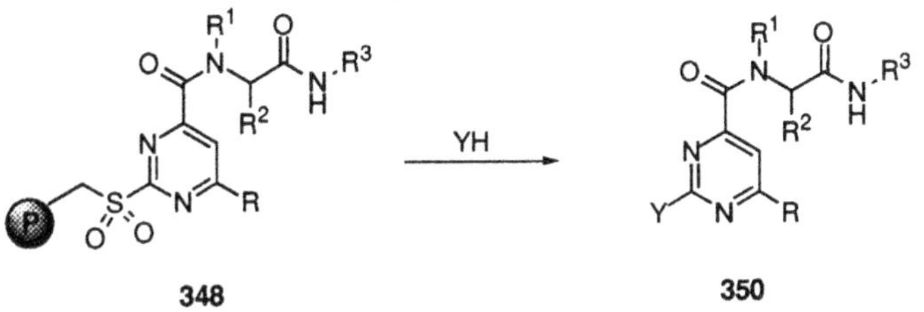

YH = R-NH_2, R^1R^2NH, R-OH, R-SH, $CH_2(COOEt)_2$; Y = CN, N_3.

One limitation encountered so far was the well known lack of reactivity of aromatic aldehydes and isonitriles under the *Ugi* conditions. The scope of this method was further extended by introduction of a third substituent at the 2-position of the pyrimidines. Thus, polymer-bound sulfones **348** were subjected to multidirectional cleavage with one equivalent of different nucleophiles affording compounds **350** in good yields and high purities[148] (*Scheme 4.6.7*).

A similar approach towards the synthesis of pyrimidines was recently described by *Suto et al.*[150] starting from polymer-bound thiol **351** as decribed in *Scheme 4.6.8*.

Scheme 4.6.8

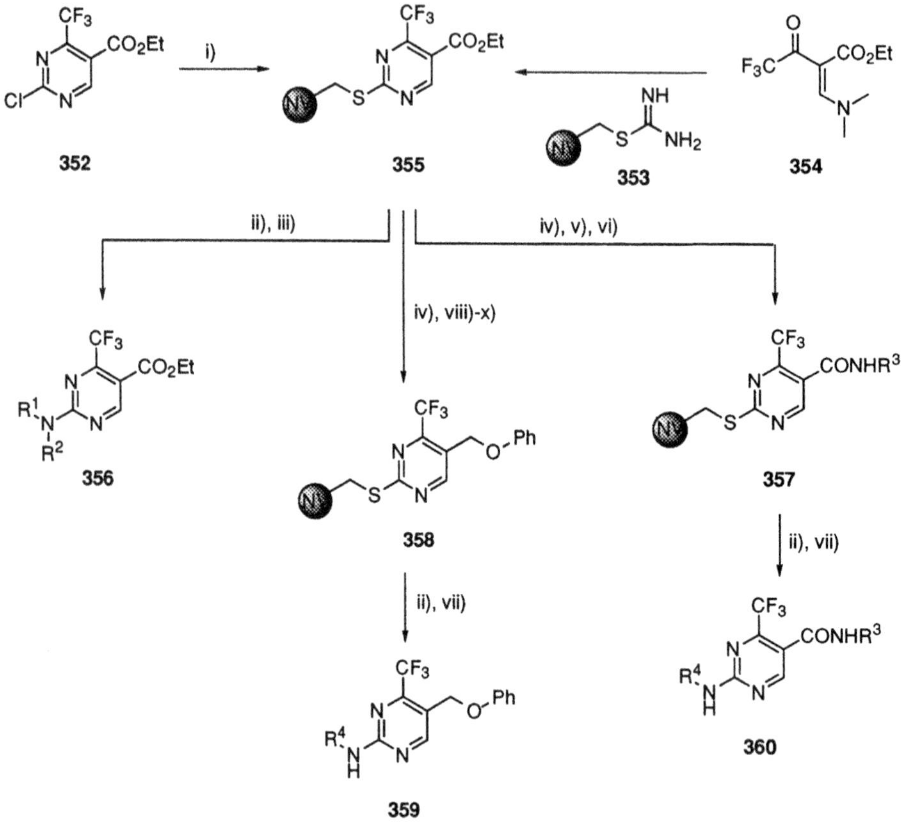

Reagents and conditions: i) NEt₃, DMF, NovaSyn® TG Thiol resin (**351**); ii) *m*-CPBA, CH₂Cl₂; iii) R¹R²NH, CH₂Cl₂; iv) NaOH, H₂O, THF, then HCl, THF; v) oxalylchloride, CH₂Cl₂; vi) R³-NH₂, CH₂Cl₂; vii) R⁴-NH₂, CH₂Cl₂; viii) NMM, ClCOOiBu, THF; ix) NaBH₄, H₂O, THF; x) PhOH, PPh₃, DEAD, NMM

Treatment of 2-chloropyrimidine **352** with NovaSyn®TG thiol resin (**351**) gave polymer-bound thiopyrimidine **355** which was in a preliminary study also obtained from isothiourea **353** and the bis-acceptor reactophor **354**. Pyrimidine **355** was cleaved from the resin after oxidation with *m*-CPBA by treatment with amines to yield 2-amino-pyrimidines **356**; or **355** was converted into the polymer-supported thiopyrimidines **357**, which after saponification, amide coupling and subsequent multidirectional cleavage gave 2-aminopyrimidines of type **360**. Alternatively, pyrimidine **355** was saponified, the resulting acid reduced to the corresponding intermediate alcohol and converted in a *Mitsunobu* coupling with phenol into polymer-bound ether **358**. Oxidation and cleavage with amines resulted in clean formation of 2-aminopyrimidines of type **359**.

Masquelin et al.[151] took advantage of a similar strategy (*Scheme 4.6.9*). Thus, polymer-bound isothiourea **338** (*Scheme 4.6.5*) was treated with the commercially available bis-acceptor building block **361** to yield cleanly polymer-bound 6-amino-5-cyano-pyrimidine **362**. This versatile building block was cleaved after oxidation with *m*-CPBA with amines to yield diamino-pyrimidines of type **363**.

Scheme 4.6.9

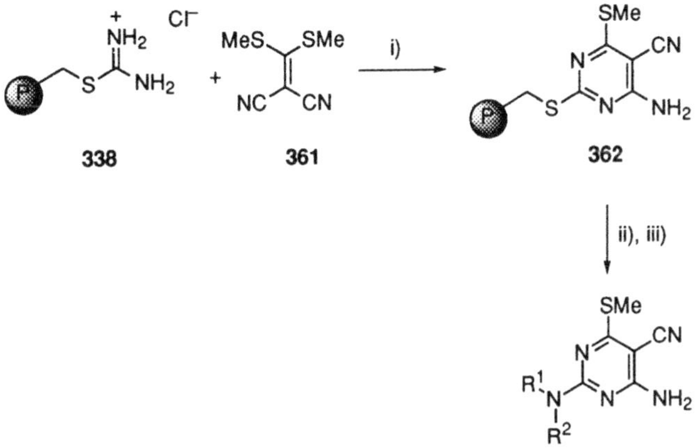

Reagents and conditions: i) DIEA, DMF, Δ; ii) *m*-CPBA, CH$_2$Cl$_2$; iii) R^{1}R^{2}NH, CH$_2$Cl$_2$ or dioxane

The strategies described in *Schemes 4.6.5-4.6.9* thus combine efficiently a novel sulfur-based "safety-catch" linker strategy, multigeneration construction of thio-linked pyrimidines with a multidirectional cleavage procedure. This approach should prove successful for the solid-phase synthesis of a whole range of nitrogen-containing heterocycles such as pyrimidines, triazines, imidazoles, benzimidazoles.

References

(1) G. B. Philips and G. P. Wei, *Tetrahedron Lett.* **1996**, *37*, 4887.

(2) B. O. Buckman and R. Mohan, *Tetrahedron Lett.* **1996**, *37*, 4439.

(3) M. G. Bock, R. M. Dipardo, B. E. Evans, K. E. Rittle, W. L. Whitter, D. F. Veber, P. S. Anderson and R. M. Freidinger, *J. Med. Chem.* **1989**, *32*, 13.

(4) D. Römer, H. H. Buschler, R. C. Hill, R. Maurer, T. J. Petcher, H. Zeugner, W. Benson, E. Finner, W. Milkowski and P. W. Thies, *Nature* **1982**, *298*, 759.

(5) E. Korneki, Y. H. Erlich and R. H. Lenox, *Science* **1984**, *226*, 1454.

(6) M. C. Hsu, A. D. Schutt, M. Holly, L. W. Slice, M. I. Sherman, D. D. Richman, M. J. Potash and D. J. Volsky, *Science* **1991**, *254*, 1799.

(7) R. Pauwels, K. Andries, J. Desmyter, D. Schols, M. J. Kukla, H. Breslin, A. Raeymaeckers and J. V. Gelder, *Nature* **1990**, *343*, 470.

(8) G. L. James, J. L. Goldstein, M. S. Brown, T. E. Rawson, T. C. Somers, R. S. McDowell, C. W. Crowley, B. K. Lucas, A. D. Levinson and J. C. Marsters, *Science* **1993**, *260*, 1937.

(9) B. A. Bunin and J. A. Ellman, *J. Am. Chem. Soc.* **1992**, *114*, 10997.

(10) L. A. Carpino, D. Sadataalaee, H. G. Chao and R. H. Deslems, *J. Am. Chem. Soc.* **1990**, *112*, 9651.

(11) M. J. Plunkett and J. A. Ellman, *J. Am. Chem. Soc.* **1995**, *117*, 3306.

(12) S. P. Hollinshead, *Tetrahedron Lett.* **1996**, *37*, 9157.

(13) M. Mergler, R. Nyfeler and J. Gosteli, *Tetrahedron Lett.* **1989**, *30*, 6741.

(14) L. A. Thompson and J. A. Ellman, *Tetrahedron Lett.* **1994**, *35*, 9333.

(15) E. K. Kick and J. A. Ellman, *J. Med. Chem.* **1995**, *38*, 1427.

(16) M. Caron, P. R. Carlier and K. B. Sharpless, *J. Org. Chem.* **1988**, *53*, 5185.

(17) A. K. Ghosh, S. P. McKee, H. Y. Lee and W. J. Thompson, *J. Chem. Soc., Chem. Commun.* **1992**, 273.

(18) P. Y. S. Lam, P. K. Jadhav, C. J. Eyermann, C. N. Hodge, Y. Ru, L. T. Bacheler, J. L. Meck, M. J. Otto, M. M. Rayner, Y. N. Wong, C.-H. Chang, P. C. Weber, D. A. Jackson, T. R. Sharpe and S. Erickson-Viitanen, *Science* **1994**, *263*, 380.

(19) A. Nefzi, J. M. Ostresh, J.-P. Meyer and R. A. Houghten, *Tetrahedron Lett.* **1997**, *38*, 931.

(20) A. C. Gaudio, A. Korolkovas and Y. Takahata, *J. Pharm. Sci.* **1994**, *83*, 1110.

(21) T. Godfraind, R. Miller and M. Wibo, *Pharmacol. Rev.* **1986**, *38*, 321.

(22) R. A. Janis, P. J. Silver and D. J. Triggle, *Adv. Drug Res.* **1987**, *16*, 309.

(23) P. P. Mager, R. A. Coburn, A. J. Solo, D. J. Triggle and H. Rothe, *Drug Des. Discovery* **1992**, *8*, 273.

(24) R. Mannhold, B. Jablonka, W. Voigdt, K. Schoenafinger and E. Schraven, *Eur. J. Med. Chem.* **1992**, *27*, 229.

(25) A. Sausins and G. Duburs, *Heterocycles* **1988**, *27*, 269.

(26) A. Galiano, *Drugs Fut.* **1995**, *20*, 231.

(27) R. Alajarín, J. J. Vaquero, J. Alvarez-Builla, M. Pastor, C. Sunkel, M. F. d. Casa-Juana, J. Priego, P. R. Statkow and J. Sanz-Aparicio, *Tetrahedron: Asymmetry* **1993**, *4*, 617.

(28) R. Alajarín, J. Alvarez-Builla, J. J. Vaquero, C. Sunkel, M. F. d. Casa-Juana, P. R. Statkow, J. Sanz-Aparicio and I. Fonseca, *J. Med. Chem.* **1995**, *38*, 830.

(29) W. A. Sannita, S. Busico, G. D. Bon, A. Ferrari and S. Riela, *Int. J. Clin. Pharmacol. Res.* **1993**, *13*, 281.

(30) Y. Uehar, Y. Kawabata, N. Ohshima, N. Hirawa, S. Takada, A. Numabe, T. Nagata, A. Goto, S. Yagi and M. Omata, *J. Cardiovasc. Pharmacol.* **1994**, *23*, 970.

(31) T. Nakagawa, Y. Yamauchi, S. Kondo, M. Fuji and N. Yokoo, *Jpn. J. Pharmacol.* **1994**, *64 (Suppl. 1)*, 260.

(32) R. Boer and V. Gekeler, *Drugs Fut.* **1995**, *20*, 499.

(33) M. Schramm, G. Thomas, R. Towart and G. Franckowiak, *Nature* **1983**, *303*, 535.

(34) C. E. Sunkel, M. F. d. Casa-Juana, L. Santos, A. G. García, C. R. Artaljero, M. Vilaroya, M. A. Gonzalez-Morales, M. G. Lopez, J. Cillero, S. Alonso and J. G. Priego, *J. Med. Chem.* **1992**, *35*, 2407.

(35) D. Vo, W. C. Matowe, M. Ramesh, M. Iqbal, M. W. Wolowyk, S. E. Howlett and K. E.E, *J. Med. Chem.* **1995**, *38*, 2851.

(36) V. Klusa, *Drugs Fut.* **1995**, *20*, 135.

(37) K. Cooper, M. J. Fray and M. J. Parry, *J. Med. Chem.* **1992**, *35*, 3115.

(38) M. F. Gordeev, D. V. Patel, J. Wu and E. M. Gordon, *Tetrahedron Lett.* **1996**, *37*, 4643.

(39) M. F. Gordeev, D. V. Patel, J. Wu and E. M. Gordon, *Tetrahedron Lett.* **1996**, *37*, 4643.

(40) F. Albericio, N. Kneib-Cordonier, S. Biancolana, L. Gera, R. I. Masada, D. Hudson and G. Barany, *J. Org. Chem.* **1990**, *55*, 3730.

(41) N. S. Gray, S. Kwon and P. G. Schultz, *Tetrahedron Lett.* **1997**, *38*, 1161.

(42) L. L. Shen, *Biochem. Pharmacol.* **1989**, *38*, 2042.

(43) T. E. Renau, J. P. Sanchez, M. A. Shapiro, J. A. Dever, S. J. Gracheck and J. M. Domagala, *J. Med. Chem.* **1995**, *38*, 2974.

(44) D. V. Caekenberghe, *J. Antimicrob. Chemother.* **1990**, *26*, 381.

(45) A. A. McDonald, S. H. DeWitt, E. M. Hogan and R. Ramage, *Tetrahedron Lett.* **1996**, *37*, 4815.

(46) R. J. Clay, T. A. Collom, G. L. Karrick and J. Wemple, *Synthesis* **1993**, 290.

(47) D. R. Cody, S. H. DeWitt, J. C. Hodges, J. S. Kiely, W. H. Moos, M. R. Pavia, B. D. Roth, M. C. Schroeder and C. J. Stankovic, US 5324483, 1994.

(48) H. J. Roth and A. Kleemann, *Pharmaceutical Chemistry*, Vol. 1: Drug Synthesis, John Wiley & Sons: New York (1988).

(49) M. M. Murphy, J. R. Schullek, E. M. Gordon and M. A. Gallop, *J. Am. Chem. Soc.* **1995**, *117*, 7029.

(50) A. Furka, F. Sebestyen, M. Asgedom and G. Dibo, *Int. J. Pept. Prot. Res.* **1991**, *37*, 487.

(51) M. F. Gordeev, H. C. Hui, E. M. Gordon and D. V. Patel, *Tetrahedron Lett.* **1997**, *38*, 1729.

(52) L. F. Tietze and A. Steinmetz, *Angew. Chem. Int. Ed. Engl.* **1996**, *35*, 651.

(53) W. D. F. Meutermans and P. F. Alewood, *Tetrahedron Lett.* **1995**, *36*, 7709.

(54) R. A. Houghten, *Proc. Natl. Acad. Sci. USA* **1985**, *82*, 5131.

(55) Y. Pei, R. A. Houghten and J. S. Kiely, *Tetrahedron Lett.* **1997**, *38*, 3349.

(56) I. Ojima, *Acc. Chem. Res.* **1995**, *28*, 383.

(57) I. Ojima and Y. Pei, *Tetrahedron Lett.* **1990**, *31*, 977.

(58) I. Ojima and Y. Pei, *Tetrahedron Lett.* **1992**, *33*, 887.

(59) M. S. Manhas, S. G. Amin and A. K. Bose, *Heterocycles* **1976**, *5*, 669.

(60) S. Kano, T. Ebata and S. Shibuya, *Chem. Pharm. Bull.* **1979**, *27*, 2450.

(61) R. A. Houghten, C. Pinilla, S. E. Blondelle, J. R. Appel, C. T. Dooley and J. H. Cuervo, *Nature* **1991**, *354*, 84.

(62) R. Sarges and J. W. Lyga, *J. Heterocycl. Chem.* **1988**, *25*, 1474.

(63) T. E. TenBrink, W. B. Im, V. H. Sethy, A. H. Tang and D. B. Carter, *J. Med. Chem.* **1994**, *37*, 758.

(64) E. J. Jacobsen, R. E. TenBrink, L. S. Stelzer, K. L. Belonga, D. B. Carter, H. K. Im, W. B. Im, V. H. Sethy, A. H. Tang, P. F. VonVoigtlander and J. D. Petke, *J. Med. Chem.* **1996**, *39*, 158.

(65) K. S. Kim, L. Quiang, J. E. Bird, K. E. Dickinson, S. Moreland, T. R. Schaeffer, T. L. Waldron, C. L. Delaney, H. N. Weller and A. V. Miller, *J. Med. Chem.* **1993**, *36*, 2335.

(66) J. Lee, W. V. Murray and R. A. Rivero, *J. Org. Chem.* **1997**, *62*, 3874.

(67) H. V. Meyers, G. J. Dilley, T. L. Durgin, T. S. Powers, N. A. Winssinger and M. R. Pavia, *Mol. Diversity* **1995**, *1*, 13.

(68) I. Maeba, M. Wakimura, Y. Ito and C. Ito, *Heterocycles* **1993**, *36*, 2591.

(69) P. M. Carabateas and L. S. Harris, *J. Med. Chem.* **1966**, *9*, 6.

(70) N. S. Cho, K. Y. Song and C. Parkanyi, *J. Heterocycl. Chem.* **1989**, *26*, 1897.

(71) R. S. McDowel, B. K. Blackburn, T. R. Gadek, L. R. McGee, T. Rawson, M. E. Reynolds, K. D. Robarge, T. C. Somers, E. D. Thorsett, M. Tischler, R. R. Webb and M. C. Venuti, *J. Am. Chem. Soc.* **1994**, *116*, 5077.

(72) D. A. Goff and R. N. Zuckermann, *J. Org. Chem.* **1995**, *60*, 5744.

(73) F. Zaragoza and S. V. Petersen, *Tetrahedron* **1996**, *52*, 10823.

(74) R. J. Clemens and J. A. Hyatt, *J. Org. Chem.* **1985**, *50*, 2431.

(75) F. Zaragoza, *Tetrahedron Lett.* **1996**, *37*, 6213.

(76) R. Laliberté and G. Médewar, *Can. J. Chem.* **1970**, *48*, 2709.

(77) R. Laliberté, H. Warwick and G. Médawar, *Can. J. Chem.* **1968**, *46*, 3643.

(78) M. J. Plunkett and J. A. Ellman, *J. Org. Chem.* **1995**, *60*, 6006.

(79) M. J. Plunkett and J. A. Ellman, *J. Am. Chem. Soc.* **1995**, *117*, 3306.

(80) P. Sieber and B. Iselin, *Helv. Chim. Acta* **1968**, *51*, 622.

(81) M. J. Plunkett and J. A. Ellman, *J. Org. Chem.* **1997**, *62*, 2885.

(82) M. R. Gowravaram and M. A. Gallop, *Tetrahedron Lett.* **1997**, *38*, 6973.

(83) J. P. Marino Jr., M. H. Osterhout and A. Padwa, *J. Org. Chem.* **1995**, *60*, 2704.

(84) A. Padwa and D. L. Hertzog, *Tetrahedron* **1993**, *49*, 2589.

(85) A. Padwa, D. L. Hertzog and R. L. Chinn, *Tetrahedron Lett.* **1989**, *30*, 4077.

(86) A. Padwa, D. L. Hertzog, W. R. Nadler, M. H. Osterhout and A. T. Price, *J. Org. Chem.* **1994**, *59*, 1418.

(87) T. Ye and M. A. McKervey, *Chem. Rev.* **1994**, *94*, 1091.

(88) G. C. Look, C. P. Holmes, J. P. Chinn and M. A. Gallop, *J. Org. Chem.* **1994**, *59*, 7588.

(89) J. A. Ellman, *The Solid-Phase Synthesis of Prostaglandins*, conference: *209th ACS National Meeting*, Anaheim, CA, 1995.

(90) C. R. Johnson and M. P. Braun, *J. Am. Chem. Soc.* **1993**, *115*, 11014.

(91) M. Suzuki, Y. Morita, H. Koyano, M. Koga and R. Noyori, *Tetrahedron* **1990**, *46*, 4809.

(92) D. Lednicer and L. A. Mitscher, *Organic Chemistry of Drug Synthesis*, Vol. 1, Wiley-Interscience: New York, USA (1976).

(93) I. Ugi, A. Domling and W. Horl, *Endeavour* **1994**, *18*, 115.

(94) C. Zhang, E. J. Moran, T. F. Woiwode, K. M. Short and A. M. M. Mjalli, *Tetrahedron Lett.* **1996**, *37*, 751.

(95) T. D. Lash, J. R. Belletini, J. A. Bastian and K. B. Couch, *Synthesis* **1994**, 170.

(96) R. Appel, R. Kleinstuck and K. Ziehn, *Angew. Chem. Int. Ed. Engl.* **1971**, *10*, 132.

(97) S. Sarshar, D. Siev and A. M. M. Mjalli, *Tetrahedron Lett.* **1996**, *37*, 835.

(98) M. F. Aly, M. I. Younes and S. A. M. Metwally, *Tetrahedron* **1994**, *50*, 3159.

(99) B. C. Hamper, D. R. Dukesherer and M. S. South, *Tetrahedron Lett.* **1996**, *37*, 3671.

(100) R. E. Lutz, P. S. Bailey, M. T. Clark, J. F. Codintong, A. J. Deinet, J. A. Freek, G. H. Harnest, N. H. Leake, T. A. Martin, R. J. Rowlett, J. M. Salsbury, N. H. Shearer, J. D. Smith and J. W. Wilson, *J. Am. Chem. Soc.* **1946**, *68*, 1813.

(101) G. A. M. Giardina, H. M. Sarau, C. Farina, A. D. Medhurst, M. Grugni, J. J. Foley, L. F. Raveglia, D. B. Schmidt, R. Rigolio, M. Vassallo, V. Vecchietti and D. W. P. Hay, *J. Med. Chem.* **1996**, *39*, 2281.

(102) A. Kar, *J. Pharm. Sci.* **1983**, *72*, 1082.

(103) O. Doebner, *Ber.* **1883**, *16*, 2357.

(104) A. Gopalsamy and P. V. Pallai, *Tetrahedron Lett.* **1997**, *38*, 907.

(105) P. Biginelli, *Gazz. Chim. Ital.* **1893**, *23*, 360.

(106) C. O. Kappe, *Tetrahedron* **1993**, *49*, 6937.

(107) P. Wipf and A. Cunningham, *Tetrahedron Lett.* **1995**, *36*, 7819.

(108) E. W. Hurst and R. Hull, *J. Med. Pharm. Chem.* **1961**, *3*, 215.

(109) H. Cho, M. Ueda, K. Shima, A. Mizuno, M. Hayashimatsu, Y. Ohnaka, Y. Takeuchi, M. Hamaguchi, K. Aisaka, T. Hidaka, M. Kawai, M. Takeda, T. Ishihara, K. Funahashi, F. Satoh, M. Morita and T. Noguchi, *J. Med. Chem.* **1989**, *32*, 2399.

(110) G. C. Rovnyak, K. S. Atwal, A. Hedberg, S. D. Kimball, S. Moreland, J. Z. Gougoutas, B. C. O'Reilly, J. Schwartz and M. F. Malley, *J. Med. Chem.* **1992**, *35*, 3254.

(111) J. Viret, J. Gabard and A. Collet, *Tetrahedron* **1987**, *43*, 891.

(112) F. Camps, J. Castells and J. Pi, *Ann. Quim.* **1974**, *70*, 848.

(113) J. P. Mayer, J. Zhang, K. Bjergarde, D. M. Lenz and J. J. Gaudino, *Tetrahedron Lett.* **1996**, *37*, 8081.

(114) G. Massimo, F. Zani, E. Coghi, A. Belloti and P. Mazza, *Farmaco* **1990**, *45*, 439.

(115) L. F. Tietze, A. Steinmetz and F. Balkenhohl, *Bioorg. Med. Chem. Lett.* **1997**, *7*, 1303.

(116) L. F. Tietze and A. Steinmetz, *Synlett* **1996**, 667.

(117) C. A. Lopez and G. G. Trigo, in: *Advances in Heterocyclic Chemistry*, C. Brown and R. M. Davidson (Ed.), Vol. 38, p. 177, Academic Press: New York (1985).

(118) B. A. Dressman, L. A. Spangle and S. W. Kaldor, *Tetrahedron Lett.* **1996**, *37*, 937.

(119) S. W. Kim, S. Y. Ahn, J. S. Koh, J. H. Lee, S. Ro and H. Y. Cho, *Tetrahedron Lett.* **1997**, *38*, 4603.

(120) S. Hanessian and R.-Y. Yang, *Tetrahedron Lett.* **1996**, *37*, 5835.

(121) A. L. Smith, C. G. Thomson and P. D. Leeson, *Bioorg. Med. Chem. Lett.* **1996**, *6*, 1483.

(122) S. T. Brennan, N. L. Colbry, R. L. Leeds, B. Leja, S. R. Priebe, M. D. Reily, H. D. H. Showalter, S. E. Uhlendorf, G. J. Atwell and W. A. Denny, *J. Heterocyclic Chem.* **1989**, *26*, 1469.

(123) D. W. Fry, A. J. Kraker, A. McMichael, L. A. Ambroso, J. M. Nelson, W. R. Leopold, R. W. Connors and A. J. Bridges, *Science* **1994**, *265*, 1093.

(124) S. Johne, *Pharmazie* **1981**, *36*, 583.

(125) S. E. Laszlo, R. S. Chang, T.-B. Cheng, K. A. Faust, W. J. Greenlee, S. D. Kivlighn, V. J. Lotti, S. S. O'Malley, T. W. Schorn, P. K. Siegl, J. Tran and G. J. Zingaro, *Bioorg. Med. Chem. Lett* **1995**, *5*, 1359.

(126) R. A. Houghten, *Linear Arrays and Combinatorial Mixtures of Quinazolinones*, conference: *Exploiting Molecular Diversity: Small Molecule Libraries For Drug Discovery*, San Diego, CA, 1996.

(127) J. Barluenga and F. Palacios, *Org. Prep. Proced. Int.* **1991**, *23*, 1.

(128) P. Molina and M. J. Vilaplana, *Synthesis* **1994**, 1197.

(129) E. C. Taylor and M. Patel, *J. Heterocyclic Chem.* **1991**, *28*, 1851.

(130) R. Frenette and R. W. Friesen, *Tetrahedron Lett.* **1994**, *35*, 9177.

(131) J. M. Villalgordo, D. Obrecht and A. Chucholowski, *Synlett* **1998**, submitted for publication.

(132) A. Chucholowski, T. Masquelin, D. Obrecht, J. Stadlwieser and J. M. Villalgordo, *Chimia* **1996**, *50*, 530.

(133) D. W. Gordon and J. Steele, *Bioorg. Med. Chem. Lett.* **1995**, *5*, 47.

(134) R. W. Armstrong, A. P. Combs, P. A. Tempest, S. D. Brown and T. A. Keating, *Acc. Chem. Res.* **1996**, *29*, 123.

(135) L. Morode, J. Lutz, F. Grams, S. Rudolph-Böhner, G. Oesapay, M. Goodman and W. Kolbeck, *Biopolymers* **1996**, *38*, 295.

(136) T. Q. Dinh and R. W. Amstrong, *Tetrahedron Lett.* **1996**, *37*, 1161.

(137) S. V. Ley, D. M. Mynett and W.-J. Koot, *Synlett* **1995**, 1017.

(138) B. J. Backes and J. A. Ellman, *J. Am. Chem. Soc.* **1994**, *116*, 11171.

(139) C. C. Leznoff, P. M. Worster and C. R. McArthur, *Angew. Chem. Int. Ed. Engl.* **1978**, *18*, 221.

(140) A. Suzuki, *Pure Appl. Chem.* **1985**, *57*, 1749.

(141) G. W. Kenner, J. R. Mc Dermott and R. C. Sheppard, *J. Chem. Soc., Chem. Commun.* **1971**, 636.

(142) B. J. Backes, A. A. Virgilio and J. A. Ellman, *J. Am. Chem. Soc.* **1996**, *118*, 3055.

(143) A. R. Katritzky and C. W. Rees, *Comprehensive Heterocyclic Chemistry*, Vol. 1, Pergamon Press: New York (1984).

(144) A. R. Katritzky and C. W. Rees, *Comprehensive Heterocyclic Chemistry*, Vol. 3, Pergamon Press: Oxford (1984).

(145) D. Obrecht, *Helv. Chim. Acta* **1989**, *72*, 447.

(146) D. Strekowsky, R. A. Harden and R. A. Watson, *Synthesis* **1988**, 70.

(147) C. J. Shishoo and K. S. Jain, *J. Heterocycl. Chem.* **1992**, *29*, 883.

(148) D. Obrecht, C. Abrecht, A. Grieder and J.-M. Villalgordo, *Helv. Chim. Acta* **1997**, *80*, 65.

(149) I. Ugi, S. Lohberger and R. Karl, in: *Comprehensive Organic Synthesis*, B. M. Trost and I. Fleming (Ed.), Vol. 2, p. 1083, Pergamon: New York (1991).

(150) L. M. Gayo and M. J. Suto, *Tetrahedron Lett.* **1997**, *38*, 211.

(151) T. Masquelin, R. Bär, F. Gerber, D. Sprenger and Y. Mercadal, *Helv. Chim. Acta* **1998**, in press.

Outlook and future directions

Combinatorial and parallel synthesis techniques will certainly have an increasing impact in drug discovery as the repertoire of synthetic methods amenable to solid-phase chemistry will expand, leading to more and more complex structures. High throughput methods, however, are by no means limited to the synthesis of compound libraries for screening and have found wide applications in process development and material sciences.[1] In process development where many parameters such as concentrations of reactants, choice of proper solvents, catalysts, reaction temperatures and reaction times have to be optimised, it is believed that combinatorial approaches will allow a simultaneous screening of parameters and thus speed up the development process. Material sciences offer a tremendous platform for combinatorial chemistry in finding novel materials for electronic, magnetic and optical devices. Efficient green, blue and red phosphors[2,3] were discovered using combinatorial techniques. Such approaches have also been used to find new iridium catalysts[4] for alkane dehydrogenation.

References

(1) E. K. Wilson, *Chem. Eng. News* **1997**, *December 8*, 24.

(2) E. Danielson, J. H. Golden, E. W. McFarland, C. M. Reaves, W. H. Weinberg and X. D. Wu, *Nature* **1997**, *389*, 944.

(3) X.-D. Sun, C. Gao, J. Wang and X.-D. Xiang, *Appl. Phys. Lett.* **1997**, *70*, 3353.

(4) W.-W. Xu, G. P. Rosini, M. Gupta, C. Jensen, W. C. Kaska, K. Krogh-Jespersen and A. S. Goldman, *J. Chem. Soc., Chem. Commun.* **1997**, 2273.

Subject Index

Acknowledgements

The authors would like to thank: Profs. Drs. Heinz Heimgartner and John A. Robinson (University of Zürich, CH) for proof-reading the manuscript; Sir Jack Baldwin (Waynflete Professor of Chemistry, Oxford, UK), Prof. Dr. Andrea Vasella (ETH Zürich, CH) and Dr. Jean-Pierre Obrecht (Polyphor Ltd., Zürich, CH) for many stimulating discussions; Dr. Alexander Chucholowski (F. Hoffmann-La Roche Ltd.) for his excellent contribution; Mr. Werner Doebelin (Prolab/LabSource, Reinach, CH) for providing *Figure 1.13* and for his technical assistance; Mrs. Liza Koltay (Koltay Design, Bottmingen, CH) for the cover-art and Dr. Cornelia Zumbrunn for the preparation of the final camera-ready manuscript.

DO would like to thank Mrs. Patricia Bir and Mrs. Jeanne Obrecht for their constant support and motivation.

339

Printed and bound by CPI Group (UK) Ltd, Croydon, CR0 4YY

03/10/2024

01040320-0009